Biomass Derived Heterogeneous and Homogeneous Catalysts

Biomass Derived Heterogeneous and Homogeneous Catalysts

Editors

José María Encinar Martín
Sergio Nogales Delgado

MDPI • Basel • Beijing • Wuhan • Barcelona • Belgrade • Manchester • Tokyo • Cluj • Tianjin

Editors
José María Encinar Martín
University of Extremadura
Spain

Sergio Nogales Delgado
University of Extremadura
Spain

Editorial Office
MDPI
St. Alban-Anlage 66
4052 Basel, Switzerland

This is a reprint of articles from the Special Issue published online in the open access journal *Catalysts* (ISSN 2073-4344) (available at: https://www.mdpi.com/journal/catalysts/special_issues/biomass_hete_homo_cata).

For citation purposes, cite each article independently as indicated on the article page online and as indicated below:

LastName, A.A.; LastName, B.B.; LastName, C.C. Article Title. *Journal Name* **Year**, *Volume Number*, Page Range.

ISBN 978-3-0365-0908-2 (Hbk)
ISBN 978-3-0365-0909-9 (PDF)

Cover image courtesy of Sergio Nogales Delgado.

© 2021 by the authors. Articles in this book are Open Access and distributed under the Creative Commons Attribution (CC BY) license, which allows users to download, copy and build upon published articles, as long as the author and publisher are properly credited, which ensures maximum dissemination and a wider impact of our publications.

The book as a whole is distributed by MDPI under the terms and conditions of the Creative Commons license CC BY-NC-ND.

Contents

About the Editors vii

José María Encinar and Sergio Nogales-Delgado
Editorial Catalysts: Special Issue on "Biomass Derived Heterogeneous and Homogeneous Catalysts"
Reprinted from: *Catalysts* **2020**, *10*, 1433, doi:10.3390/catal10121433 1

Dinesh Kumar, Chan Hee Park and Cheol Sang Kim
One-Pot Solvent-Free Synthesis of N,N-Bis(2-Hydroxyethyl) Alkylamide from Triglycerides Using Zinc-Doped Calcium Oxide Nanospheroids as a Heterogeneous Catalyst
Reprinted from: *Catalysts* **2019**, *9*, 774, doi:10.3390/catal9090774 5

José María Encinar, Sergio Nogales-Delgado, Nuria Sánchez and Juan Félix González
Biolubricants from Rapeseed and Castor Oil Transesterification by Using Titanium Isopropoxide as a Catalyst: Production and Characterization
Reprinted from: *Catalysts* **2020**, *10*, 366, doi:10.3390/catal10040366 19

Nuria Sánchez, José María Encinar, Sergio Nogales and Juan Félix González
Biodiesel Production from Castor oil by Two-Step Catalytic Transesterification: Optimization of the Process and Economic Assessment
Reprinted from: *Catalysts* **2019**, *9*, 864, doi:10.3390/catal9100864 31

Alberto Navajas, Inés Reyero, Elena Jiménez-Barrera, Francisca Romero-Sarria, Jordi Llorca and Luis M. Gandía
Catalytic Performance of Bulk and Al_2O_3-Supported Molybdenum Oxide for the Production of Biodiesel from Oil with High Free Fatty Acids Content
Reprinted from: *Catalysts* **2020**, *10*, 158, doi:10.3390/catal10020158 47

Hui Chen, Zihui Shang, Huanjie Cai and Yan Zhu
Irrigation Combined with Aeration Promoted Soil Respiration through Increasing Soil Microbes, Enzymes, and Crop Growth in Tomato Fields
Reprinted from: *Catalysts* **2019**, *9*, 945, doi:10.3390/catal9110945 61

Sohaib Hameed, Lu Lin, Aiqin Wang and Wenhao Luo
Recent Developments in Metal-Based Catalysts for the Catalytic Aerobic Oxidation of 5-Hydroxymethyl-Furfural to 2,5-Furandicarboxylic Acid
Reprinted from: *Catalysts* **2020**, *10*, 120, doi:10.3390/catal10010120 77

Huiyuan Xue, Xingxing Gong, Jingjing Xu and Rongrong Hu
Performance of a Ni-Cu-Co/Al_2O_3 Catalyst on In-Situ Hydrodeoxygenation of Bio-derived Phenol
Reprinted from: *Catalysts* **2019**, *9*, 952, doi:10.3390/catal9110952 103

Aa'ishah Abdul Gafar, Mohd Ezuan Khayat, Siti Aqlima Ahmad, Nur Adeela Yasid and Mohd Yunus Shukor
Response Surface Methodology for the Optimization of Keratinase Production in Culture Medium Containing Feathers by *Bacillus* sp. UPM-AAG1
Reprinted from: *Catalysts* **2020**, *10*, 848, doi:10.3390/catal10080848 121

Samkelo Malgas, Shaunita H. Rose, Willem H. van Zyl and Brett I. Pletschke
Enzymatic Hydrolysis of Softwood Derived Paper Sludge by an In Vitro Recombinant Cellulase Cocktail for the Production of Fermentable Sugars
Reprinted from: *Catalysts* **2020**, *10*, 775, doi:10.3390/catal10070775 . . . . **139**

Mohan Turup Pandurangan and Krishnan Kanny
Study of Curing Characteristics of Cellulose Nanofiber-Filled Epoxy Nanocomposites
Reprinted from: *Catalysts* **2020**, *10*, 831, doi:10.3390/catal10080831 . . . . **157**

Annu Rusanen, Riikka Kupila, Katja Lappalainen, Johanna Kärkkäinen, Tao Hu and Ulla Lassi
Conversion of Xylose to Furfural over Lignin-Based Activated Carbon-Supported Iron Catalysts
Reprinted from: *Catalysts* **2020**, *10*, 821, doi:10.3390/catal10080821 . . . . **171**

About the Editors

José María Encinar Martín was born on September 16, 1952, in Cáceres, Spain. He received a Bachelor of Science degree from the Science Faculty of Extremadura University (Spain) and a Ph.D degree from the Extremadura University, in 1977 and 1984, respectively, both in chemical science. He has been working for the Extremadura University since 1977, where he is currently a Professor in the Department of Chemical Engineering and Physical Chemistry. His current research interests include combustion; pyrolysis; gasification; thermochemical processes; the energy use of the biomass residues; active carbons; and production of biodiesel, bioethanol, and biolubricants.

Sergio Nogales Delgado was born on March 16, 1984, in Badajoz, Spain. He received a Bachelor of Science degree and a Ph.D. degree from the University of Extremadura, in 2007 and 2016, respectively, both in chemical science. He worked as a technologist dealing with subjects like minimally processed vegetable and fruits, biomass pelletizing and emissions, and biodiesel and biolubricant production, among others. His current research interests include combustion, pyrolysis, gasification, thermochemical processes, the energy use of biomass wastes, and biodiesel and biolubricant production and characterization (including viscosity, oxidative stability, flash and combustion points, and cold filter plugging point).

Editorial

Editorial Catalysts: Special Issue on "Biomass Derived Heterogeneous and Homogeneous Catalysts"

José María Encinar and Sergio Nogales-Delgado *

Department of Chemical Engineering and Physical-Chemistry, University of Extremadura, Av. De Elvas s/n, 06006 Badajoz, Spain; jencinar@unex.es
* Correspondence: senogalesd@unex.es

Received: 27 November 2020; Accepted: 4 December 2020; Published: 8 December 2020

The replacement of petrol products for environmentally-friendly ones is a reality today, as many governments and international organizations are promoting the implementation of renewable energy sources and natural feedstocks in industrial activity. The multiple advantages of using bio-based products instead of their equivalent material of mineral origin are well known, like lower associated emissions, higher biodegradability, and a sustainable economic development of poor regions or developing countries, among others. In addition, the performance of these products, compared to the traditional ones, is acceptable, even showing some competitive advantages in certain cases. However, the implementation of technologies for environmentally-friendly production is sometimes worrisome, because it usually requires high costs. In these cases, the use of catalysts in order to improve the yield during production is vital to make green processes more competitive from an economic point of view.

In this special issue, we have tried to focus on the performance of homogeneous and heterogeneous catalysts applied in biomass processing, paying attention to the main advantages and challenges related to each kind of catalyst. Indeed, these challenges are opportunities to develop new research lines which can be fruitful in the near future. Thus, the use of homogeneous catalysts tends to be really useful to obtain acceptable product yields, whereas their management (including separation, among other steps) is generally difficult. On the other hand, heterogeneous catalysts can prevent this kind of problem, but their effectiveness compared to homogeneous ones is lower. That is, the yield is usually lower and therefore higher temperature or longer reaction times are required. In addition, their reusability is generally poor, not allowing many cycles.

Thus, we can note, in this special issue, the wide range of uses (or approaches) of homogeneous and heterogeneous catalysts, from the production of fatty acid amides [1] to the production of fatty acid esters for biodiesel or biolubricant use through different techniques and catalysts [2–4]. This fact points out that the use of homogeneous and, especially, heterogeneous catalysts has a wide range of possibilities in biomass processing, which will be a promising research line in the medium and long term.

Indeed, the role of enzymes as biocatalysts is important, even taking part in pre-harvest conditions of biomass. This way, soil respiration was studied under different irrigation levels (with and without aeration) in a greenhouse tomato system, showing differences in root length as well as dry biomass of leaf, stem or fruit [5].

This special issue has not only focused on the production of typical biofuels like biodiesel. As abovementioned, there were different and specific processes where the use of catalyst was vital to make the process feasible in a large-scale industry. For instance, the production of fatty acid amides from natural triglycerides through amidation was considered. Fatty acid amides have multiple uses (like surfactants, lubricants, and detergents) in various industries (in cosmetic industry, in biodiesel development technology, etc.), and the use of heterogeneous catalysts (Zn-doped CaO nanospheroids) for its production was studied by Kumar et al. [1], obtaining high efficiency and excellent reusability without losing much catalytic activity (one of the main drawbacks related to the use of heterogeneous catalysts or enzymes).

On the other hand, Hameed et al. reviewed the use of metal-based catalysts for the catalytic aerobic oxidation of biomass to produce 2,5-furandicarboxilic acid (FDCA), which has multiple uses (especially the replacement for terephthalic acid, PTA). The main challenges are the improvement of selectivity for FDCA when non-noble metals were used, and the performance of the catalysts was dependent on properties like the support, the active phase, or particle size, among others [6].

Another application of catalysts in biomass could be upgrading bio-oils in order to improve their performance in engine fuels or as fuel additives. Xue et al. assessed the performance of an active trimetallic Ni-Cu-Co/Al_2O_3 catalyst for hydrodeoxigenation of bio-derived phenol, showing a high catalytic activity compared to their monometallic and bimetallic equivalents and obtaining high cyclohexane yields (over 98%) [7].

On the other hand, the use of enzymes in many processes like food, bioremediation, and industrial biotechnology is also important. Gafar et al. studied the optimization of the production of keratinase using feathers as the only source of carbon and nitrogen, being the starting point for the treatment of other wastes like waste bagasse or palm mill oil effluent, among others [8].

This way, another waste such as paper sludge is an attractive biomass feedstock for bioconversion to ethanol, and Malgas et al. used a recombinant cellulose cocktail for the saccharification of this waste to produce fermentable sugars. Thus, the enzymatic cocktail was optimized, and its performance was comparable to commercial preparations for paper sludge saccharification [9].

Searching natural replacements for synthetic nanofillers in order to reinforce polymeric matrices could be another important aspect to be covered. Pandurangan et al. studied the effect of cellulose nanofiber (CNF) from banana fibers on curing characteristics, structure, thermal, and mechanical properties in epoxy polymer matrix. This way, CNF could be a promising green nanofiller for the epoxy matrix, possibly acting as a curing catalyst during epoxy gelation [10].

Last, but not least, furfural is an intermediate step for the generation of many pharmaceutical and chemical products, mainly obtained from xylose in agricultural wastes. The use of lignin-based activated carbon-supported iron catalysts for that purpose was studied by Rusanen et al., being a feasible alternative for $FeCl_3$, with furfural yields up to 57%, although their reusability should be improved [11].

As a conclusion, the use of homogeneous and heterogeneous catalysts contributed to the improvement of the performance of many different processes, in order to produce bio-fuels or bio-based materials. Indeed, new trends were observed, like the use of natural feedstocks to take part in the catalytic process. In any case, the improvement of the performance of these green processes seems to be a promising research line in the medium or long term. Thus, as pointed out in Figure 1, catalysts can play an important role in the replacement of fossil fuels and its derivatives for natural feedstocks (like the ones covered in this special issue).

Figure 1. Main prospects for the implementation of green technologies.

Thus, catalysts can contribute to the implementation of new feedstocks used in biorefineries in order to obtain environmentally-friendly products. Moreover, one of the main challenges would be the replacement of artificial catalysts for bio-based catalysts, which would make currents technologies "greener", if possible.

Funding: This research received no external funding.

Acknowledgments: The guest editors would like to thank the interest of the authors who have contributed to this special issue, and we are especially grateful for the assistance of Zerlinda Tian during this interesting and rewarding experience.

Conflicts of Interest: The authors declare no conflict of interest.

References

1. Kumar, D.; Park, C.H.; Kim, C.S. One-pot solvent-free synthesis of N,N-Bis(2-hydroxyethyl) alkylamide from triglycerides using zinc-doped calcium oxide nanospheroids as a heterogeneous catalyst. *Catalysts* **2019**, *9*, 774. [CrossRef]
2. Encinar, J.M.; Nogales-Delgado, S.; Sánchez, N.; González, J.F. Biolubricants from rapeseed and castor oil transesterification by using titanium isopropoxide as a catalyst: Production and characterization. *Catalysts* **2020**, *10*, 366. [CrossRef]
3. Sánchez, N.; Encinar, J.M.; Nogales, S.; González, J.F. Biodiesel production from castor oil by two-step catalytic transesterification: Optimization of the process and economic assessment. *Catalysts* **2019**, *9*, 864. [CrossRef]
4. Navajas, A.; Reyero, I.; Jiménez-Barrera, E.; Romero-Sarria, F.; Llorca, J.; Gandía, L.M. Catalytic Performance of Bulk and Al_2O_3-Supported Molybdenum Oxide for the Production of Biodiesel from Oil with High Free Fatty Acids Content. *Catalysts* **2020**, *10*, 158. [CrossRef]
5. Chen, H.; Shang, Z.; Cai, H.; Zhu, Y. Irrigation combined with aeration promoted soil respiration through increasing soil microbes, enzymes, and crop growth in tomato fields. *Catalysts* **2019**, *9*, 945. [CrossRef]
6. Hameed, S.; Lin, L.; Wang, A.; Luo, W. Recent Developments in Metal-Based Catalysts for the Catalytic Aerobic Oxidation of 5-Hydroxymethyl-Furfural to 2, 5-Furandicarboxylic Acid. *Catalysts* **2020**, *10*, 120. [CrossRef]
7. Xue, H.; Gong, X.; Xu, J.; Hu, R. Performance of a Ni-Cu-Co/Al_2O_3 Catalyst on in-situ Hydrodeoxygenation of Bio-derived Phenol. *Catalysts* **2019**, *9*, 952. [CrossRef]
8. Gafar, A.A.; Khayat, M.E.; Ahmad, S.A.; Yasid, N.A.; Shukor, M.Y. Response surface methodology for the optimization of keratinase production in culture medium containing feathers by *Bacillus* sp. Upm-aag1. *Catalysts* **2020**, *10*, 848. [CrossRef]
9. Malgas, S.; Rose, S.H.; van Zyl, W.H.; Pletschke, B.I. Enzymatic hydrolysis of softwood derived paper sludge by an in vitro recombinant cellulase cocktail for the production of fermentable sugars. *Catalysts* **2020**, *10*, 775. [CrossRef]
10. Pandurangan, M.T.; Kanny, K. Study of curing characteristics of cellulose nanofiber-filled epoxy nanocomposites. *Catalysts* **2020**, *10*, 831. [CrossRef]
11. Rusanen, A.; Kupila, R.; Lappalainen, K.; Kärkkäinen, J.; Hu, T.; Lassi, U. Conversion of xylose to furfural over lignin-based activated carbon-supported iron catalysts. *Catalysts* **2020**, *10*, 821. [CrossRef]

Publisher's Note: MDPI stays neutral with regard to jurisdictional claims in published maps and institutional affiliations.

© 2020 by the authors. Licensee MDPI, Basel, Switzerland. This article is an open access article distributed under the terms and conditions of the Creative Commons Attribution (CC BY) license (http://creativecommons.org/licenses/by/4.0/).

Communication

One-Pot Solvent-Free Synthesis of N,N-Bis(2-Hydroxyethyl) Alkylamide from Triglycerides Using Zinc-Doped Calcium Oxide Nanospheroids as a Heterogeneous Catalyst

Dinesh Kumar [1,*], Chan Hee Park [1,2,*] and Cheol Sang Kim [1,*]

1 Department of Bionanosystem Engineering, Chonbuk National University, Jeonju 561756, Korea
2 Department of Mechanical Design Engineering, Chonbuk National University, Jeonju 561756, Korea
* Correspondence: dinesh.tu@gmail.com (D.K.); biochan@jbnu.ac.kr (C.H.P.); chskim@jbnu.ac.kr (C.S.K.); Tel.: +82-63-270-4283 (D.K.); +82-63-270-4284 (C.H.P.); +82-63-270-4284 (C.S.K.)

Received: 1 September 2019; Accepted: 13 September 2019; Published: 14 September 2019

Abstract: N,N-Bis(2-hydroxyethyl) alkylamide or fatty acid diethanolamides (FADs) were prepared from a variety of triglycerides using diethanolamine in the presence of different transition metal-doped CaO nanocrystalline heterogeneous catalysts. The Zn-doped Cao nanospheroids were found to be the most efficient heterogeneous catalyst, with complete conversion of natural triglycerides to fatty acid diethanolamide in 30 min at 90 °C. The Zn/CaO nanoparticles were recyclable for up to six reaction cycles and showed complete conversion even at room temperature. The amidation reaction of natural triglycerides was found to follow the pseudo-first-order kinetic model, and the first-order rate constant was calculated as 0.171 min^{-1} for jatropha oil aminolysis. The activation energy (Ea) and pre-exponential factor (A) for the same reaction were found to be 47.8 kJ mol^{-1} and 4.75×10^8 min^{-1}, respectively.

Keywords: nanospheroids; zinc-doped CaO; natural triglycerides; aminolysis; heterogeneous catalyst; recyclability

---

1. Introduction

The rapid depletion of fossil fuel resources and global warming has motivated researchers to expand different technologies that utilize renewable energy sources [1–3]. Natural triglycerides [4,5] and lignocellulosic biomass [6–8] have been converted in various platforms to value-added products. Triglycerides in vegetable oils and animal fats have been used in industries as a feedstock for the preparation of fatty amides, nitriles, amines, and alcohols, which in turn are useful for the preparation of various commodity chemicals such as surfactants and different types of polymers [9].

Fatty acid amides have a wide range of applications, viz. in surfactants, cosmetics, fungicides, lubricants, foam-control agents, water repellents, shampoos, detergents, corrosion inhibitors, and antiblocking agents, in plastics processing technologies [5,10,11]. Fatty acid amides possess better ignition properties than simple esters and hence are more useful in biodiesel development technology [12]. Moreover, fatty acid amide derivatives of natural triglycerides (vegetable oils or animal fats) or fatty acids have been found to be free from sulfur or any other aromatic compounds and thus help to lessen the greenhouse effect, showing an improvement in cetane number and cold flow properties and a beneficial effect on particulate matter emissions [12]. The industrial synthesis of fatty acid amides involves a two-step process: first, the conversion of triglycerides into fatty acid methyl/ethyl esters, followed by a high-temperature treatment to prepare fatty acid amides [5].

Due to their great importance, some methodologies to prepare fatty acid amides from fatty acids or fatty acid alkyl esters or triglycerides through treatments with different amines have been proposed previously [13,14]. At the industrial level, homogeneous catalysts such as sodium ethoxide [15], sodium methoxide [16], and calcium chloride [17] are used for the preparation of fatty acid amides. Enzymes have been frequently reported [18,19] as a heterogeneous catalyst for amidation reactions, though they require a longer reaction duration. In addition, solvent-free conditions [20,21], Sm^{III} complexes [22], Sn^{IV} complexes [23], the Deoxo-Fluor reagent [bis(2-methoxyethyl)amino-sulfur trifluoride] [24], and other chemicals [25] have been utilized to obtain the desired amide. However, a few drawbacks are associated with these methods, viz. low product yields, a longer reaction duration, difficult product separation, a lack of catalyst reusability, a large molar excess of reactants, contamination of the product, and the creation of stoichiometric amounts of undesired products. Therefore, it is necessary to extend new, efficient, environmentally friendly, ecologically correct, and reusable catalytic methods for the amidation of fatty acids and triglycerides [26–28]. In the recent past, heterogeneous catalysts have attracted considerable attention, as they are nonhazardous, have good selectivity and recyclability, and are easy to separate from reaction medium [29–31].

In the present report, zinc-doped CaO, MgO, and ZnO were prepared in nanoparticle form using an incipient-wetness impregnation method and were used as heterogeneous catalysts for solvent-free direct amidation of natural triglycerides. The effect of transition metal ion impregnation on CaO activity and calcination temperature was also studied by preparing a series of catalysts with Fe, Co, Cu, Zn, and Cd doped on CaO.

2. Results and Discussion

2.1. Brunauer–Emmett–Teller (BET) Surface Area and Hammett Indicator Test

The basic strength (pK_{BH+}) and surface area of prepared catalysts were analyzed using the Hammett indicator test and Brunauer–Emmett–Teller (BET) surface area measurement, respectively. The basic strength of CaO was found to be increased from 9.8–10.1 to 11.1–15.0 after 2 wt% doping of Zn, which further increased to a maximum of 18.4 after calcination at 400 °C. However, a further increase in calcination temperature decreased the basic strength. The improvement of the basic strength of CaO after zinc ion doping with an increase in calcination temperature up to 400 °C could have been due to the partial dehydration and strong increase in the surface area [32]. An increase in the Zn ion concentration did not improve the basic strength. However, after Zn doping on MgO and ZnO, the basic strength was increased to the range of 15.0–18.4 (Table 1). The BET surface area was another critical factor that had a direct impact on catalytic efficiency. Bare CaO had a surface area of 3.56 m^2/g, which improved to 16.87 m^2/g after 2 wt% doping of Zn along with calcination at 400 °C. However, calcination at high temperatures, viz. 600 °C and 800 °C, caused a reduction in surface area to 10.12 m^2/g and 5.25 m^2/g, respectively, which could have been due to the sintering of material at high temperatures. The doping of the Zn ion on MgO and ZnO also caused an increase in surface area from 10.4 m^2/g to 14.89 m^2/g and 4.72 m^2/g to 12.13 m^2/g, respectively (Table 1). Hence, the doping of the Zn ion and calcination were critical factors responsible for the increase in basic strength and surface area.

2.2. Structural Analysis of Catalyst

The effect of calcination temperature, Zn ion concentration, and different Zn doping on different metal oxides was studied by powder XRD analysis (Figure 1 and Figure S1 in Supporting Information). The presence of the cubic phase of CaO was confirmed by peaks at 2θ values of 32.27°, 37.47°, 53.89°, and 67.39° (JCPDS (Joint Committee on Powder Diffraction Standards) card no. 821691), as shown in Figure 1a. After zinc doping (2 wt%) and calcination at 200 °C (2-Zn/CaO-200), the cubic phase of CaO was converted into a hexagonal form of $Ca(OH)_2$, as supported by the peaks at 18.05°, 28.6°, 34.17°, 47.17°, 50.78°, and 62.57° (JCPDS 84-1276), as shown in Figure S1a. A further increase in the calcination temperature to 400 °C showed the coexistence of both the cubic and hexagonal phases, which might have been the reason for the abrupt enhancement in the basic strength of 2-Zn/CaO-400. However, a further increase in calcination temperature to 600 °C showed the presence of only the cubic phase. The increase of the Zn ion concentration had no impact on the structure of Zn/CaO-400 (Figure S1b).

Table 1. Effect of calcination temperature, Zn ion concentration, and different oxides on the basic strength, surface area, and crystallite size. BET: Brunauer–Emmett–Teller.

Catalyst Type	Basic Strength (pK_{BH+})	BET Surface Area (m^2/g)	Average Crystallite Size (nm)*
CaO	$9.8 < pK_{BH+} < 10.1$	3.56	108
2-Zn/CaO-100	$11.1 < pK_{BH+} < 15.0$	4.51	35
2-Zn/CaO-200	$15.0 < pK_{BH+} < 18.4$	6.03	37
2-Zn/CaO-400	$18.4 < pK_{BH+}$	16.87	33
2-Zn/CaO-600	$15.0 < pK_{BH+} < 18.4$	10.12	35
2-Zn/CaO-800	$11.1 < pK_{BH+} < 15.0$	5.25	55
1-Zn/CaO-400	$11.1 < pK_{BH+} < 15.0$	11.72	37
3-Zn/CaO-400	$15.0 < pK_{BH+} < 18.4$	16.24	34
4-Zn/CaO-400	$15.0 < pK_{BH+} < 18.4$	16.54	39
5-Zn/CaO-400	$15.0 < pK_{BH+} < 18.4$	14.36	35
2-Zn/MgO-400	$15.0 < pK_{BH+} < 18.4$	14.89	35
2-Zn/ZnO-400	$15.0 < pK_{BH+} < 18.4$	12.13	38

* on (200) plane by Debye–Scherrer method [33].

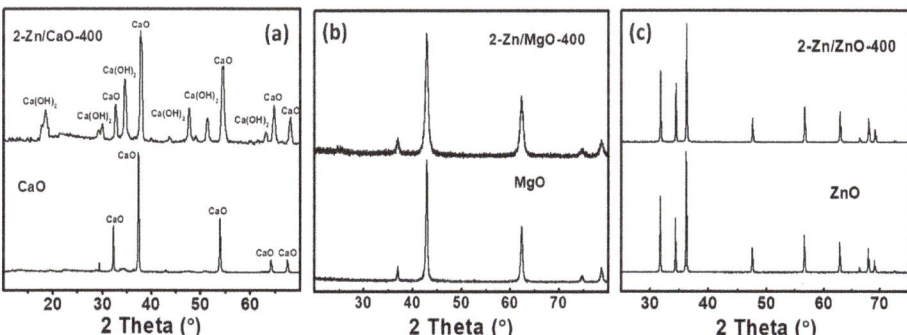

Figure 1. Comparative powder XRD patterns of (a) CaO with 2-Zn/CaO-400, (b) MgO with 2-Zn/MgO-400, and (c) ZnO with 2-Zn/ZnO-400.

XRD patterns for the 2-Zn/MgO-400 catalysts showed sharp diffraction peaks at 2θ = 36.89°, 42.92°, 62.28°, 74.58°, and 78.53°, which were attributed to the crystalline phase of the MgO (JCPDS 4-829) (Figure S1b). The XRD analysis of 2-Zn/ZnO-400 showed the presence of a hexagonal phase of ZnO (JCPDS: 80-0075), as indicated by the peaks at 31.79°, 34.51°, 36.2°, 47.55°, 56.65°, 62.84°, 66.31°, 67.9°, and 69.03° (Figure 1c). The diffraction pattern of Zn was not observed in any XRD spectrum, which might have been due to its high degree of dispersion on CaO/MgO/ZnO or it being below the detection limit of XRD. The particle size of prepared nanoparticles was also calculated from powder XRD analysis data using the Debye–Scherrer method [33]. Bare CaO particles were found to have a 108-nm size, and after Zn doping, it was found to decrease to 33–39 nm in range (Table 1). However, a change in Zn ion concentration and calcination temperature was not found to alter the particle size significantly.

Dynamic light scattering analysis (DLS) was performed for the measurement of the particle size distribution of CaO and 2-Zn/CaO-400 and showed that the average particle size of CaO and 2-Zn/CaO-400 was 115 nm and 35 nm, respectively (Figure 2a,b). The particle size and surface morphology of the prepared catalyst were analyzed through field emission scanning electron microscopy (FESEM), and the average particle size was observed in the range of 100–200 nm, with irregular surface morphology (Figure 2c). The same particles were analyzed using transmission electron microscopy (TEM) for clear observation of the particle size, and it was found that the 2-Zn/CaO-400 nanoparticles had an average particle size of ~30 nm with an oblate spherical shape (Figure 2d).

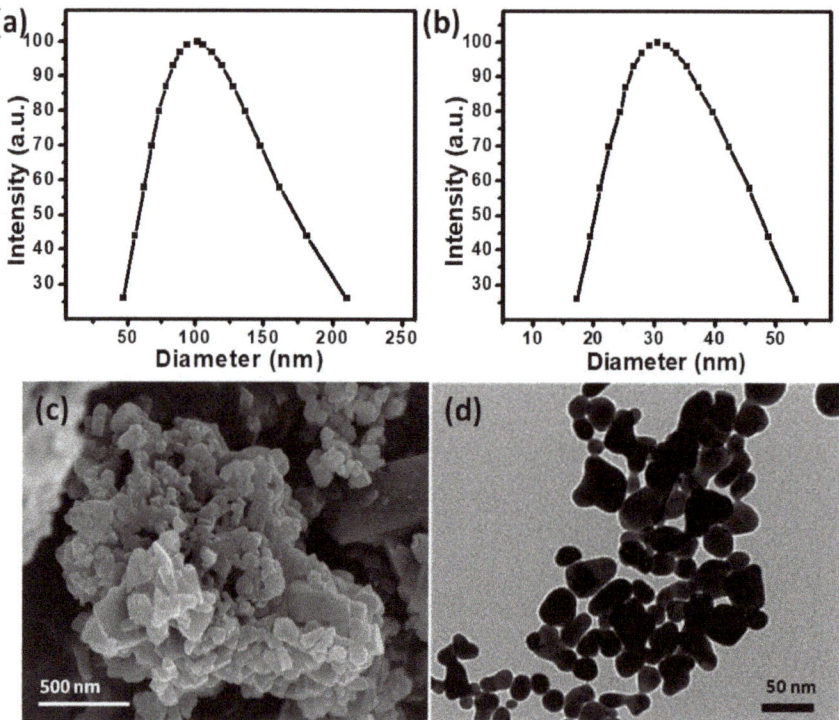

Figure 2. Particle size distributions of (**a**) CaO and (**b**) 2-Zn/CaO-400: (**c**) field emission scanning electron microscopy (FESEM) and (**d**) transmission electron microscopy (TEM) images of 2-Zn/CaO-400.

2.3. Aminolysis Reaction

The amidation of a variety of natural triglycerides, such as virgin soybean oil (VSO), waste soybean oil (WSO), jatropha oil (JO), animal fat (AF), Karanja oil (KO), and fatty acid methyl esters (FAMEs) derived from these oils, as well as methyl laurate (ML) with different molar concentrations of diethanolamine and different catalyst amounts, was performed at 90 °C. All amidation reactions for studying various reaction parameters were carried out with diethanolamine in the presence of 4 wt% of the 2-Zn/CaO-400 catalyst at 90 °C for 0.5 h (Scheme 1). The schematic for the FAMEs derivation from vegetable oils was performed with methanol (9:1 methanol/oil molar ratio) by using 5 wt% of the same catalyst at 65 °C (Scheme S1, Supplementary Materials).

The progress of the amidation reaction was monitored by taking out samples from the reaction mixture and analyzing them with FTIR and ^{1}H-NMR (Nuclear Magnetic Resonance) techniques. The catalyst nanoparticles were removed by simple centrifugation of the final reaction mixture at 7000 rpm, and the organic layer was then washed with distilled water and dried over sodium sulfate. The amide derivatives thus obtained were further analyzed by FTIR (Figure 3A) and ^{1}H-NMR (Figure 1B) analysis techniques. The final reaction product obtained from the methyl laurate amidation reaction was also characterized by mass spectrometry (Figure S4, Supporting Information) along with the ^{1}H-NMR and FTIR studies. A shifting of the ester carbonyl peak to the ester amide peak from 1739 cm^{-1} to

1617 cm^{-1} indicated the formation of fatty acid diethanolamine (FAD) (Figure 3A). The formation of a diethanolamide derivative was also supported by the presence an –OH group peak at 3406 cm^{-1}.

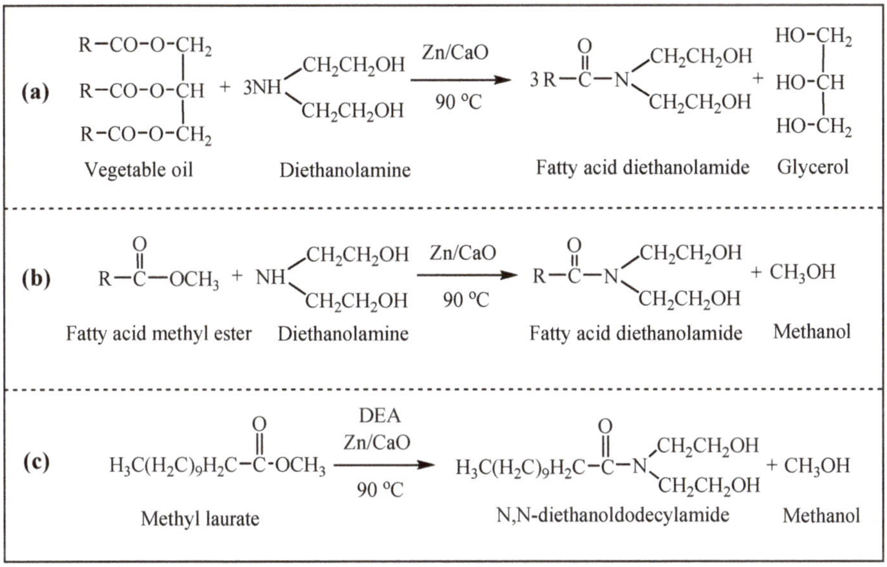

Scheme 1. (a) Waste soybean oil or jatropha oil; (b) vegetable oil-derived fatty acid methyl esters (FAMEs); and (c) methyl laurate-derived fatty acid diethanolamide preparation in the presence of a Zn/CaO solid catalyst.

On the other hand, in the ^{1}H-NMR spectrum, the appearance of a multiplet at 3.48 and 3.78 ppm due to -NCH$_2$- and -CH$_2$OH protons (Figure 3Bb) and the disappearance of characteristic glyceridic proton signals at 4.13 and 4.30 ppm (Figure 3Bi) supported the conversion of triglyceride to corresponding FAD. Furthermore, in the case of a FAD derivative of JO-derived FAMEs and methyl laurate, the disappearance of a methyl ester proton signal at 3.65 ppm (Figure 3Biii,v) and the appearance of amide proton signals at 3.48 and 3.78 ppm (Figure 3Bd,f) (corresponding to -NCH$_2$- and -CH$_2$OH protons) confirmed the formation of respective FAD.

2.3.1. Optimization of Different Parameters

The prepared nanocrystalline catalysts with different transition metals and different metal oxides were utilized for the amidation of natural triglycerides with diethanolamine. However, jatropha oil (JO) was selected for the optimization of parameters, as it has a high level of free fatty acid contents (8.2 wt % free fatty acids).

A series of transition metals was used for the doping in CaO to test their impact on the catalytic activity of CaO for amidation, and it was found that Mn/CaO, Fe/CaO, Co/CaO, Ni/CaO, Cu/CaO, Zn/CaO, and Cd/CaO showed 18%, 24%, 16%, 58%, 15%, 99%, and 25% FADs, which were yielded in 0.5 h at 90 °C. Bare CaO-400 was also tested and was found to have a 10% FAD yield (Figure 4a). Among all the prepared catalysts, Zn/CaO was found to be the most efficient, and it was selected for further optimization studies. Further, for the selection of a metal oxide as a base material, Zn was doped in CaO, MgO, and ZnO, as all of these metal oxides have been extensively reported to be efficient catalysts for different reactions. For the amidation of JO, Zn/CaO was found to be the most effective, with a 99% conversion yield for FADs, whereas Zn/MgO and Zn/ZnO also showed significant conversion rates, with 84% and 76% FAD yields, respectively (Figure 4b). The high surface area was the deciding factor for the catalytic activity.

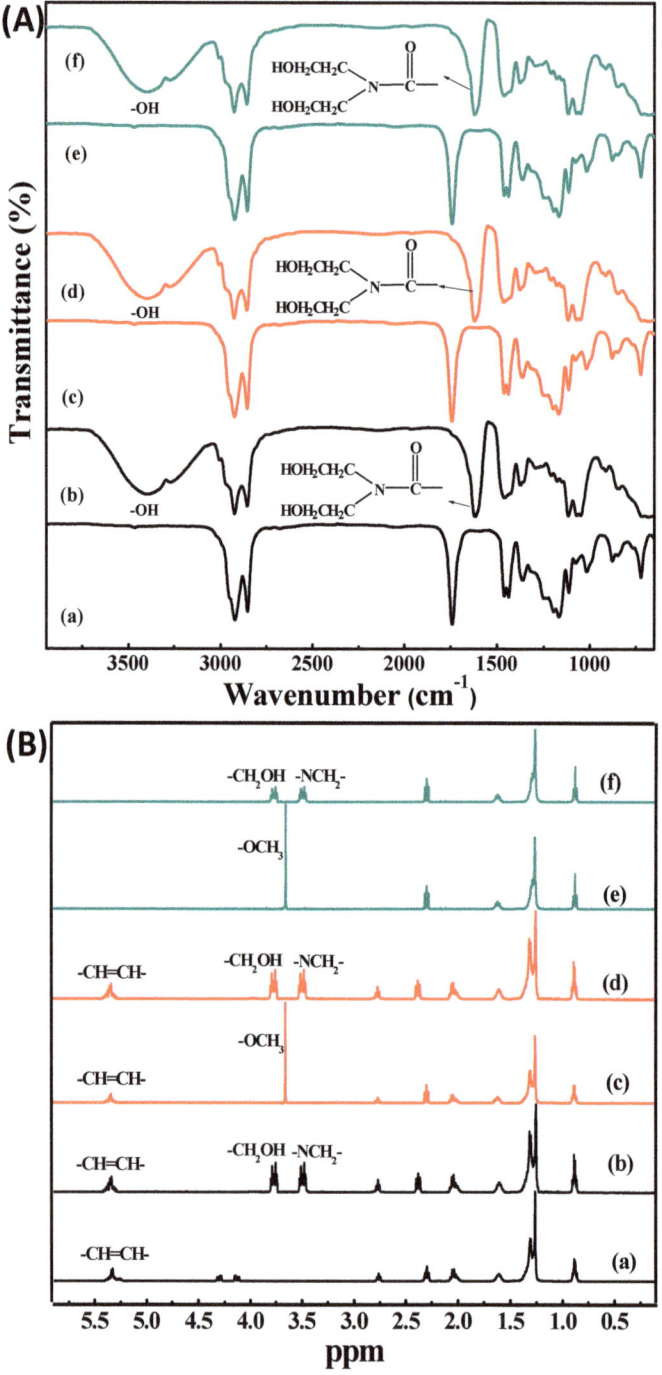

Figure 3. Comparative (**A**) FTIR spectrum and (**B**) ^{1}H-NMR spectrum of (**a**) waste cotton seed oil, (**b**) fatty acid amide of waste cotton seed oil, (**c**) waste soybean oil (WSO)-derived FAMEs, (**d**) a fatty acid amide of WSO-derived FAMEs, (**e**) methyl laurate, and (**f**) an amide derivative of methyl laurate.

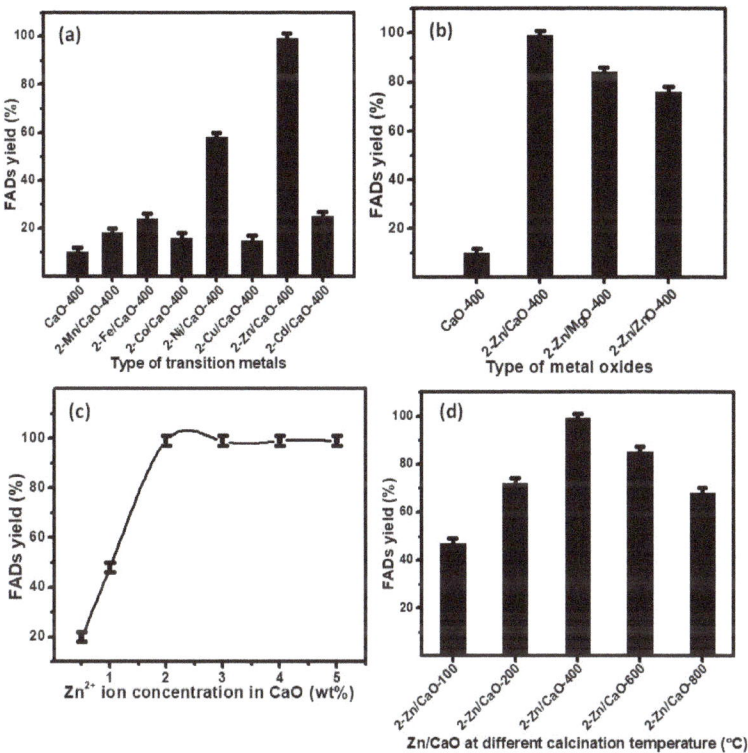

Figure 4. Effect of (**a**) different transition metal doping on CaO, (**b**) Zn ion doping on different metal oxides, (**c**) Zn ion concentration in CaO, and (**d**) calcination temperature in Zn/CaO on the amidation of jatropha oil (JO).

To optimize the concentration of Zn ions for a higher reaction rate, a series of aminolysis reactions was carried out by varying the Zn^{2+} concentration from 0.5 to 5 wt%. There was a significant increase in the FAD yield from 20% to 99%, as the Zn^{2+} concentration increased from 0.5 to 2 wt %, respectively. A further increase in Zn ion concentration had no effect on the reaction rate, and hence 2 wt% was chosen as an optimized amount of Zn for maximum efficiency (Figure 4c). The Zn/CaO nanoparticles prepared at different calcination temperatures were also used for the amidation of JO, and it was observed that the FAD yield was enhanced from 47% to 99% as the calcination temperature increased from 100 to 400 °C, respectively. Interestingly, calcination at a higher temperature such as 600 °C and 800 °C caused a reduction in the reaction rate as the FAD yield lowered to 85% and 68%, respectively. This could have been due to the fact that high temperatures caused the sintering of particles, which in turn decreased the surface area and basic strength (Table 1).

Different catalyst amounts from 1 to 10 wt % and a range of reaction temperatures from 30 to 130 °C were tried to figure out the optimized catalyst amount and reaction temperature. The FAD yield was increased from 25% to 99% when the catalyst amount was enhanced from 1 to 4 wt%. A further increase in the catalyst amount had no impact on the reaction rate (Figure 5a). Similarly, when the reaction temperature was increased from 30 °C to 90 °C, the FAD yield increased significantly from 24% to 99%, respectively. A higher reaction temperature did not show any effect on the reaction rate (Figure 5b). The optimization of the reaction temperature parameter was carried out by taking out samples at 5-, 15-, and 30-min time intervals and analyzing them through FTIR and ^{1}H-NMR analysis.

After 90 °C, there was no change in the reaction rate, and hence 90 °C was used as the optimized reaction temperature.

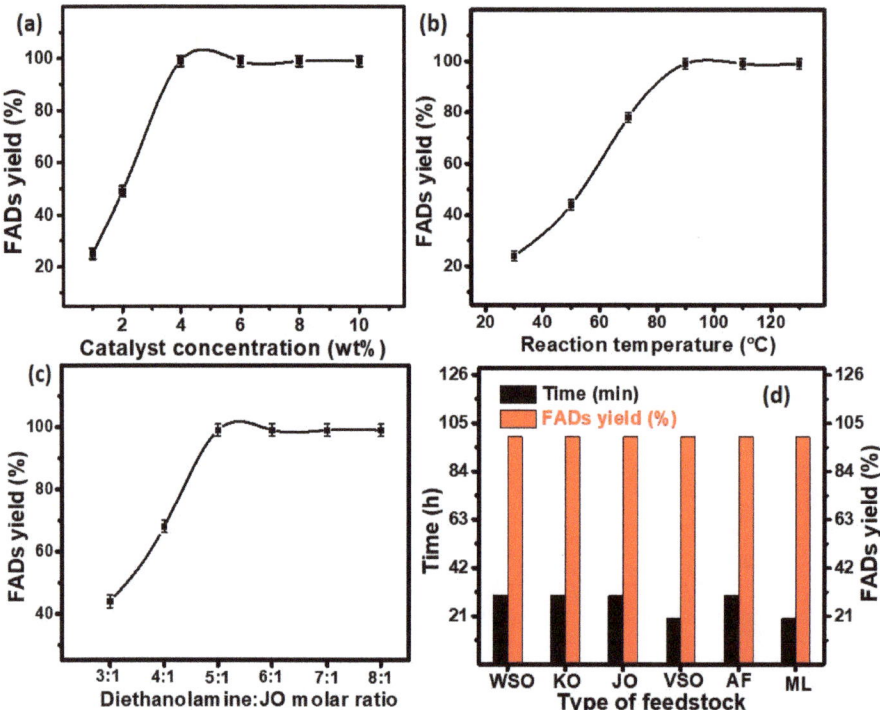

Figure 5. Effect of (**a**) catalyst concentration, (**b**) reaction temperature, and (**c**) the diethanolamine (DEA)/JO molar ratio on the complete aminolysis of used cotton seed oil (reaction time = 0.5 h). (**d**) The effect of different feedstock on the time required and the fatty acid diethanolamine (FAD) yield for the amidation reaction. Reaction conditions: diethanolamine/feedstock = 5:1 (m/m), catalyst amount = 4 wt % of feedstock, temperature = 90 °C.

Six different molar ratios of diethanolamine and JO were used to optimize the diethanolamine (DEA) amount for maximum conversion in the minimum time. As the molar ratio increased from 3:1 to 5:1, the FAD yield also increased from 44% to 99%, respectively, in 0.5 h (Figure 5c). A diethanolamine/JO molar ratio of 5:1, a 4 wt% catalyst amount, and a 90 °C reaction temperature were the final optimum reaction conditions for the complete conversion of JO to fatty acid diethanolamides in the minimum possible time (0.5 h). However, the Zn/CaO nanospheroids were found to convert JO to FADs completely, with a 3:1 diethanolamine/JO molar ratio and a 1 wt% catalyst amount at room temperature (35 °C), but the reaction time increased to 4 h.

A variety of triglycerides, which included natural triglycerides as well as methyl laurate, were tested with an amidation reaction to check the efficiency of the prepared Zn/CaO nanospheroids. Zn/CaO was found to convert all of the triglycerides to FADs, where it took 30 min to complete the reaction in the case of JO, KO, WSO, and AF; and it took only 20 min for VSO and ML. The low free fatty acid (FFA) content was the reason for the lower reaction time for VSO (FFAs = 0.2%) and ML, whereas Zn/CaO was found to be highly efficient for high FFAs containing feedstock, viz. AF (1.4), WSO (2.1), KO (4.4), and JO (8.2). The high FFAs (free fatty acids) caused the partial deactivation of the catalytic sites of Zn/CaO.

To examine the reusability of Zn/CaO nanospheroids, after being recovered from the final reaction mixture through centrifugation, Zn/CaO was washed with hexane and dried at 400 °C. The recovered catalyst was tested for the six catalytic runs under the same reaction conditions and regeneration technique. The recycled and regenerated catalyst was also found to complete (>99 %, m/m) the amidation of JO, though it required 35 min for the second catalytic recycle and 70 min for the sixth catalytic run (Figure S2, Supplementary Materials). The partial loss of catalytic activity could have been due to the loss of Zn/CaO particles during successive centrifugation and the partial leaching of active species.

2.3.2. Kinetic Study

To calculate the reaction rate, the samples from the Zn/CaO-mediated amidation of JO were withdrawn regularly every 10 min and then centrifuged to remove the catalyst; the rotary evaporated was used to remove any remaining amount of diethanolamine; and the samples were finally subjected to FTIR studies to analyze the fatty acid diethanolamide yield. As the reaction progressed, the intensity of the ester carbonyl peak at 1739 cm^{-1} decreased regularly, and the intensity of the fatty acid amide carbonyl band at 1617 cm^{-1} increased due to the conversion of triglyceride to corresponding amide (Figure S3, Supplementary Materials). The amidation of JO in the presence of Zn/CaO was found to follow pseudo-first-order kinetics, and the reaction rate constant (k) could be given as

$$k = -\ln\{(1 - X_{FAD})/t\}, \quad (1)$$

where X_{FAD} is the fatty acid diethanolamide yield at time t.

The kinetics of the Zn/CaO-catalyzed amidation of JO were studied at a 5:1 diethanolamine/JO molar ratio in the temperature range of 30–90 °C. The linear nature of the $-\ln(1 - X_{FAD})$ versus t (time) plots (Figure 6a) supported the idea that the reaction followed pseudo-first-order kinetics. The rate constant values were calculated as 0.171, 0.043, 0.019, and 0.011 min^{-1} at 90, 70, 50, and 30 °C, respectively.

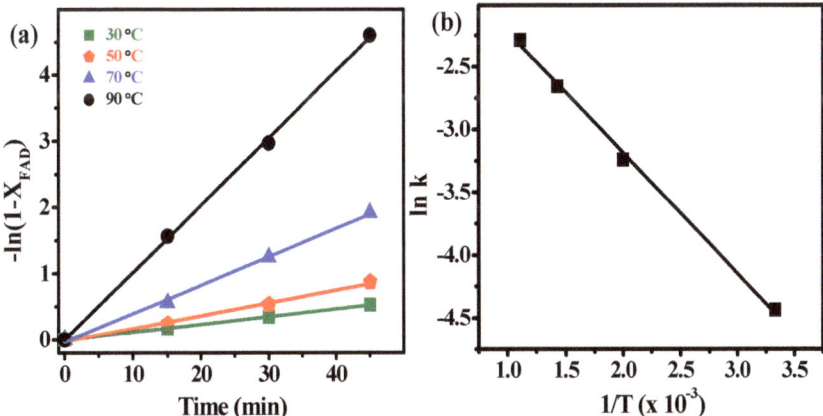

Figure 6. (a) Plot of $-\ln(1 - X_{FAD})$ versus kt at different reaction temperatures for the 2-Zn/CaO-400-catalyzed aminolysis of JO. (b) The Arrhenius equation curve for the aminolysis of JO. Reaction conditions: diethanolamine/JO = 5:1 (m/m), catalyst amount = 4 wt%, temperature = 90 °C.

To calculate the activation energy (E_a) and the pre-exponential factor (A) for the amidation reaction, an Arrhenius model was used, as given in Equation (2):

$$\ln k = -E_a/RT + \ln A, \quad (2)$$

where R = the gas constant (8.31 J K^{-1} mol^{-1}), and T is the reaction process temperature in kelvin.

The values of E_a and A calculated from the graph between $1/T$ and $\ln k$ were found to be 47.8 kJ mol^{-1} and 4.75×10^8 min^{-1}, respectively (Figure 6b). The resulting activation energy (47.8 kJ mol^{-1}) for the amidation of JO was observed within the range for the heterogeneous catalysis (33–84 kJ mol^{-1}).

3. Materials and Methods

All chemical reagents were purchased from Sigma-Aldrich (St. Louis, MO, USA) and were used without further purification. Transition metal-doped CaO, MgO, and ZnO catalysts were synthesized in nanoparticle form by following a modified incipient-wetness impregnation method.

3.1. General

Nitrates of Mn, Fe, Co, Cu, Ni, Zn, and Cd; zinc acetate; CaO; methanol (99.8%); methyl laurate (99%); and diethanolamine (99%) were purchased from Sigma-Aldrich, USA, and were used as such without further purification. Waste soybean oil was collected from local restaurants located in Jeonju. Scanning electron microscopy (SEM) was performed on a JEOL JSM 6510LV (JEOL Ltd., Akishima, Tokyo, Japan) to collect the SEM images, and transmission electron microscopy (TEM) was performed on a HITACHI 7500 to record TEM images. Scanning microscopy-energy-dispersive X-ray analysis (SEM-EDX) was performed for the qualitative analysis of the catalysts. Fourier-transform–nuclear magnetic resonance (FT–NMR) spectra of vegetable oils, fatty acid methyl esters (FAMEs), and fatty acid amides were recorded on a Bruker Avance-II (400 MHz) spectrophotometer (Bruker Corporation, Billerica, MA, USA). The presence of amide functional groups was supported with the help of FTIR spectra recorded on a Thermo Scientific Nicolet iS10 spectrometer (Thermo Fisher Scientific, Waltham, MA, USA). Mass spectra of methyl laurate and the amide derivative of methyl laurate were recorded on a Waters Micromass Q-ToF Micro mass spectrophotometer (Waters Corporation, Milford, MA, USA) equipped with electrospray ionization (ESI) and atmospheric pressure chemical ionization (APcI) sources with a mass range of 4000 amu in quadruple and 20,000 amu in ToF (Time-Of-Flight). The free fatty acids (FFA), saponification, iodine values, and moisture content of the virgin soybean oil (VSO), animal fat (AF), waste soybean oil (WSO), Karanja oil (KO), and jatropha oil (JO) were determined by following methods reported in the literature [34] (Table S1).

3.2. Experimental Section

3.2.1. Preparation of Catalyst

Transition metal-doped CaO, MgO, and ZnO catalysts were synthesized in nanoparticle form by following a modified incipient-wetness impregnation method [35]. In a typical preparation, metal oxide (CaO or MgO or ZnO) slurry (10 mg/40 mL ethanol) was sonicated for 1 h, and then 10 mL of transition metal (Fe or Co or Cu or Zn or Cd) solution in ethanol of a desired concentration was added dropwise into the metal oxide slurry and stirred moderately for 3 h at 25 °C. The resulting mixture was then dried and calcined at varying temperatures from 100 to 800 °C for 12 h. The solid then obtained was characterized by BET surface area measurement ((ASAP 2010, Micromeritics, USA)), a Hammett indicator test, powder-XRD, DLS (Brookhaven Instruments Corporation, Austin, TX, United States), FESEM, and TEM.

3.2.2. Aminolysis Reaction

Aminolysis reactions of a variety of feedstock (viz. WSO, AF, VSO, KO, and JO), FAMEs derived from them, and methyl laurate (ML) with varying molar concentrations of diethanolamine and catalyst amounts were performed at 90° C. All reactions were carried out until the completion of the reaction by varying one parameter at a time in order to establish the reaction conditions required for complete aminolysis in the minimum possible time. The amide derivative produced during the aminolysis reaction was characterized by FTIR and proton NMR and quantified by FTIR spectroscopy. The amide derivative of methyl laurate was additionally analyzed using a mass spectroscopic technique. FAMEs prepared through the transesterification of a variety of vegetable oils and mutton fat were characterized and quantified by proton NMR spectroscopy.

Fatty acid amide derivative of WSO: yield > 99%. FTIR (cm^{-1}): 3406 (ν_{OH}), 1617 ($\nu_{C=O}$); ^{1}H-NMR (CDCl$_3$, δ ppm): 5.3 (m, -CH=CH-), 3.77 (m, -CH$_2$OH), 3.46 (m, -NCH$_2$-), 2.7 (m, -CH=CH-CH$_2$-CH=CH-), 2.31 (m, -CH$_2$-CO-), 2.0 (m, -CH$_2$-(CH$_2$)$_n$-CH-), 1.6-1.25 (m, -(CH$_2$)$_n$-), 0.95 (m, -CH=CH-CH$_3$), 0.87 (m, -CH$_2$-CH$_3$).

Fatty acid amide derivative of fatty acid methyl ester of WSO: yield > 99%. FTIR (cm^{-1}): 3406 (ν_{OH}), 1617 ($\nu_{C=O}$); ^{1}H-NMR (CDCl$_3$, δ ppm): 5.3 (m, -CH=CH-), 3.78 (m, -CH$_2$OH), 3.48 (m, -NCH$_2$-), 2.7 (m, -CH=CH-CH$_2$-CH=CH-), 2.3 (m, -CH$_2$-CO-), 2.0 (m, -CH$_2$-(CH$_2$)$_n$-), 1.6-1.25 (m, -(CH$_2$)$_n$-), 0.95 (m, -CH=CH-CH$_3$), 0.87 (m, -CH$_2$-CH$_3$).

Fatty acid amide derivative of animal fat: yield > 99%. FTIR (cm^{-1}): 3406 (ν_{OH}), 1617 ($\nu_{C=O}$); ^{1}H-NMR (CDCl$_3$, δ ppm): 5.34 (m, -CH=CH-), 3.77 (m, -CH$_2$OH), 3.46 (m, -NCH$_2$-), 2.3 (m, -CH$_2$-CO-), 2.0 (m, -CH$_2$-(CH$_2$)$_n$-), 1.6-1.25 (m, -(CH$_2$)$_n$-), 0.87 (m, -CH$_2$-CH$_3$).

Fatty acid amide derivative of fatty acid methyl ester of animal fat: yield > 99%. FTIR (cm^{-1}): 3406 (ν_{OH}), 1617 ($\nu_{C=O}$); ^{1}H-NMR (CDCl$_3$, δ ppm): 5.3 (m, -CH=CH-), 3.75 (m, -CH$_2$OH), 3.47 (m, -NCH$_2$-), 2.3 (m, -CH$_2$-CO-), 2.0 (m, -CH$_2$-(CH$_2$)$_n$-), 1.6-1.25 (m, -(CH$_2$)$_n$-), 0.87 (m, -CH$_2$-CH$_3$).

Fatty acid amide derivative of Karanja oil: yield > 99%. FTIR (cm^{-1}): 3406 (ν_{OH}), 1617 ($\nu_{C=O}$); ^{1}H-NMR (CDCl$_3$, δ ppm): 5.3 (m, -CH=CH-), 3.77 (m, -CH$_2$OH), 3.46 (m, -NCH$_2$-), 2.7 (m, -CH=CH-CH$_2$-CH=CH-), 2.31 (m, -CH$_2$-CO-), 2.0 (m, -CH$_2$-(CH$_2$)$_n$-CH-), 1.6-1.25 (m, -(CH$_2$)$_n$-), 0.95 (m, -CH=CH-CH$_3$)0.87 (m, -CH$_2$-CH$_3$).

Fatty acid amide derivative of fatty acid methyl ester of Karanja oil: yield > 99%. FTIR (cm^{-1}): 3406 (ν_{OH}), 1617 ($\nu_{C=O}$); ^{1}H-NMR (CDCl$_3$, δ ppm): 5.3 (m, -CH=CH-), 3.78 (m, -CH$_2$OH), 3.48 (m, -NCH$_2$-), 2.7 (m, -CH=CH-CH$_2$-CH=CH-), 2.3 (m, -CH$_2$-CO-), 2.0 (m, -CH$_2$-(CH$_2$)$_n$-), 1.6-1.25 (m, -(CH$_2$)$_n$-), 0.95 (m, -CH=CH-CH$_3$), 0.87 (m, -CH$_2$-CH$_3$).

Fatty acid amide derivative of jatropha oil: yield > 99%. FTIR (cm^{-1}): 3406 (ν_{OH}), 1617 ($\nu_{C=O}$); ^{1}H-NMR (CDCl$_3$, δ ppm): 5.3 (m, -CH=CH-), 3.77 (m, -CH$_2$OH), 3.46 (m, -NCH$_2$-), 2.7 (m, -CH=CH-CH$_2$-CH=CH-), 2.31 (m, -CH$_2$-CO-), 2.0 (m, -CH$_2$-(CH$_2$)$_n$-CH-), 1.6-1.25 (m, -(CH$_2$)$_n$-), 0.95 (m, -CH=CH-CH$_3$)0.87 (m, -CH$_2$-CH$_3$).

Fatty acid amide derivative of fatty acid methyl ester of jatropha oil: yield > 99%. FTIR (cm^{-1}): 3406 (ν_{OH}), 1617 ($\nu_{C=O}$); ^{1}H-NMR (CDCl$_3$, δ ppm): 5.3 (m, -CH=CH-), 3.78 (m, -CH$_2$OH), 3.48 (m, -NCH$_2$-), 2.7 (m, -CH=CH-CH$_2$-CH=CH-), 2.3 (m, -CH$_2$-CO-), 2.0 (m, -CH$_2$-(CH$_2$)$_n$-), 1.6-1.25 (m, -(CH$_2$)$_n$-), 0.95 (m, -CH=CH-CH$_3$), 0.87 (m, -CH$_2$-CH$_3$).

Fatty acid amide derivative of methyl laurate: yield > 99%. FTIR (cm^{-1}): 3406 (ν_{OH}), 1617 ($\nu_{amide-C=O}$); ^{1}H-NMR (CDCl$_3$, δ ppm): 3.8 (m, -CH$_2$OH), 3.5 (m, -NCH$_2$-), 2.3 (m, -CH$_2$-CO-), 1.6-1.25 (m, -(CH$_2$)$_n$-), 0.87 (m, -CH$_2$-CH$_3$); EI-MS (electron ionization mass spectrometer) (m/z) (intensity (%), fragment): 287.2 (3, M), 270.3 (100, M-H$_2$O), 227.22 (10, M-CH$_3$(CH$_2$)$_3$), 175.2 (10, M-CH$_3$(CH$_2$)$_7$), 132.2 (12, M-CH$_3$(CH$_2$)$_{10}$), 114.2 (5, M-CH$_3$(CH$_2$)$_{10}$OH).

3.2.3. Synthesis of Fatty Acid Methyl Esters

In order to yield FAMEs for aminolysis, the same catalyst was also utilized for the transesterification of a variety of triglycerides (waste cotton seed oil, jatropha oil, and animal fat) with methanol, as shown in Scheme 2. All transesterification reactions were carried out in a refluxing unit consisting of a two-necked round-bottom flask (100 mL) fitted with a water-cooled condenser, an oil bath, and a magnetic stirrer. In a typical transesterification process, vegetable oil or animal fat (triglyceride) was mixed with methanol (in a 9:1 molar ratio with respect to triglyceride) with 5 wt % of Zn/CaO and heated at 65 °C until the completion of the reaction.

The catalyst was removed from the reaction mixture by centrifugation (at 8000 rpm) after the completion of the reaction, the rotary evaporated was used to recover the excess methanol, and then everything was kept in a separate funnel for 12 h to separate the FAMEs from the glycerol. The FAMEs were thus obtained, further analyzed, and quantified through methods reported in the literature [32] using ^{1}H-NMR data.

The FAMEs were synthesized through the transesterification of a variety of triglycerides with methanol (9:1 methanol/oil molar ratio) by using 5 wt % of the same catalyst at 65 °C (Scheme 2).

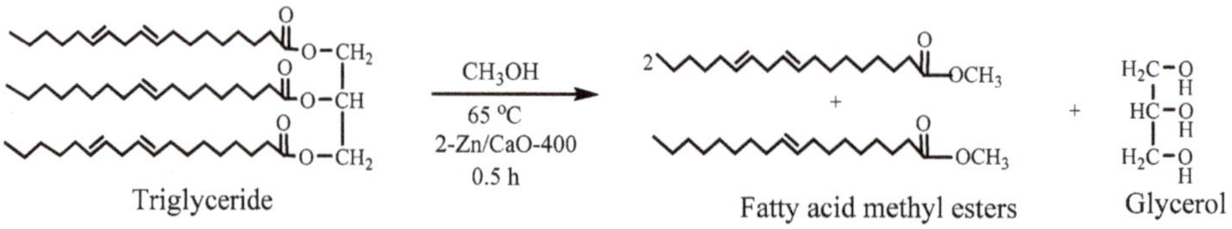

Scheme 2. Transesterification of triglycerides using the 2-Zn/CaO-400 nanocatalyst.

4. Conclusions

Zn-doped CaO nanospheroids were prepared by utilizing a simple method and were used as a heterogeneous catalyst for the amidation of a variety of natural triglycerides. Zn/CaO-400 nanospheroids were found to have a ~30 nm size, more than an 18.4 in basic strength, and 16.87 m^2/g of surface area. All of these factors made it a highly efficient catalyst in amidation, as it took only a 4 wt % catalyst amount for the complete conversion of high FFAs containing JO triglyceride, with a 5:1 molar ratio of DEA/JO at 90 °C. The Zn/CaO nanocatalyst was also found to be efficient at room temperature (35 °C) for amidation reactions of JO. The Zn/CaO-400 nanospheroids were found to be most efficient when compared to other transition metal-doped CaO nanomaterials (Mn, Fe, Co, Ni, Cu, and Cd). In addition, CaO was found to be the most effective support material compared to MgO and ZnO. The presence of both CaO and Ca(OH)$_2$ phases in Zn/CaO-400 made it the most efficient heterogeneous catalyst for the aminolysis of various triglycerides and FAMEs derived from them. Zn/CaO-400 was also found to have excellent recyclability, as it was used for six consecutive reaction cycles without losing much catalytic activity. The high reaction rate of 0.171 min^{-1} with 47.8 kJ mol^{-1} of activation energy and 4.75 X 10^8 pre-exponential factors made it a highly efficient heterogeneous catalyst.

Supplementary Materials: The following are available online at http://www.mdpi.com/2073-4344/9/9/774/s1 (FTIR and NMR data), Scheme S1: Transesterification of triglycerides using 2-Zn/CaO-400 nanocatalyst, Figure S1: Comparative XRD spectrum of (a) CaO with Zn/CaO calcined at different temperatures and (b) Zn/CaO with different doping percentages of zinc ion; Figure S2: Recyclability studies of the catalyst in the aminolysis of JO; Figure S3: Progress of aminolysis using FTIR; Figure S4: Mass spectra of fatty acid amide derived from methyl laurate; Table S1: The chemical analysis of vegetable oils.

Author Contributions: D.K. was involved in the discovery and development of the photocatalyst platform. D.K., C.H.P., and C.S.K. analyzed the data and contributed to designing the experiments. D.K., C.H.P., and C.S.K. gave approval for the final version of the manuscript.

Funding: This work was supported by the National Research Foundation of Korea (NRF-2016R1D1A1B03934226 and 2019R1A2B5B02070092) Project. We are thankful to CURF, Chonbuk National University, South Korea, for the FESEM, XRD, FTIR, TEM, NMR, and mass spectroscopic analysis.

Conflicts of Interest: The authors declare no conflicts of interest.

References

1. Kordulis, C.; Bourikas, K.; Gousi, M.; Kordouli, E.; Lycourghiotis, A. Development of nickel based catalysts for the transformation of natural triglycerides and related compounds into green diesel: A critical review. *Appl. Catal. B Environ.* **2016**, *181*, 156–196. [CrossRef]
2. Peng, B.; Yao, Y.; Zhao, C.; Lercher, J.A. Towards quantitative conversion of microalgae oil to diesel-range alkanes with bifunctional catalysts. *Angew. Chem. Int. Ed.* **2012**, *51*, 2072–2075. [CrossRef] [PubMed]
3. Sheldon, R.A. Green chemistry, catalysis and valorization of waste biomass. *J. Mol. Catal. A Chem.* **2016**, *422*, 3–12. [CrossRef]

4. Besson, M.; Gallezot, P.; Pinel, C. Conversion of biomass into chemicals over metal catalysts. *Chem. Rev.* **2014**, *114*, 1827–1870. [CrossRef] [PubMed]
5. Pelckmans, M.; Renders, T.; Van de Vyver, S.; Sels, B.F. Bio-based amines through sustainable heterogeneous catalysis. *Green Chem.* **2017**, *19*, 5303–5331. [CrossRef]
6. Esposito, D.; Antonietti, M. Redefining biorefinery: The search for unconventional building blocks for materials. *Chem. Soc. Rev.* **2015**, *44*, 5821–5835. [CrossRef]
7. Luterbacher, J.S.; Martin Alonso, D.; Dumesic, J.A. Targeted chemical upgrading of lignocellulosic biomass to platform molecules. *Green Chem.* **2014**, *16*, 4816–4838. [CrossRef]
8. Tuck, C.O.; Pérez, E.; Horváth, I.T.; Sheldon, R.A.; Poliakoff, M. Valorization of biomass: Deriving more value from waste. *Science* **2012**, *337*, 695–699. [CrossRef]
9. Foley, P.; Kermanshahi pour, A.; Beach, E.S.; Zimmerman, J.B. Derivation and synthesis of renewable surfactants. *Chem. Soc. Rev.* **2012**, *41*, 1499–1518. [CrossRef]
10. Biermann, U.; Friedt, W.; Lang, S.; Lühs, W.; Machmüller, G.; Metzger, J.O.; Rüsch gen. Klaas, M.; Schäfer, H.J.; Schneider, M.P. New syntheses with oils and fats as renewable raw materials for the chemical industry. *Angew. Chem. Int. Ed.* **2000**, *39*, 2206–2224. [CrossRef]
11. Jamil, M.A.R.; Siddiki, S.M.A.H.; Touchy, A.S.; Rashed, M.N.; Poly, S.S.; Jing, Y.; Ting, K.W.; Toyao, T.; Maeno, Z.; Shimizu, K.-i. Selective transformations of triglycerides into fatty amines, amides, and nitriles by using heterogeneous catalysis. *ChemSusChem* **2019**, *12*, 3115–3125. [CrossRef] [PubMed]
12. Serdari, A.; Lois, E.; Stournas, S. Tertiary fatty amides as diesel fuel substitutes. *Int. J. Energy Res.* **2000**, *24*, 455–466. [CrossRef]
13. Awasthi, N.P.; Singh, R.P. Microwave-assisted facile and convenient synthesis of fatty acid amide (erucamide): Chemical-catalyzed rapid method. *Eur. J. Lipid Sci. Technol.* **2009**, *111*, 202–206. [CrossRef]
14. Rawlins, J.; Pramanik, M.; Mendon, S. Synthesis and characterization of soyamide ferulate. *J. Am. Oil Chem. Soc.* **2008**, *85*, 783–789. [CrossRef]
15. Al-Mulla, E.A.J.; Yunus, W.M.Z.W.; Ibrahim, N.A.B.; Rahman, M.Z.A. Synthesis and characterization of n,n′-carbonyl difatty amides from palm oil. *J. Oleo Sci.* **2009**, *58*, 467–471. [CrossRef] [PubMed]
16. Yapa Mudiyanselage, A.; Yao, H.; Viamajala, S.; Varanasi, S.; Yamamoto, K. Efficient production of alkanolamides from microalgae. *Ind. Eng. Chem. Res.* **2015**, *54*, 4060–4065. [CrossRef]
17. Bundesmann, M.W.; Coffey, S.B.; Wright, S.W. Amidation of esters assisted by $Mg(OCH_3)_2$ or $CaCl_2$. *Tetrahedron Lett.* **2010**, *51*, 3879–3882. [CrossRef]
18. Litjens, M.J.J.; Sha, M.; Straathof, A.J.J.; Jongejan, J.A.; Heijnen, J.J. Competitive lipase-catalyzed ester hydrolysis and ammoniolysis in organic solvents; equilibrium model of a solid–liquid–vapor system. *Biotechnol. Bioeng.* **1999**, *65*, 347–356. [CrossRef]
19. Wang, X.; Wang, X.; Wang, T. Synthesis of oleoylethanolamide using lipase. *J. Agric. Food Chem.* **2012**, *60*, 451–457. [CrossRef]
20. Karis, N.D.; Loughlin, W.A.; Jenkins, I.D. A facile and efficient method for the synthesis of novel pyridone analogues by aminolysis of an ester under solvent-free conditions. *Tetrahedron* **2007**, *63*, 12303–12309. [CrossRef]
21. Sabot, C.; Kumar, K.A.; Meunier, S.; Mioskowski, C. A convenient aminolysis of esters catalyzed by 1,5,7-triazabicyclo[4.4.0]dec-5-ene (tbd) under solvent-free conditions. *Tetrahedron Lett.* **2007**, *48*, 3863–3866. [CrossRef]
22. Ishii, Y.; Takeno, M.; Kawasaki, Y.; Muromachi, A.; Nishiyama, Y.; Sakaguchi, S. Acylation of alcohols and amines with vinyl acetates catalyzed by cp*2sm(thf)2. *J. Org. Chem.* **1996**, *61*, 3088–3092. [CrossRef] [PubMed]
23. Chisholm, M.H.; Delbridge, E.E.; Gallucci, J.C. Modeling the catalyst resting state in aryl tin(iv) polymerizations of lactide and estimating the relative rates of transamidation, transesterification and chain transfer. *New J. Chem.* **2004**, *28*, 145–152. [CrossRef]
24. Kangani, C.O.; Kelley, D.E. One pot direct synthesis of amides or oxazolines from carboxylic acids using deoxo-fluor reagent. *Tetrahedron Lett.* **2005**, *46*, 8917–8920. [CrossRef] [PubMed]
25. Kumar, K.N.; Sreeramamurthy, K.; Palle, S.; Mukkanti, K.; Das, P. Dithiocarbamate and dbu-promoted amide bond formation under microwave condition. *Tetrahedron Lett.* **2010**, *51*, 899–902. [CrossRef]
26. Pan, J.; Devarie-Baez, N.O.; Xian, M. Facile amide formation via s-nitrosothioacids. *Org. Lett.* **2011**, *13*, 1092–1094. [CrossRef] [PubMed]

27. Sasaki, K.; Crich, D. Facile amide bond formation from carboxylic acids and isocyanates. *Org. Lett.* **2011**, *13*, 2256–2259. [CrossRef]
28. Sathishkumar, M.; Shanmugavelan, P.; Nagarajan, S.; Maheswari, M.; Dinesh, M.; Ponnuswamy, A. Solvent-free protocol for amide bond formation via trapping of nascent phosphazenes with carboxylic acids. *Tetrahedron Lett.* **2011**, *52*, 2830–2833. [CrossRef]
29. Hayyan, A.; Hashim, M.A.; Hayyan, M. Application of a novel catalyst in the esterification of mixed industrial palm oil for biodiesel production. *BioEnergy Res.* **2014**, *8*, 459–463. [CrossRef]
30. Kumar, D.; Kim, S.M.; Ali, A. One step synthesis of fatty acid diethanolamides and methyl esters from triglycerides using sodium doped calcium hydroxide as a nanocrystalline heterogeneous catalyst. *New J. Chem.* **2015**, *39*, 7097–7104. [CrossRef]
31. Lee, A.F.; Bennett, J.A.; Manayil, J.C.; Wilson, K. Heterogeneous catalysis for sustainable biodiesel production via esterification and transesterification. *Chem. Soc. Rev.* **2014**, *43*, 7887–7916. [CrossRef] [PubMed]
32. Kumar, D.; Ali, A. Transesterification of low-quality triglycerides over a Zn/CaO heterogeneous catalyst: Kinetics and reusability studies. *Energy Fuels* **2013**, *27*, 3758–3768. [CrossRef]
33. Kumar, D.; Kim, S.M.; Ali, A. Solvent-free one step aminolysis and alcoholysis of low-quality triglycerides using sodium modified cao nanoparticles as a solid catalyst. *RSC Adv.* **2016**, *6*, 55800–55808. [CrossRef]
34. Kumar, D.; Ali, A. Direct synthesis of fatty acid alkanolamides and fatty acid alkyl esters from high free fatty acid containing triglycerides as lubricity improvers using heterogeneous catalyst. *Fuel* **2015**, *159*, 845–853. [CrossRef]
35. Degirmenbasi, N.; Boz, N.; Kalyon, D.M. Biofuel production via transesterification using sepiolite-supported alkaline catalysts. *Appl. Catal. B Environ.* **2014**, *150–151*, 147–156. [CrossRef]

© 2019 by the authors. Licensee MDPI, Basel, Switzerland. This article is an open access article distributed under the terms and conditions of the Creative Commons Attribution (CC BY) license (http://creativecommons.org/licenses/by/4.0/).

Article

Biolubricants from Rapeseed and Castor Oil Transesterification by Using Titanium Isopropoxide as a Catalyst: Production and Characterization

José María Encinar [1], Sergio Nogales-Delgado [1],*, Nuria Sánchez [1] and Juan Félix González [2]

[1] Department of Chemical Engineering and Physical-Chemistry, University of Extremadura, 06006 Badajoz, Spain; jencinar@unex.es (J.M.E.); nuriass@unex.es (N.S.)
[2] Department of Applied Physics, University of Extremadura, 06006 Badajoz, Spain; jfelixgg@unex.es
* Correspondence: senogalesd@unex.es

Received: 4 March 2020; Accepted: 26 March 2020; Published: 29 March 2020

Abstract: The transesterification of rapeseed and castor oil methyl esters with different alcohols (2-ethyl-1-hexanol, 1-heptanol and 4-methyl-2-pentanol) and titanium isopropoxide as a catalyst, to produce biolubricants, was carried out. Parameters such as temperature, alcohol/methyl ester molar ratio, and catalyst concentration were studied to optimize the process. The reaction evolution was monitored with the decrease in FAME concentration by gas chromatography. In general, the reaction was almost complete in two hours, obtaining over 93% conversions. All the variables studied influenced on the reaction yields. Once the optimum conditions for the maximum conversion and minimum costs were selected, a characterization of the biolubricants obtained, along with the study of the influence of the kind of alcohol used, was carried out. The biolubricants had some properties that were better than mineral lubricants (flash points between 222 and 271 °C), needing the use of additives when they do not comply with the standards (low viscosity for rapeseed biolubricant, for instance). There was a clear influence of fatty acids of raw materials (oleic and ricinoleic acids as majority fatty acids in rapeseed and castor oil, respectively) and the structure of the alcohol used on the final features of the biolubricants.

Keywords: fatty acid methyl esters; 2-ethyl-1-hexanol; 1-heptanol; 4-methyl-2-pentanol; viscosity; flash and combustion points; methyl oleate; methyl ricinoleate

1. Introduction

The main objective of lubrication is the protection of surfaces in close proximity and moving relative to each other. This protection is carried out by interposing a substance (lubricant) between the abovementioned surfaces. This way, the use of lubricants is important to reduce wear, avoiding corrosion and reducing oxidation in surfaces [1]. Consequently, lubricants are essential in industrial processes. However, petro-based lubricants, which are extensively used, contribute to many environmental and sustainability problems [2]. In addition, the scarcity of crude oil reserves provokes the increase in prices, which makes sustainability and development of poor areas more difficult. This way, the search for alternative, sustainable, and biodegradable products is becoming more and more necessary. Thus, biolubricants (mainly derived from vegetable oils) are gaining in importance, as they can be used in many applications, being environmental friendly (that is, biodegradable, with low ecotoxicity and not contributing to volatile organic chemicals) [1–3]. Moreover, the fatty acid component of biolubricants can form layers and a stable film on the surfaces of rubbing zones, avoiding the contact between surfaces and therefore, corrosion and wear [4]. Many raw materials (such as rapeseed or castor) could contribute to the sustainable development, being an alternative to rotate typical crops in many regions and being an alternative to petroleum fuels or products and, consequently, a key for the development of local economies [5–8]. Indeed, they have been used as raw materials for biolubricant production from different ways [1,9,10]. However, the low temperature properties and oxidation stability of these biolubricants make extensive use difficult [11]. In general, vegetable oils are mainly composed of fatty acids. Thus, oleic, linoleic, linolenic, or ricinoleic acid play an important role in some typical vegetable oils, such as rapeseed, safflower, sunflower, corn, or castor oil. Thus, the specific composition of these vegetable oils can vary depending on the kind of oil or pre-harvest conditions such as soil, climate, etc [2]. Consequently, the characteristics of these vegetable oils (and their derivatives, such as fatty acid methyl esters), especially some properties such as viscosity or oxidative stability, depend on the percentages of these components [1,5,12–14]. Oleic acid, with one unsaturation in its molecular structure, promotes high oxidative stabilities, whereas ricinoleic acid increases viscosity due to the hydroxyl group in its structure. As a result, many properties of these fatty acids depend on the stereochemistry of the molecular chains, the length and degree of branching or unsaturation, among other factors [2,15–17]. Concerning biolubricant production, there are plenty of chemical reactions to obtain them, especially epoxidation and double transesterification (from triglycerides to fatty acid complex esters, through fatty acid methyl esters) [1,13,18]. Thus, transesterification is a chemical reaction between an ester and an alcohol to transform the former into another ester by interchanging alkyl groups. It can be classified into acid or base-catalyzed reactions, depending on the catalyst used. The complexity of the final ester obtained (and their characteristics as a biolubricant) is dependent on the nature of the alcohol used for the chemical reaction [1]. Thus, the use of catalysts is necessary, especially when the alcohol used in this transesterification is complex (especially when biolubricants are produced), which makes transesterification difficult. The aim of this research work was to optimize temperature, alcohol/FAME ratio, and catalyst concentration for the suitable production of biolubricants from two typical vegetable oils in Spain, assessing the effect of the raw material and the kind of alcohol used in biolubricant features. This research work could lead to a sustainable production of biolubricants from local feedstocks.

2. Results and Discussion

2.1. Raw Material Characterization

In order to determine the biolubricant yield, gas chromatography coupled to FID detection was carried out. In this case, the decrease in total FAME content over time was used to determine the yield of biolubricant production. FAME distribution, after the first transesterification (see Figure 1), was similar to previous studies with the same oils, observing slight differences [6,12,16,19]. For rapeseed oil, the majority ester was methyl oleate (63.07%), followed by methyl linoleate (21.15%), and linolenate (8.71%). On the other hand, other FAMEs (methyl myristate, palmitoleate, and erucate) hardly constitute 0.5%. Initially, total FAME content was 96.59%. On the other hand, castor oil presented high proportions of methyl ricinoleate (over 88%), and the other FAMEs were scarce, with methyl linoleate being the second most abundant one (4.5%).

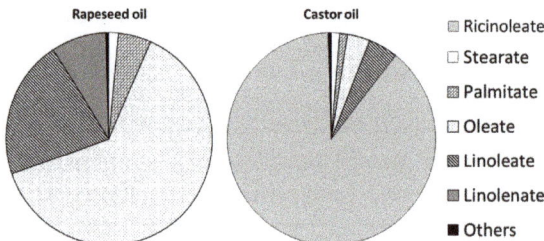

Figure 1. FAME profile of biodiesel from rapeseed and castor oils.

In general, fatty acid methyl ester distribution, as it occurred in biodiesel composition, could play an important role in the behavior of the subsequent biolubricants [14,16,20]. This is important in the case of castor oil where, as indicated, there was a notorious majority in methyl ricinoleate. Effectively, according to Table 1, there were clear differences between the samples, especially concerning viscosity. For rapeseed oil methyl esters, similar values for methyl oleate (between 3.73 and 4.51 cSt) were found in the literature. It must be pointed out that methyl oleate is the main FAME in rapeseed oil biodiesel, and many of its physical and chemical characteristics could influence rapeseed biodiesel. In addition, methyl linoleate, with a double unsaturation, could contribute to low viscosity values [16]. On the other hand, castor oil FAMEs had an extremely high viscosity. This fact could be explained by the structure of the main FAME after the first transesterification. That way, viscosity increases proportionally with the presence of hydroxyl groups, which is the distinguishing characteristic of ricinoleic acid. As it is the main fatty acid in castor oil, this could explain the high viscosity found for the corresponding FAMES of this oil [16].

Table 1. Characteristics of FAMES from rapeseed and castor oils.

Parameter	Rapeseed FAME	Castor oil FAME
Density (kg/m3 at 15 °C)	878.23	916.20
Viscosity (cSt at 40 °C)	5.32	15.8
Acid number (mg KOH/g)	0.175	0.875

2.2. Influence of Temperature

Concerning the influence of temperature on biolubricant yield (Figure 2), it can be seen that the worst performance was related to 150 °C, obtaining 89% at the end of the experiment. For 160 °C and 170 °C, over 90% of the reaction was completed after 60 min, reaching in both cases similar final yields (94.26% and 95.38%, respectively), albeit the reaction at 170 °C was kinetically faster. Moreover, after 60 min, an asymptotic behavior was observed for these experiments. This way, a clear influence of temperature was observed, not only on final yield, but also on kinetics, especially if temperatures were too low. Accordingly, from an economic point of view, 160 °C was the recommended temperature for further studies.

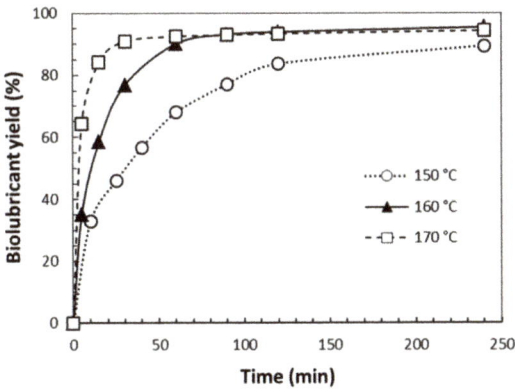

Figure 2. Biolubricant production depending on temperature ([Catalyst] = 1% w/w; alcohol = 2-ethyl-1-hexanol; molar ratio = 3:1).

2.3. Influence of Catalyst Concentration

The influence of catalyst concentration (see Figure 3) was determined at 150 °C, considering that at low temperatures, the concentration of catalyst would be more decisive. From a kinetic point of view, the concentration of catalyst exerted a positive effect on the reaction rate. Taking into account the biolubricant yield, with 0.5% catalyst (w/w), the final conversion was too poor (under 80%) compared to 1% and 2% (89.40% and 90.18%, respectively). For 1% and 2% of catalyst, the reaction was almost complete after two hours, reaching an asymptotic curve from then on. There seemed to be a maximum catalyst concentration, from which no improvement in conversion was observed. On this basis, the most suitable catalyst concentration was 1% w/w.

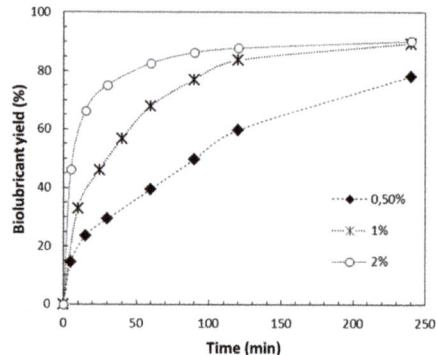

Figure 3. Biolubricant production depending on catalyst concentration. (Temperature = 150 °C; alcohol = 2-ethyl-1-hexanol; molar ratio = 3:1).

2.4. Influence of Alcohol/FAME Molar Ratio

With respect to 2-ethyl-1-hexanol/FAME molar ratio, there were differences, especially for 1:1 molar ratio (Figure 4). With this value, hardly 50% conversion was achieved, whereas for 2:1 and 3:1, the yield was better (81.71% and 89.40%, respectively). These results confirmed that the concentration of alcohol strongly affected the equilibrium, displacing the reaction towards the formation of the biolubricant. Concerning kinetics, the influence of alcohol concentration was positive, increasing the formation rate of the final product. For this reason, some authors recommend, for transesterification reactions using FAME as substrate, an amount of alcohol slightly higher than 1:1 molar ratio [21]. In this case, the most suitable molar ratio was 3:1.

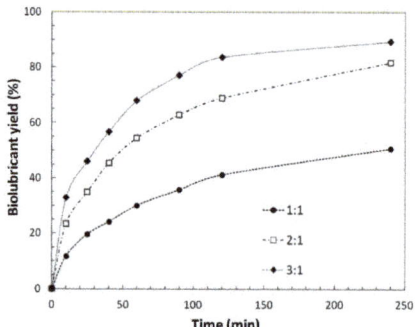

Figure 4. Biolubricant production depending on alcohol/FAME molar ratio. ([Catalyst] = 1% *w/w*; alcohol = 2-ethyl-1-hexanol; temperature: 150 °C).

In summary, and taking into account the above-mentioned discussion and the relationship of commitment between yield and economic criteria, the selected values for the final biolubricant production were the following: 160 °C, 1% catalyst (w/w), and 3:1 molar ratio.

2.5. Biolubricant Characterization

Once the final biolubricant was produced under these conditions (160 °C, 1% catalyst (w/w), 2-ethyl-1-hexanol and 3:1 molar ratio), a thorough characterization was carried out, obtaining for rapeseed oil biolubricant the following values: saponification number, 148.5 mg KOH/g; iodine number, 55.5% w/w; moisture, 0.026% w/w; and cold filter plugging point, 0 °C. The low value for iodine number could imply a better stability of this biolubricant during storage, as the amount of unsaturations (that is, reactive points to increase oxidation) is considerably lower to those found in the literature for FAME [5].

2.6. Influence of Raw Material on Biolubricants

For the rest of parameters, a comparison between rapeseed (RB), castor (CB) and two commercial lubricants (L1 and L2) was carried out; the results are shown in Table 2. The commercial lubricants were mainly composed of aliphatic hydrocarbons, and their main use was for hydraulic systems, turbines, and gears. As can be observed, both biolubricants (obtained by transesterification of FAMEs with 2-ethyl-1-hexanol) were obtained with adequate yields (94–97%) and similar density values, close to the values typically found in the literature and the commercial biolubricants. There were considerable differences in viscosity values, as mentioned earlier for the respective biofuels (especially for CB, much higher than the rest of the samples). For rapeseed oil, a viscosity of 11 cSt at 40 °C was found, which does not meet most requirements of ISO norms, whereas castor oil amply complies with these requirements with 208.25 cSt at 40 °C, exceeding the lower limit of the ISO VG100 norm (90.0 cSt at the same temperature). These values were similar to those found in the literature for other vegetable oils used as biolubricants or derivatives [18,22]. It was possibly due to the fact that the fatty acids of biolubricants derived from rapeseed and castor oils kept a similar chemical structure, especially paying attention to the hydroxyl group in ricinoleate (majority in castor oil biolubricant), which increases its polarity and, therefore, its viscosity both at 40 °C and 100 °C [16,17]. Concerning flash and combustion points, they were high in both biolubricants, especially for castor oil. This fact implies that they are safe when it comes to storage and management, exceeding most requirements for this parameter in the case of rapeseed oil, and improving the results of the commercial lubricants [18]. Regarding the viscosity index, it could be said that both biolubricants showed multigrade characteristics (exceeding 140, being suitable for industrial purposes), and therefore, viscosity is less variable over temperature for these products, compared to the commercial lubricants, which showed lower viscosity indexes (below 100) [23,24].

Table 2. Comparison of rapeseed and castor biolubricants *.

Parameter	RB	CB	L1	L2
Yield (%)	96.59	93.80	-	-
Density (kg/m^3 at 15 °C)	873	930	845	860
Viscosity (cSt at 40 °C)	10.04	208.25	7.0	13.8
Viscosity (cSt at 100 °C)	4.09	26.74	2.1	3.4
Viscosity index	377	163	97	95
Pour point (°C)	<−10	<−16	−30	−18
CFPP (°C)	0	-	−27	−18
Flash point (°C)	222	271	150	180
Combustion point (°C)	236	285	-	-
Acid number (mg KOH/g)	0.39	0.45	0.38	0.50
Oxidative stability (h)	0.94	-	-	-

* Experimental conditions: Temperature = 160 °C; [Catalyst] = 1% w/w; 2-ethyl-1-hexanol/FAME ratio = 3:1

For CFPP and pour point, it could be said that, in general, the commercial lubricants showed lower (and better) values compared to the biolubricants obtained, with the values of the latter being comparable to other biolubricants [18]. The oxidative stability (data available for rapeseed biolubricant) was low, not exceeding one hour. Nevertheless, according to the literature, this value is in a similar order of magnitude to other biolubricants [18,24].

2.7. Influence of Alcohol Structure on Biolubricants

Finally, and comparing the same raw material for the production of biolubricants with different alcohols for transesterification, castor FAMEs were used with 2-ethyl-1-hexanol, 1-heptanol, and 4-methyl-2-pentanol (Table 3). In this case, there were clear differences between the biolubricant obtained by the reaction of castor FAME with 2-ethyl-1 hexanol and the rest of the alcohols. It has already been reported that the number of carbons or branching in the alcohol chain structure could also influence properties such as viscosity [17,25]. However, some authors do not consider this fact as a determinant, finding some functional groups (such as phenyl) more influential [10]. Thus, the complexity of 2-ethyl-1-hexanol structure is able to promote more intermolecular bonds, increasing viscosity.

Table 3. Comparison of biolubricants from castor oil FAMEs by using 2-ethyl-1-hexanol (A), 1-heptanol (B), and 4-methyl-2-pentanol (C).

Parameter	A	B	C
Density (mg/ml at 15 °C)	0.930	0.908	0.912
Viscosity (cSt at 40 °C)	208.25	39.96	34.47
Viscosity (cSt at 100 °C)	26.74	7.11	6.13

* Experimental conditions: Temperature = 160 °C, 150 °C, and 120 °C for A, B, and C, respectively; [Catalyst] = 1% w/w; alcohol/FAME ratio = 3:1

3. Materials and Methods

3.1. Raw Materials and Experimental Design

The raw materials, rapeseed and castor oil, were provided by the Research Center "La Orden-Valdesequera" (Badajoz-Spain) Section of Non-Food (CICYTEX). These vegetable oils were the starting point of the experimental design, being quite different in physical and chemical features. An overview of the experimental design is shown in Figure 5.

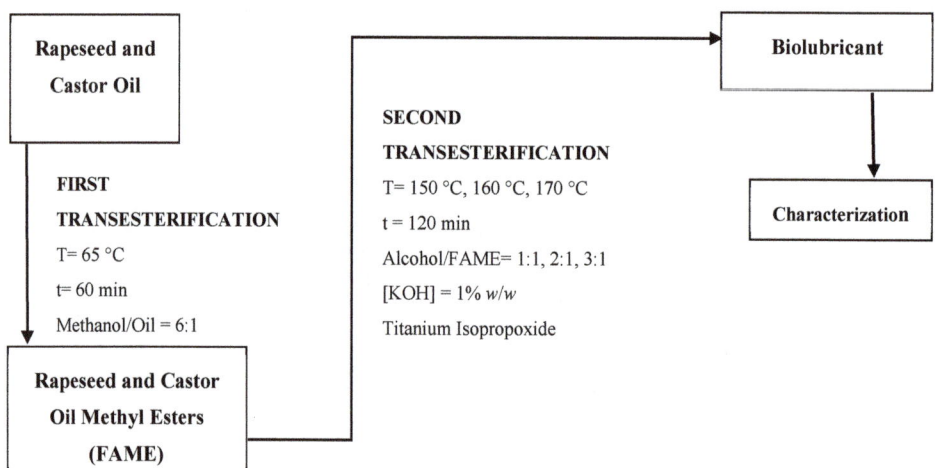

Figure 5. Experimental design.

The first transesterification produced fatty acid methyl esters (FAMEs or biodiesel), which were obtained by using optimum parameters. Afterwards, the second transesterification was carried out to produce fatty acid complex esters (biolubricants), assessing different temperature values, catalyst percentages, alcohol/FAME ratios, and kinds of alcohol. The characterization of the biolubricant obtained was carried out, paying attention to viscosity.

3.2. First Transesterification

As an intermediate stage, fatty acid methyl esters (FAME) were produced from rapeseed and castor oil. In this reaction, the triglycerides that constitute the vegetable oils react with three moles of methanol to produce three moles FAME and glycerol.

This reaction is reversible, and therefore parameters such as temperature, catalyst concentration, and molar ratio are important to obtain better FAME yields.

For this purpose, a reactor coupled to a condenser and supplied with temperature and stirring rate control was used. The reaction conditions, optimized in previous studies [26], were as follows: a reaction temperature of around 65 °C was used (to avoid excessive methanol evaporation); the reaction took place for 60 min (to make sure that the FAME proportion in the final product was over 96.5%); the methanol/oil ratio was 6:1 (according to previous studies, additional methanol was not necessary); the catalyst (potassium hydroxide) added was 1% w/w (of the total reaction medium). Following these steps, the FAME content of the product was over 96.5%, complying with the standard and making it a suitable reagent for the next step (the second transesterification for biolubricant production). Details of the experimental procedure, methodology, and analytical methods can be found in previous works [6,26].

3.3. Second Transesterification Reaction (Biolubricant Production)

Concerning biolubricant production, different alcohols (such as 2-ethyl-1-hexanol, 1-heptanol, and 4-methyl-2-pentanol; see Table 4), FAME (obtained earlier), and titanium isopropoxide (as a catalyst) were used for the second transesterification reaction (again, a reversible reaction; see Figure 2). These alcohols were chosen on account of their different structure, with different branching levels (see Table 4), in order to obtain different viscosity values for the products obtained. On the other hand, titanium isopropoxide was chosen due to its effectiveness and the use of titanium catalysts in industry, especially as the precursor of other catalysts such as TiO_2 [27–29].

Table 4. Structure of the different alcohols used for the second transesterification.

Alcohol	Structure	Boiling Point °C
2-ethyl-1-hexanol	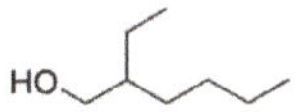	180
1-heptanol		176
4-methyl-2-pentanol	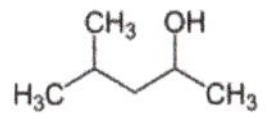	131.6

As can be observed in Figure 6, methanol was released along with the biolubricant and its removal contributes to better biolubricant production.

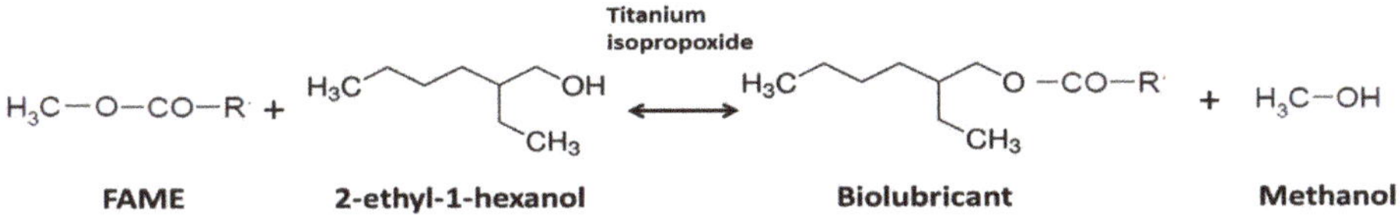

Figure 6. Second transesterification.

As was observed in Figure 7, the experimental facility was similar to that described for the FAME production, except for the Dean Stark trap, which collects (and therefore removes from the reaction medium, improving the yield of the biolubricant) the methanol that is evolved during the second transesterification process, and the sampling point (with the aim of taking samples to analyze the FAME content evolution by gas chromatography).

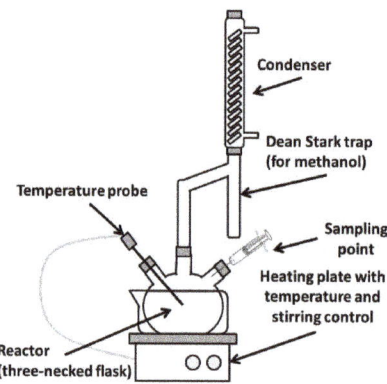

Figure 7. Experimental arrangement for biolubricant production.

In the second transesterification reaction, temperature, catalyst concentration, and molar ratio, in addition to the alcohol type, were studied for biolubricant yield optimization. Table 5 shows the range for these variables. The second transesterification process was carried out near the corresponding boiling point for each alcohol. This way, the temperature was as high as possible, not exceeding each boiling point, so that the alcohol is conserved in the reaction medium and not evaporated. In order to remove the surplus alcohol after the reaction, vacuum distillation was carried out. Figure 8 shows the experimental setup for this purpose.

Table 5. Experimental design for the optimization of biolubricant production.

Experiment	Temperature (°C)	Catalyst Concentration (%)	Alcohol */FAME Molar Ratio
Influence of temperature	150, 160 and 170	1	3:1
Influence of catalyst concentration	150	0.5, 1 and 2	3:1
Influence of molar ratio	150	1	1:1, 2:1 and 3:1

* For variable optimization, 2-ethyl-1-hexanol was used.

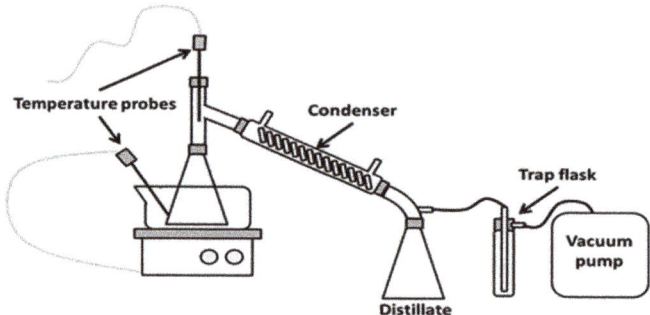

Figure 8. Biolubricant purification.

Thus, the installation was composed of heating and stirring systems (where the lubricant is heated to remove surplus alcohol at approximately 120–150 °C), temperature probes, a condenser, the corresponding Erlenmeyer flask to collect the alcohol, a trap flask (to protect the pump), and a vacuum pump.

To measure the decrease in FAMEs during the second transesterification reaction (implying the increase in biolubricant yield), a VARIAN 3900 chromatograph, provided with a FID, and a silica capillary column (30 m length, 0.32 mm ID, and 0.25 mm film thickness) was used. The carrier gas was helium (0.7 mL/min flow rate), and heptane was used as a solvent. The injector temperature was kept at 270 °C and the detector temperature at 300 °C.

Temperature ramp started at 20 °C, and then went 20 °C/min up to 220 °C. A calibration curve was done for each FAME, using its corresponding standard (Sigma-Aldrich). The calibration was carried out by using an internal standard (methyl heptadecanoate).

3.4. Biolubricant Characterization

Once the biolubricant was obtained and purified, it was characterized. For this purpose, several determinations were carried out. Density determination was obtained by using a pycnometer. Viscosity is determined following the ISO 3104:1994 standard [30] with an Ostwald viscometer. For viscosity index, the ASTM D2270 standard was used [31]. Pour point was measured according to ATSM D-97 standard [32]. To determine cold filter plugging point (CFPP), the EN 116 standard was consulted [33]. For flash and combustion point determination, the Cleveland open-cup method was used (UNE 51-023-90) [34]. For moisture, a Metrohm 870 trinitro plus equipment was used, using the Karl-Fischer method (UNE-EN ISO 12937:2000) [35]. The saponification number was determined following the UNE-EN 55012 standard [36]. The acid number was determined according to the UNE-EN 12634:1999 standard [37]. For Iodine number, the UNE-EN 14111:2003 standard was followed [38]. The oxidative stability was obtained for biolubricant, according to the Rancimat test [39]. Thus, three grams of the sample was placed in a test tube, bubbling air (10 L/h) into the sample, and heating it at 110 °C. The resulting stream of air, after passing through the sample, bubbled 50 ml of deionized water. To monitor the whole process, the conductivity of this deionized water was measured. As the sample was oxidized, some by-products were developed and dissolved into the deionized water, increasing the conductivity of the latter. Thus, the induction point was determined at the time when the conductivity increased considerably.

4. Conclusions

The production of biolubricants (kinetics and conversion) was influenced by temperature, concentration of catalyst (titanium isopropoxide), and alcohol/FAME molar ratio. Thus, temperature slightly increased the final yield, with notorious kinetic changes at higher temperatures. Equally, the final conversion was higher, as catalyst concentration increased from low values to intermediate ones, apparently reaching a point (1% w/w) where it was not dependent on concentration. Finally, as the molar ratio alcohol/FAME was higher, the yield was better. In order to meet both yield and economic factors, intermediate values for temperature and catalyst concentration (T = 160 °C, [Catalyst] = 1% w/w) and the highest alcohol/FAME ratio (alcohol:FAME = 3:1) were considered to be suitable for biolubricant production. Once the main parameters were optimized for biolubricant production, the yield obtained for rapeseed and castor oils was acceptable, exceeding 93% in both cases.

The raw materials studied, rapeseed and castor oils, presented different fatty acid profiles. For the former, the majority fatty acid was oleic acid, whereas for the latter, it was ricinoleic acid. Therefore, the kind of seed selected, among other factors such as pre-harvest conditions, make this initial characterization necessary. Consequently, the proportion of fatty acids (especially the majority one), plays an important role in biolubricant features. Thus, the structure of these fatty acids, including unsaturations, length of the chain, and hydroxyl groups, influenced many parameters of the final biolubricant. For example, the presence of a hydroxyl group in ricinoleic acid could explain the high viscosity of castor oil and its corresponding biodiesel and biolubricant, compared to rapeseed oil. Both biolubricants showed (in general) suitable properties, sometimes better than in the case of the commercial lubricants that were studied.

The kind of alcohol used in biolubricant production seemed to influence some of its characteristics, especially viscosity. The longer and more branched the molecule chain, the higher the viscosity. Consequently, the right choice of the alcohol could optimize the performance of the biolubricant, especially concerning tribology.

For further studies on this subject, analyses to complete characterization of the final biolubricant produced are necessary, paying special attention to IR, mass spectrometry, etc.

Author Contributions: Conceptualization, J.M.E., S.N.-D., and N.S.; methodology, S.N.-D. and N.S.; formal analysis, S.N.-D. and N.S.; investigation, J.M.E., S.N.-D. and N.S.; resources, J.M.E. and J.F.G.; data curation, S.N.-D. and N.S.; writing—original draft preparation, S.N.-D.; writing—review and editing, J.M.E., N.S. and J.F.G.; visualization, J.M.E. and J.F.G.; supervision, J.M.E. and J.F.G.; project administration, J.M.E. and J.F.G.; funding acquisition, J.M.E. and J.F.G. All authors have read and agreed to the published version of the manuscript.

Funding: This research was funded by Junta de Extremadura and FEDER "Fondos Europeos de Desarrollo Regional", grant numbers GR18150 and IB18028.

Acknowledgments: We would like to thank Junta de Extremadura and FEDER "Fondos Europeos de Desarrollo Regional" for the financial support, and the Research Center "La Orden-Valdesequera" (from Cicytex) for the raw material provided for this research work.

Conflicts of Interest: The authors declare no conflict of interest.

References

1. Panchal, T.M.; Patel, A.; Chauhan, D.D.; Thomas, M.; Patel, J.V. A methodological review on bio-lubricants from vegetable oil based resources. *Renew. Sustain. Energy Rev.* **2017**, *70*, 65–70. [CrossRef]
2. Salih, N.; Salimon, J.; Abdullah, B.M.; Yousif, E. Thermo-oxidation, friction-reducing and physicochemical properties of ricinoleic acid based-diester biolubricants. *Arab. J. Chem.* **2017**, *10*, S2273–S2280. [CrossRef]
3. Kania, D.; Yunus, R.; Omar, R.; Abdul Rashid, S.; Mohamad Jan, B. A review of biolubricants in drilling fluids: Recent research, performance, and applications. *J. Pet. Sci. Eng.* **2015**, *135*, 177–184. [CrossRef]
4. Sudan Reddy Dandu, M.; Nanthagopal, K. Tribological aspects of biofuels—A review. *Fuel* **2019**, *258*, 116066. [CrossRef]
5. Martínez, G.; Sánchez, N.; Encinar, J.M.; González, J.F. Fuel properties of biodiesel from vegetable oils and oil mixtures. Influence of methyl esters distribution. *Biomass Bioenergy* **2014**, *63*, 22–32. [CrossRef]
6. Encinar, J.M.; Pardal, A.; Sánchez, N. An improvement to the transesterification process by the use of co-solvents to produce biodiesel. *Fuel* **2016**, *166*, 51–58. [CrossRef]
7. Dehghani Soufi, M.; Ghobadian, B.; Atashgaran, M.; Mousavi, S.M.; Najafi, G. Biolubricant production from edible and novel indigenous vegetable oils: Mainstream methodology, and prospects and challenges in Iran. *Biofuels Bioprod. Biorefining* **2019**, *13*, 838–849. [CrossRef]
8. Chen, J.; Bian, X.; Rapp, G.; Lang, J.; Montoya, A.; Trethowan, R.; Bouyssiere, B.; Portha, J.F.; Jaubert, J.N.; Pratt, P.; et al. From ethyl biodiesel to biolubricants: Options for an Indian mustard integrated biorefinery toward a green and circular economy. *Ind. Crops Prod.* **2019**, *137*, 597–614. [CrossRef]
9. Hajar, M.; Vahabzadeh, F. Modeling the kinetics of biolubricant production from castor oil using Novozym 435 in a fluidized-bed reactor. *Ind. Crops Prod.* **2014**, *59*, 252–259. [CrossRef]
10. Saboya, R.M.A.; Cecilia, J.A.; García-Sancho, C.; Sales, A.V.; de Luna, F.M.T.; Rodríguez-Castellón, E.; Cavalcante, C.L. Synthesis of biolubricants by the esterification of free fatty acids from castor oil with branched alcohols using cationic exchange resins as catalysts. *Ind. Crops Prod.* **2017**, *104*, 52–61. [CrossRef]
11. Fox, N.J.; Stachowiak, G.W. Vegetable oil-based lubricants—A review of oxidation. *Tribol. Int.* **2007**, *40*, 1035–1046. [CrossRef]
12. Hamdan, S.H.; Chong, W.W.F.; Ng, J.H.; Ghazali, M.J.; Wood, R.J.K. Influence of fatty acid methyl ester composition on tribological properties of vegetable oils and duck fat derived biodiesel. *Tribol. Int.* **2017**, *113*, 76–82. [CrossRef]
13. Heikal, E.K.; Elmelawy, M.S.; Khalil, S.A.; Elbasuny, N.M. Manufacturing of environment friendly biolubricants from vegetable oils. *Egypt. J. Pet.* **2017**, *26*, 53–59. [CrossRef]

14. Reeves, C.J.; Menezes, P.L.; Jen, T.C.; Lovell, M.R. The influence of fatty acids on tribological and thermal properties of natural oils as sustainable biolubricants. *Tribol. Int.* **2015**, *90*, 123–134. [CrossRef]
15. Salimon, J.; Salih, N.; Yousif, E. Biolubricant basestocks from chemically modified ricinoleic acid. *J. King Saud Univ. Sci.* **2012**, *24*, 11–17. [CrossRef]
16. Knothe, G.; Razon, L.F. Biodiesel fuels. *Prog. Energy Combust. Sci.* **2017**, *58*, 36–59. [CrossRef]
17. Cecilia, J.A.; Plata, D.B.; Maria, R.; Saboya, A.; Murilo, F.; De Luna, T.; Cavalcante, C.L.; Rodríguez-castellón, E. An Overview of the Biolubricant Production Process: Challenges and Future Perspectives. *Processes* **2020**, *8*, 257. [CrossRef]
18. McNutt, J.; He, Q.S. Development of biolubricants from vegetable oils via chemical modification. *J. Ind. Eng. Chem.* **2016**, *36*, 1–12. [CrossRef]
19. Mazanov, S.V.; Gabitova, A.R.; Usmanov, R.A.; Gumerov, F.M.; Labidi, S.; Ben Amar, M.; Passarello, J.P.; Kanaev, A.; Volle, F.; Neindre, B. Le Continuous production of biodiesel from rapeseed oil by ultrasonic assist transesterification in supercritical ethanol. *J. Supercrit. Fluids* **2016**, *118*, 107–118. [CrossRef]
20. Jose, T.K.; Anand, K. Effects of biodiesel composition on its long term storage stability. *Fuel* **2016**, *177*, 190–196. [CrossRef]
21. Kleinaite, E.; Jaška, V.; Tvaska, B.; Matijošyte, I. A cleaner approach for biolubricant production using biodiesel as a starting material. *J. Clean. Prod.* **2014**, *75*, 40–44. [CrossRef]
22. Madankar, C.S.; Dalai, A.K.; Naik, S.N. Green synthesis of biolubricant base stock from canola oil. *Ind. Crops Prod.* **2013**, *44*, 139–144. [CrossRef]
23. Verdier, S.; Coutinho, J.A.P.; Silva, A.M.S.; Alkilde, O.F.; Hansen, J.A. A critical approach to viscosity index. *Fuel* **2009**, *88*, 2199–2206. [CrossRef]
24. Greco-Duarte, J.; Cavalcanti-Oliveira, E.D.; Da Silva, J.A.C.; Fernandez-Lafuente, R.; Freire, D.M.G. Two-step enzymatic production of environmentally friendly biolubricants using castor oil: Enzyme selection and product characterization. *Fuel* **2017**, *202*, 196–205. [CrossRef]
25. Gryglewicz, S.; Muszyński, M.; Nowicki, J. Enzymatic synthesis of rapeseed oil-based lubricants. *Ind. Crops Prod.* **2013**, *45*, 25–29. [CrossRef]
26. Encinar, J.M.; González, J.F.; Pardal, A.; Martínez, G. Rape oil transesterification over heterogeneous catalysts. *Fuel Process. Technol.* **2010**, *91*, 1530–1536. [CrossRef]
27. Fischer, K.; Schulz, P.; Atanasov, I.; Latif, A.A.; Thomas, I.; Kühnert, M.; Prager, A.; Griebel, J.; Schulze, A. Synthesis of high crystalline tio2 nanoparticles on a polymer membrane to degrade pollutants from water. *Catalysts* **2018**, *8*, 376. [CrossRef]
28. Gao, X.; Wang, C.; Xu, Q.; Lv, H.; Chen, T.; Liu, C.; Xi, X. N-doped K3Ti5Nbo14@TiO2 core-shell structure for enhanced visible-light-driven photocatalytic activity in environmental remediation. *Catalysts* **2019**, *9*, 106. [CrossRef]
29. Fischer, K.; Gawel, A.; Rosen, D.; Krause, M.; Latif, A.A.; Griebel, J.; Prager, A.; Schulze, A. Low-temperature synthesis of anatase/rutile/brookite TiO2 nanoparticles on a polymer membrane for photocatalysis. *Catalysts* **2017**, *7*, 209. [CrossRef]
30. UNE-EN ISO 3104/AC:1999. Petroleum products. Transparent and opaque liquids. In *Determination of Kinematic Viscosity and Calculation of Dynamic Viscosity*; ISO 3104:1994; Asociación Española de Normalización: Madrid, Spain, 1999.
31. ASTM-D2270-10. *Standard Practice for Calculating Viscosity Index from Kinematic Viscosity at 40 °C and 100 °C*; ATSM International: West Conshohocken, PA, USA, 2016.
32. ASTM-D97-17b. *Standard Test Method for Pour Point of Petroleum Products*; ATSM International: West Conshohocken, PA, USA, 2017.
33. UNE-EN 116:2015. *Diesel and Domestic Heating Fuels—Determination of Cold Filter Plugging Point- Stepwise Cooling Bath Method*; Asociación Española de Normalización: Madrid, Spain, 2015.
34. UNE-EN 51023:1990. Petroleum products. Determination of flash and fire points. In *Cleveland Open Cup Method*; Asociación Española de Normalización: Madrid, Spain, 1990.
35. UNE-EN-ISO-12937:2000. Productos petrolíferos. Determinación de agua. In *Método de Karl Fischer por Valoración Culombimétrica*; Asociación Española de Normalización: Madrid, Spain, 2001.
36. UNE-EN-55012. Vehículos, embarcaciones y dispositivos propulsados por motores de combustión interna. Características de las perturbaciones radioeléctricas. In *Límites y Métodos de Medición para la Protección de Receptores Externos*; Asociación Española de Normalización: Madrid, Spain, 2008.

37. UNE-EN-12634:1999. Productos petrolíferos y lubricantes. Determinación del índice de ácido. In *Método de Valoración Potenciométrica en un Medio no Acuoso*; Asociación Española de Normalización: Madrid, Spain, 1999.
38. UNE-EN 14111:2003. Fat and oil derivatives. Fatty Acid Methyl Esters (FAME). In *Determination of Iodine Value*; Asociación Española de Normalización: Madrid, Spain, 2003.
39. Focke, W.W.; Van Der Westhuizen, I.; Oosthuysen, X. Biodiesel oxidative stability from Rancimat data. *Thermochim. Acta* **2016**, *633*, 116–121. [CrossRef]

© 2020 by the authors. Licensee MDPI, Basel, Switzerland. This article is an open access article distributed under the terms and conditions of the Creative Commons Attribution (CC BY) license (http://creativecommons.org/licenses/by/4.0/).

Article

Biodiesel Production from Castor oil by Two-Step Catalytic Transesterification: Optimization of the Process and Economic Assessment

Nuria Sánchez [1], José María Encinar [1], Sergio Nogales [1,*] and Juan Félix González [2]

[1] Department of Chemical Engineering and Physical Chemistry, University of Extremadura, Avda. de Elvas s/n, 06006 Badajoz, Spain; nuriass@unex.es (N.S.); jencinar@unex.es (J.M.E.)

[2] Department of Applied Physics, University of Extremadura, Avda. de Elvas s/n, 06006 Badajoz, Spain; jfelixgg@unex.es

* Correspondence: senogalesd@unex.es; Tel.: +34 924289672

Received: 22 September 2019; Accepted: 16 October 2019; Published: 17 October 2019

Abstract: The use of biodiesel and the requirement of improving its production in a more efficient and sustainable way are becoming more and more important. In this research work, castor oil was demonstrated to be an alternative feedstock for obtaining biodiesel. The production of biodiesel was optimized by the use of a two-step process. In this process, methanol and KOH (as a catalyst) were added in each step, and the glycerol produced during the first stage was removed before the second reaction. The reaction conditions were optimized, considering catalyst concentration and methanol/oil molar ratio for both steps. A mathematical model was obtained to predict the final ester content of the biodiesel. Optimal conditions (0.08 mol·L^{-1} and 0.01 mol·L^{-1} as catalyst concentration, 5.25:1 and 3:1 as methanol/oil molar ratio for first and second step, respectively) were established, taking into account the biodiesel quality and an economic analysis. This type of process allowed cost saving, since the amounts of methanol and catalyst were significantly reduced. An estimation of the final manufacturing cost of biodiesel production was carried out.

Keywords: catalyst; sodium hydroxide; fatty acid methyl ester; central composite rotatable design; operational conditions

1. Introduction

Nowadays, fossil fuel depletion and the increase in atmospheric carbon dioxide concentration are two of the main reasons to promote sustainable alternatives for petroleum products. Biodiesel can be considered as a real alternative for diesel because of its renewable character and its use in any compression ignition engine [1]. The most commonly used route to obtain biodiesel is through transesterification of vegetable oils (or other sources such as animal fats) with methanol as an alcohol and NaOH, KOH, CH$_3$ONa, or CH$_3$OK as catalysts [1–3]. Non-edible vegetable oils such as castor oil could be considered as appropriate raw material because it is not used in human diet, its plants can grow in agronomically poor soils, and its oil yield is higher than in the case of other energy crops [4]. The main component of this oil is the triglyceride formed of the unsaturated hydroxyl-fatty acid, ricinoleic acid [(9Z, 12R)-12-hydroxy-9-octadecenoicacid]. This compound is the main cause of the high viscosity and polarity of castor oil [4]. Such properties would limit its use as a biodiesel; nevertheless, as is usually done for other biodiesel samples [5], mixtures with castor oil biodiesel show good properties when mixed with conventional diesel or other less viscous biodiesels [6], and can also be used in mixtures as oil [7]. In addition, castor oil shows high solubility in alcohols, which favors transesterification [8].

Biodiesel production from castor oil was studied in mixtures with soybean oil. Nevertheless, nonsignificant substrate preference was observed [9]. On the other hand, the use of co-solvents was an additional method for improving castor oil biodiesel yield [10]. In this case, hexane was used as a co-solvent and biodiesel yield was not significantly affected by the presence of this compound. A high alcohol/oil molar ratio, 20:1, was also necessary. The transesterification of castor oil using methanol was done with ultrasound; the highest ester content was 93.3% [11]. Solid catalysis was also tested in castor oil transesterification [12]. The catalyst was composed of Ag salts and 29:1 methanol/oil molar ratio, 60 °C, and a reaction time of 3 h were necessary to reach 90% biodiesel yield. As seen in previous works, the effort to enhance the results of the conventional method did not lead to completely satisfactory conclusions. The highest ester content, 97%, was obtained with homogeneous basic catalysis and conventional heating. However, 18.8:1 methanol/oil molar ratio was necessary [8].

The transesterification is a reversible reaction. When enough catalyst is present in the reaction medium, chemical equilibrium is reached. Methanol is usually added in higher ratios than the stoichiometric ratio (3:1) in order to shift the equilibrium position of the reaction towards the product side. However, this fact strongly increases the final cost of the process due to the fact that methanol expenses are higher. Therefore, the optimization of the process is vital to reduce environmental impacts and costs [13]. In this research work, a process with two steps was proposed to obtain castor oil biodiesel with high methyl ester content and decrease costs. In this way, two reactions were carried out and, before the second one, the glycerol produced in the first reaction was removed. The removal of this product promoted a change in the equilibrium position towards the products. The aim of this work was to assess the best transesterification conditions to improve the process and reduce production costs. The operation conditions were optimized to obtain high methyl ester yields and the best conditions were economically evaluated. Some economic assessments have been carried out for industrial plants of biodiesel production [14–18]. However, there was little information when castor oil was used as a feedstock [19]. Therefore, the global process of biodiesel production from castor oil in an industrial plant was evaluated, and the final cost of biodiesel production was calculated for the analyzed plant.

2. Results and Discussion

2.1. Raw Material

Oil properties and its corresponding fatty acid content are shown in Table 1. The oil content of the feedstock used was equivalent to the composition of a typical castor oil: 90% ricinoleic acid, 4.5% linoleic acid, and 3.6% oleic acid [20]. Ricinoleic acid, with a hydroxyl group, shows very different properties compared to other fatty acids, that is, regarding density and viscosity, it was highly hygroscopic and had a low iodine value and high solubility in alcohols. The latter is the most interesting characteristic considering transesterification to obtain biodiesel because it promotes this chemical reaction at low temperatures [21,22]. Compared with other vegetable oils, there were clear differences in fatty acid profile, with oleic acid being the majority fatty acid for rapeseed and sunflower oils. This difference in fatty acid composition could explain the difference in observed properties.

The low acid value of this oil made the use of basic catalysis possible for transesterification [23]. This way, potassium hydroxide was selected, as basic catalysts are suitable for oils with low acid values [21,24].

Table 1. Castor oil fatty acid profile and properties and comparison with other biodiesel from vegetable oils.

Oil	Castor	Rapeseed [25]	Sunflower [25]
Fatty acid profile, %			
C16:0 palmitic	1.30	4.92	4.88
C18:0 stearic	1.22	1.63	4.78
C18:1 oleic	3.61	66.59	67.66
C18:2 linoleic	4.58	17.08	21.26
C18:3 linolenic	0.39	7.75	0.09
C18:1–OH ricinoleic	88.9	N.D.	N.D.
Physical and chemical properties			
Density at 15 °C, kg·m^{-3}	961	919	918
Viscosity at 40 °C, cSt	262	38.5	38.3
Water content, %	0.31	0.06	0.06
Acid value, mgKOH·g^{-1}	1.19	0.71	1.90
Acid number, %	0.55	0.36	0.95
Iodine value, gI$_2$·(100 g)$^{-1}$	80.5	101.1	93.5
Saponification value, mgKOH·g^{-1}	179	193.2	184.0

N.D. = not detected.

2.2. Reaction Conditions and Variables of the Design

In previous works, castor biodiesel was obtained using a one-step reaction process, achieving 97% methyl esters [8]. This ester content was suitable for its use as biodiesel; however, a very high concentration of alcohol was necessary. The methanol/oil ratio was 18.8:1. In addition, the used catalyst was CH$_3$OK, which means high costs and preventive measures to avoid contact with atmospheric moisture. In this work, two serial transesterification reactions were proposed to decrease MeOH concentration and to avoid the use of CH$_3$OK as a catalyst because it would involve higher costs in an industrial process [14,16,19,26]. In this case, KOH was used as a catalyst because it is cheaper and easier to use.

The transesterification reaction has five important operational conditions: catalyst percentage, methanol/oil molar ratio, temperature, time, and stirring speed. The high solubility between castor oil and methanol was to avoid mass transfer problems. A stirring speed of 700 rpm was maintained in order to ensure thermal homogeneity, and based on previous works with the same system [8,25–28]. Regarding reaction temperature, this parameter was maintained at 45 °C as this was the optimal temperature in previous works with castor oil, and its variation showed just slight effects in transesterification [8,29,30]. In the literature, the reaction time is usually 1–2 h; however, the equilibrium is normally reached during the first minutes of reaction [29,31,32]. In addition, previous work carried out with castor oil showed 10 min was a suitable reaction time [8]. Hence, 10 min was chosen for the first and second step in this work.

On the other hand, catalyst and methanol concentrations have been the most influencing factors in transesterification, and their effects are related each other [8,33,34]. Therefore, these variables for the first and second stages were considered in the experimental design. The ranges of these variables were established based on previous reactions. In the first stage, 0.02–0.10 mol·L^{-1} KOH and 3:1–6:1 molar ratio of CH$_3$OH/oil were used. Regarding second stage, the ranges of operation variables were 0.01–0.05 mol·L^{-1} KOH and 1:1–5:1 molar ratio of CH$_3$OH/oil.

2.3. Regression Model Development

The experimental conditions of the runs by the coded levels of the variables are shown in the Materials and Methods section. As previously mentioned, the studied variables were catalyst concentration in the first step (A), CH$_3$OH/oil molar ratio in the first step (B), catalyst concentration in the second step (C) and CH_3OH/oil molar ratio in the second step (D). These variables were analyzed by central composite rotatable design and the experimental conditions of the runs by the coded levels of the variables are shown in Table 2. The response variable was the biodiesel ester content achieved for each reaction, and these data were also collected in the table. The central conditions of the design produced biodiesel with average ester content of 93.0%. The results were analyzed through multiple regressions, testing various models such as linear, two-factor interaction, three-factor interaction, two and three factor interaction, cubic, quadratic, and cubic plus quadratic models, with the quadratic one best fitting real data as was seen for the transesterification reaction in previous works [8,33,35]. Equation (1) shows the estimated response model equation for methyl ester content of biodiesel (related to original factors).

$$\text{Ester content (wt\%)} = 92.961 + 3.159 \cdot A + 2.940 \cdot B + 1.046 \cdot C + 2.024 \cdot D - 0.894 \cdot AB - 0.644 \cdot AC \\ + 0.459 \cdot AD - 0.989 \cdot BC - 1.046 \cdot BD - 0.119 \cdot CD - 0.470 \cdot AA - 0.704 \cdot BB - 0.150 \cdot CC - 1.133 \cdot DD \quad (1)$$

As can be seen in Equation (1), linear terms showed positive coefficient values, quadratic terms showed negative coefficients, and some cross-product terms were positive and some of them negative. For this reason, the equation of the model will describe a response surface where the maximum ester yield can be observed. The ANOVA test of the response surface is shown in Table 2. The effect of the factors in the response variable followed this order: catalyst concentration first step > CH_3OH/oil molar ratio first step > CH_3OH/oil molar ratio second step > catalyst concentration second step. The determination coefficient pointed to the suitability of the model (0.966). The P-value of the model was lower than 0.05, implying a statistical relation between the response surface and the variables at a confidence level of 95%. In addition, the p-value for the parameter lack of fit was 0.0941, greater than 0.05; then, the model was appropriate to fit the actual data, there was no significant lack of fit. Most terms of the model were significant. In conclusion, the model fits the experimental data faithfully and can be used to predict experimental data.

Table 2. Analysis of variance table for response surface quadratic model.

Source	Sum of Squares	DF	Mean Square	F-Value	P-Value
Model	676.867	14	48.348	32.190	0.0000
A	239.528	1	239.528	159.476	0.0000
B	207.446	1	207.446	138.116	0.0000
C	26.250	1	26.250	17.477	0.0007
D	98.334	1	98.334	65.470	0.0000
AB	12.781	1	12.781	8.509	0.0101
AC	6.631	1	6.631	4.415	0.0518
AD	3.367	1	3.367	2.242	0.1538
BC	15.642	1	15.642	10.414	0.0053
BD	17.514	1	17.514	11.661	0.0035
CD	0.226	1	0.226	0.150	0.7034
AA	6.326	1	6.326	4.212	0.0569
BB	14.177	1	14.177	9.439	0.0073
CC	0.646	1	0.646	0.430	0.5211
DD	36.699	1	36.699	24.434	0.0001
Error	24.031	16	1.502		
Lack of fit	20.055	10	2.006	3.026	0.0941
Pure error	3.976	6	0.663		
Total error	24.031	16	1.502		
		$R^2 = 0.966$			

In the Materials and Methods section, the predicted values from the model and the measured values under the same experimental conditions are shown. As seen, the predicted values agreed with the observed ones in these operating conditions. On the other hand, the residuals were randomly

2.4. Response Surface Graphs

Response surface graphs are one of the most usual ways to show the regression equation in the RSM. When the model considers more than two variables, two of them can be plotted, keeping the remaining constant. In Figure 1a, the effect of catalyst concentration and CH_3OH/oil molar ratio in the first step and their interaction are shown. The conditions for second reaction were kept at the central values of the model (0.03 mol·L^{-1} and 3:1 as KOH concentration and CH_3OH/oil molar ratio, respectively). As seen, the increase of catalyst and methanol concentrations led to a significant increase of the ester content of the biodiesel. The stoichiometric molar ratio between castor oil and methanol is 3:1; however, this alcohol ratio was not enough to reach ester content greater than 95%. As seen in the figure, when CH_3OH/oil molar ratio was lower, the increase in catalyst concentration in the first step would lead to higher ester contents, and vice versa. This behavior has been observed by other authors when these variables were studied in one-step processes [5,34]. On the other hand, Figure 1b shows the response surface of methyl ester yield when the catalyst proportion and methanol/oil molar ratio in the second step were varied. In this case, the higher the catalyst or methanol concentrations, the higher the obtained ester content. However, the effect of catalyst concentration on the second step was less significant than in the case of the first step and, in general, changes in conditions of the second step had less effect on the final result. According to this figure, the most suitable reaction conditions were 0.05 mol·L^{-1} of KOH and 4:1 CH_3OH/oil ratio. Then, these conditions were kept constant and the response surface of Figure 1c was plotted, where catalyst and methanol concentration in the first step were varied. In this case, the maximum ester content was achieved with high catalyst concentration and low CH_3OH/oil molar ratio in the first step. The 3:1 ratio would be enough to reach high conversions, in contrast to the results plotted in Figure 1a. Finally, the condition in the first step which maximized the ester content in Figure 1a were considered (0.10 mol·L^{-1} KOH and 6:1 CH_3OH/oil molar ratio), and the response surface of Figure 1d was drawn. The optimal results differed with the previous figure once again. Therefore, there is a need to reach the condition of equilibrium between both reactions to determine the most suitable conditions.

2.5. Process Optimization

The most interesting aspect of the response surface graphs was their wide area of high ester yield. This implies stability, being a desirable effect because high ester contents can be obtained under various experimental conditions. In particular, it is possible to find lots of reaction conditions which lead to an ester content greater than 96.5%, the minimum value specified by the European Standard UNE-EN 14214. According to Equation (1), the maximum ester content would exceed 100%; using 0.10 mol·L^{-1} KOH and 3:1 CH_3OH/oil molar ratio in the first step and 0.05 mol·L^{-1} KOH and 5:1 CH_3OH/oil ratio in the second step, the predicted ester content would be 101.2% (Table 3). However, experimental ester content higher than 98% was not obtained in any reaction. It was expected that the empirical ester content would be close to 98% when the predicted ester content was higher than 98%. This hypothesis was supported by the first reaction in Table 3, whose conditions led to predicted ester contents higher than 100%, and the measured one was close to 98%. On the other hand, the second reaction of this table was carried out under conditions which led to ester content over 96.5%, and the measured one was also higher than 96.5%, so this biodiesel would be within the European standard.

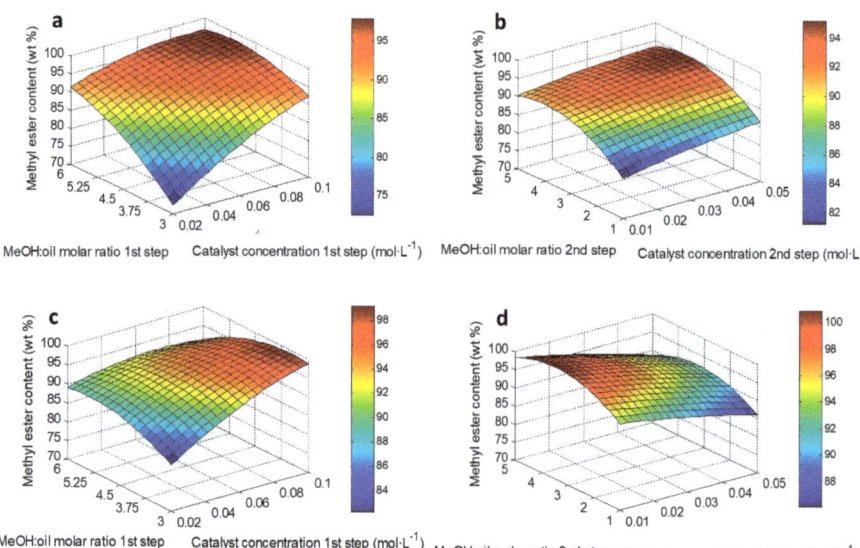

Figure 1. Response surface plots of ester content: (**a**) Catalyst concentration vs. MeOH/oil molar ratio first step; catalyst concentration second step: 0.03 mol·L^{-1}, MeOH/oil molar ratio second step: 3:1; (**b**) Catalyst concentration vs. MeOH/oil molar ratio second step; catalyst concentration first step: 0.06 mol·L^{-1}, MeOH/oil molar ratio first step: 4.5:1; (**c**) Catalyst concentration vs. MeOH/oil molar ratio first step; catalyst concentration second step: 0.05 mol·L^{-1}, MeOH/oil molar ratio second step: 4:1; (**d**) Catalyst concentration vs. MeOH/oil molar ratio second step; catalyst concentration first step: 0.10 mol·L^{-1}, MeOH/oil molar ratio first step: 6:1.

Table 3. Optimization of the process.

A	B	C	D	Predicted Ester Content (wt %)	Experimental Ester Content (wt %)	Relative Error (%)
0.10 mol·L^{-1} (A = 2)	3.00:1 (B = −2)	0.05 mol·L^{-1} (C = 2)	5:1 (D = 2)	101.2	97.6	3.6
0.08 mol·L^{-1} (A = 1)	5.25:1 (B = 1)	0.04 mol·L^{-1} (C = 1)	4:1 (D = 1)	96.7	96.9	−0.2

Since the experimental conditions to obtain biodiesel from castor oil were optimized, an economic evaluation was carried out. For this economic evaluation, the conditions assuming that the predicted ester content was higher than 96.5% were considered. These conditions are collected in Table S1 of Supplementary Material. The main variable costs of the process, such as the consumed methanol, catalyst, and neutralizer, were determined for each condition. To simplify, only the levels of the factors integrated in the model were considered to this calculation, although similar conditions would be expected to achieve similar results and there would be infinite options. Since four factors and five levels were considered, 54 alternatives were evaluated. Among them, the ester content was predicted to be greater than 96.5% under 74 conditions.

Firstly, a biodiesel plant which uses 50,000 tons of castor oil per year was considered. Since biodiesel yield is usually close to 100%, this yield was assumed to the following calculations of this section [19,26]. The process is composed by two heated series reactors. The biodiesel and glycerol phase of the product of the first reactor would be separated, and biodiesel phase would be transferred to the second reactor. Fresh alcohol catalyst solution would also be added. From the products of the reaction, methanol would be recovered in about 90% of unreacted alcohol [2]. The neutralizer was

H_3PO_4, so an input from the sale of K_3PO_4 to the industry of fertilizers was added. The prices to purchase one kilogram of CH_3OH, KOH, and H_3PO_4 were $0.47, $1.87, and $0.40, respectively. The price of selling one kilogram of K_3PO_4 was $0.64. These data were obtained from local companies and the Methanex Methanol Price Sheet [36]. According to these values, the annual cost of methanol, catalyst, and neutralizer in the aforementioned biodiesel plant were evaluated. These expenses were calculated for each condition (Table S1 of Supplementary Material) and they were plotted in Figure 2 as variable cost per liter of biodiesel. The cost of castor oil was not considered because it would be the same for all conditions. The numbers in the x-axis represent the experimental conditions considered to calculate the cost. According to the model, all of these conditions will lead to an ester content greater than 96.5%. Among them, the reaction conditions which showed the cheapest processes in terms of feedstock cost would be the numbers 57, 58, 39, and 65. These numbers represent the conditions collected in Table 4. As an example, the number 57 represents 0.06 mol·L^{-1} as catalyst concentration and 6:1 as MeOH/oil molar ratio for the first step and 0.01 mol·L^{-1} as catalyst concentration and 3:1 as MeOH/oil molar ratio for the second step. When these conditions were used in a biodiesel plant of 50,000 tons of castor oil, a cost saving close to $400,000 could be obtained in comparison to the most unfavorable conditions collected in Figure 2. On the other hand, the optimal conditions as shown in a previous work, where one-step process was used, were 0.064 mol·L^{-1} CH_3OK and 18.8:1 as catalyst concentration and MeOH/oil molar ratio, respectively [8]. Considering these conditions and the same biodiesel plant, close to $800,000 per year could be saved if the process with two steps were used. Therefore, the use of two-step transesterification for this process will suppose important saving costs for reagents.

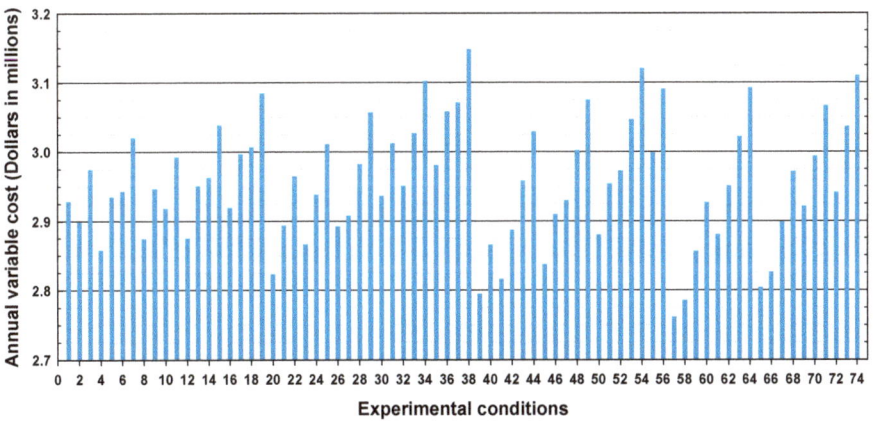

Figure 2. Annual variable cost in millions of dollars for the selected conditions.

Table 4. Experimental conditions with the lowest variable costs per liter of biodiesel.

Run	Catalyst Concentration First Step	CH_3OH/Oil Molar Ratio First Step	Catalyst Concentration Second Step	CH_3OH/Oil Molar Ratio Second Step
57	0.06 mol·L^{-1}	6.00:1	0.01 mol·L^{-1}	3:1
58	0.08 mol·L^{-1}	6.00:1	0.01 mol·L^{-1}	2:1
39	0.08 mol·L^{-1}	5.25:1	0.01 mol·L^{-1}	3:1
65	0.06 mol·L^{-1}	6.00:1	0.02 mol·L^{-1}	3:1

2.6. Process Simulation

Experimental conditions related to run 39 were considered to simulate the process. Due to the high solubility between methanol and castor biodiesel, phase separation was extremely slow when the conditions of run 57 and 58 were used. Therefore, the third condition with lower cost was taken into account. The simulation of a plant of 50,000 tons·year^{-1} of castor oil was carried out. A continuous process was considered because it is common in industry, especially in plants with high capacities [2]. The process flowsheet is shown in Figure 3. The simulation was carried out with the software UniSim Design, and the properties of the main streams of the process were collected in Tables 5 and 6. The process could be improved by energy integration and a study in-depth of the pumping system.

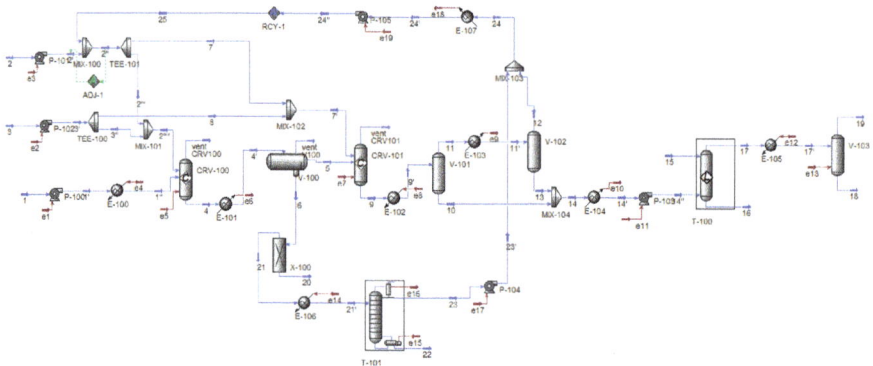

Figure 3. Process flowsheet.

Table 5. Properties of the main streams (part I).

Stream	1"	2	3	4	5	6	7'	9
Temperature, °C	45.0	20.0	20.0	45.0	25.0	25.0	20.0	45.0
Pressure, kPa	250	101	101	250	250	250	250	250
Molar flow, kmol·h^{-1}	6.69	20.5	3.88	42.9	30.5	12.5	20.3	50.8
Mass flow, kg·h^{-1}	6250	56	132	7418	6449	970	652	7100
Component mass fraction								
Ricinolein	1.000	0.000	0.000	0.091	0.078	0.177	0.000	0.016
Methanol	0.000	1.000	0.700	0.075	0.057	0.200	0.993	0.137
NaOH	0.000	0.000	0.300	0.005	0.004	0.009	0.006	0.004
Methyl ricinoleate	0.000	0.000	0.000	0.755	0.859	0.060	0.000	0.836
Glycerol	0.000	0.000	0.000	0.074	0.002	0.554	0.000	0.007
H$_2$O	0.000	0.000	0.000	0.000	0.000	0.000	0.000	0.000

Table 6. Properties of the main streams (part II).

Stream	12	15	17	18	19	22	23	25
Temperature, °C	50	40.0	40.1	150.0	150.0	138.8	47.5	20.0
Pressure, kPa	50	101	101	20	20	60	50	250
Molar flow, kmol·h^{-1}	26.9	49.4	19.3	19.2	0.0285	6.74	5.50	32.2
Mass flow, kg·h^{-1}	863	890	6017	6016	0.942	784	176	1031
Component mass fraction								
Ricinolein	0.001	0.000	0.019	0.019	0.032	0.273	0.000	0.000
Methanol	0.999	0.000	0.001	0.001	0.964	0.080	1.000	1.000
NaOH	0.000	0.000	0.000	0.000	0.000	0.000	0.000	0.000
Methyl ricinoleate	0.000	0.000	0.980	0.980	0.002	0.027	0.000	0.000
Glycerol	0.000	0.000	0.000	0.000	0.000	0.864	0.000	0.000
H$_2$O	0.000	1.000	0.000	0.000	0.003	0.000	0.000	0.000

Firstly, the chemical components were defined for the simulation process. Methanol, glycerol, NaOH (instead of KOH) and water were available in the software component library. The castor oil feedstock and biodiesel were defined as the triglyceride of ricinoleic acid and methyl ricinoleate, respectively. Both compounds were added as hypothetical components [19,37]. Due to the presence of highly polar components, non-random two liquid (extended NRTL) was recommended as thermodynamic model. In addition, liquid–liquid equilibrium data for the system of methanol–glycerol–methyl ricinoleate were faithfully provided by the model [38,39].

Plant capacity was established as 50,000 tons·year^{-1} of castor oil transformed to biodiesel; therefore, 6.25 tonnes·hour^{-1} were considered (8,000 annual operating hours) [10]. As shown in Figure 3, castor oil and fresh methanol were fed to the process by stream 1 and 2, respectively. The stream 3 was a 30% catalyst solution in methanol. Conditions and compositions of these streams are shown in Tables 5 and 6. Methanol was added in excess, so the surplus was recovered and purified, and 89.2% of unreacted methanol was recycled. In the first reactor, the reaction was carried out with a 5.25:1 methanol/oil molar ratio, 0.08 mol·L^{-1} NaOH, 45 °C, and 10 min as residence time. In the second reactor, composition and flow of the streams were regulated to use a 3:1 methanol/oil molar ratio, 0.01 mol·L^{-1} NaOH, 45 °C, and 10 min as residence time. These conditions were the established in run 39, Table 4.

The reactors used in the simulation were conversion reactor models with 89% oil conversion in the first reactor (CRV-100) and 77.5% remain oil conversion in the second reactor (CRV-101). As previously checked by other authors, the presence of the theoretical reaction intermediates, diacylglycerols and monoacylglycerols, was only observed in the initial stages of the reaction, due to high methanol to oil ratios [40].

In the process flowsheet (Figure 3), the product of the first reactor was led to a liquid–liquid separator (V-100). In the process design, a high-speed disc bowl centrifuge was considered because this separation is very slow for castor oil biodiesel and glycerol [41]. Biodiesel rich phase was led to the second reactor and fresh catalyst and methanol were also added. After this reaction, the product was a monophasic mixture. Methanol was recovered by two in series vacuum distillation (V-101 and V-102) at 50 kPa. The temperatures of these operations were 150 °C in the first separator and 50 °C in the second one. The recycled stream had 99.9% methanol.

The streams of biodiesel from the separators were washed in a water washing column (T-100) with hot water (40 °C). It was a column with four theoretical stages at atmospheric pressure [2,40], and in this step, the total removal of the remaining catalyst, methanol, and glycerol was achieved.

Final biodiesel refinement was conducted through vacuum distillation, in order to obtain biodiesel which was within the EN 14214:2013 standard. According to these specifications, water and methanol contents were lower than 0.05% and 0.20%, respectively. The final amount of biodiesel obtained in the process was 6016 kg·h^{-1}, with a yield of 96.3% based on the initial oil.

Glycerol rich phase was neutralized with H_3PO_4 and this operation was simulated by the tool "Component Splitter" (X-100). This tool allowed for the separation of NaOH. The real operation requires a reactor where H_3PO_4 reacts with KOH and a following step for the separation of the synthesized salts [2]. Finally, a distillation column (T-101) was used for methanol recovery and glycerol purification. The design of this column was carried out according to previous works [40,42] and using the "Short Cut Distillation" software. The achieved glycerol purity was 86.4%, but depending on its desired use, additional purification could be necessary.

2.7. Cost Evaluation

Once the production process was established, a cost evaluation was carried out. Firstly, the main units were identified and their sizes calculated. To determine the volume of the transesterification reactors, the volumetric flow of reagents and their residence time were considered; the reactor was a stirred tank with 0.5 as a fill factor. The size of the high-speed disc bowl centrifuge was estimated based on the flow of product that had to be separated. Flash distiller sizes were obtained according to the guidelines established by Silla [43]. The evaporator V-101 was a vertical vessel with cylindrical shape, 4.5 m^3 as total volume and 3 as the length/diameter ratio. The design of V-102 was carried out considering that the inlet stream was mainly composed of vapor. In this case, the length to diameter ratio was 2 and its volume was 0.70 m^3. The most suitable shape of the distiller V-103 was a horizontal cylinder with 50% as liquid level and 10 min as residence time. Under these assumptions, the volume of the vessel was 2.7 m^3.

The cost of the water washing column and the neutralization and salt removal units was estimated based on the inlet flow rate [16,40]. Regarding the distillation column for glycerol purification (T-101), design criteria for this type of unit were used [44]. A packed column was chosen instead of a plate column, because of its short dimensions. The packing material was INTALOX®, 1". The diameter and length of the column were 0.30 and 5.2 m, respectively, considering 65% efficiency [43–45]. The cost of the reboiler and condenser were considered in the heat exchanger section.

The costs of the processing units were calculated based on its size and previous data. In addition, the exponential rule of economy of scale was applied, following Equation (2), where S_0 and C_{S0} are the capacity and cost of the known unit, respectively, and S and C_S are the capacity and cost of the unit of which the cost is unknown. Exponent, δ, is characteristic of each technology and $\delta = 0.65$ for the estimation of tanks, reactors, and columns, and $\delta = 0.80$ for the pumps [46].

$$C_S = C_{S0} \left(\frac{S}{S_0}\right)^\delta \quad (2)$$

This rule was also applied to determine the cost of the equipment in an additional plant. A plant which used 16,000 tons·year^{-1} of castor oil was also evaluated. This plant was considered because the production of castor oil could be smaller in some areas. Fifty thousand tons of castor oil and about one-third of this amount were considered. The cost of the equipment has to be updated, following changes in the value of money due to inflation and deflation according to the chemical engineering plant cost index. Equation (3) was employed, where C_A and C_B are the current capital cost and the cost in the base period, respectively, and $CEPCI_A$ and $CEPCI_B$ are the index published in *Chemical Engineering Journal*. The costs of the major processing units are presented in Table 7.

$$C_A = C_B \frac{CEPCI_A}{CEPCI_B} \quad (3)$$

Table 7. Estimation of equipment costs.

Equipment	Cost, Thousands of Dollars	
	50,000 t·year^{-1}	16,000 t·year^{-1}
Reactor (CRV–100)	163.13	77.78
Reactor (CRV–101)	158.06	75.37
Centrifuge (V–100)	31.34	14.94
Flash distiller (V–101)	89.41	42.63
Flash distiller (V–102)	52.84	25.19
Flash distiller (V–103)	530.80	253.09
Washing column (T–100)	31.10	14.83
Neutralization and removal of the catalyst (X–100)	138.35	65.97
Distillation column (T–101)	123.00	58.65
Heat exchangers	410.71	195.83
Pumps and valves	112.86	53.81
Tanks	2056.71	980.66
Total equipment cost	3898.30	1858.76

The total capital cost of the plant was estimated by the method of factors developed by Lang and improved by Peter and Timmerhaus [47]. This method was based on the cost of the major processing units. In Table 8, factors and capital costs of the plant were collected. As seen in this table, the cost of the plant increased because of the increase of capacity, however, one plant was more than three times bigger than the other, while its cost was only double.

Table 8. Estimated fixed capital cost (year 2015).

Concept	Factor, %	Cost, Thousands of Dollars	
		50,000 t·year^{-1}	16,000 t·year^{-1}
Equipment costs	100	3898.30	1858.76
Installation	47	1832.20	873.62
Instrumentation and controls	18	701.69	334.58
Piping	66	2572.88	1226.78
Electrical systems	11	428.81	204.46
Buildings	18	701.69	334.58
Civil & structure	10	389.83	185.88
Service facilities	70	2728.81	1301.13
Total direct cost		13,254.23	6319.78
Engineering supervision	33	1286.44	613.39
Construction costs	41	1598.30	762.09
Total indirect cost		2884.74	1375.48
Legal expenses and contractors fee (about 5% of direct and indirect costs)	21	818.64	390.34
Contingency (about 10% of direct and indirect costs)	42	1637.29	780.68
Total capital cost (TCC)		18,594.91	8866.28

The price of castor oil was considered as an average price of this oil for the European region. However, this price strongly depends on Indian producers, because this country accounts for more than 60% of the global yield [48]. The prices of the rest of the raw materials were also obtained. As shown in Table 9, the costs of the utilities, and fixed costs were collected to obtain the final manufacturing cost of biodiesel in plants with capacity of 50,000 and 16,000 tonnes·year^{-1}.

Table 9. Annual manufacturing costs.

Concept	Price, $·kg^{-1}	Cost, $·kg^{-1} Biodiesel	
		50,000 t·year^{-1}	16,000 t·year^{-1}
Castor oil	0.948	0.9849	
Methanol	0.393	0.0490	
KOH	1.87	0.0203	
H$_3$PO$_4$	0.40	0.0006	
Water	0.0017	0.0002	
Raw material costs		0.5023	
K$_3$PO$_4$	0.64	0.0017	
By-product		0.0017	
Waste treatment	0.21	0.0433	
Energetic streams	0.024	0.0144	
Electricity	0.157	0.0295	
Energy costs		0.0439	
Variable costs, $·kg^{-1} biodiesel		1.1405	
Depreciation	0.10·TCC	0.0386	0.0576
Repair	0.03·TCC	0.0116	0.0173
Administrative costs	0.03·TCC	0.0116	0.0173
Personal	530,000 $·year^{-1}	0.0111	0.0344
Fixed cost, $·kg^{-1} biodiesel		0.0729	0.1265
Total manufacturing cost, $·kg^{-1} biodiesel		1.2134	1.2670
Total manufacturing cost, $·L^{-1} biodiesel		1.1163	1.1657

As seen in Table 9, vegetable oil represents 81% of the biodiesel costs in the plant of 50,000 tons·year^{-1} and 78% in the smallest plant. Castor oil can be a promising raw material for biodiesel production on account of the low production requirements. Castor bean can be grown in poor or low fertile lands with low rain indexes, making it a good option for poor regions. However, this oil shows high price on the international market, possibly due to its dependence on the Indian market and its use as lubricant and in the chemical industry. The issue of this type of biodiesel is its high viscosity, which make its direct use in injection engines difficult. However, there are some works where castor oil is used to obtain biodiesel because it has good behavior in engines when it is mixed with other biofuels or diesel [4,49].

3. Materials and Methods

3.1. Materials

Castor oil was supplied by INTERFAT (Barcelona, Spain), and transesterification was carried out by using methanol (MeOH 99.6%, Panreac, Barcelona, Spain), and the catalyst used was potassium hydroxide (85%, Alfa Aesar, Kandel, Germany). Sulfuric acid (95%–98%, Panreac, Barcelona, Spain) was used to neutralize the catalyst. For oil characterization, all the reagents (Panreac, Barcelona, Spain) were of recognized analytical grade. Methyl ester standards from Merck (Darmstadt, Germany) were used in chromatographic analysis.

3.2. Transesterification Reaction

Transesterification reactions were carried out in a 500 mL spherical glass reactor connected to a condenser, which had a sampling outlet and stirring, heating, and temperature control systems. The process was composed of two successive transesterification reactions which were carried out in the same experimental setup. For the first step, oil was preheated to 45 °C, the reaction temperature, and a solution with the desired amount of methanol and KOH was added to the reactor. For the second step, after 10 min, the mixture of reaction was separated in a decantation funnel, removing glycerol. Biodiesel phase was put into the reactor again, it was heated (45 °C) and a new solution of catalyst and alcohol was added. After the reaction time (10 min), the catalyst was neutralized with sulfuric acid, glycerol and methanol were removed, and the biodiesel was washed with distilled water. The remaining water was removed by heating at 110 °C.

To optimize the process, catalyst concentration and methanol/oil molar ratio were studied, whereas temperature and time were set at 45 °C and 10 min, respectively, for both steps and every reaction.

3.3. Experimental Design and Statistical Analysis

Central composite rotatable design (that is, CCD) was used to evaluate the influence of operational conditions on methyl ester yield. There were five levels of points and four factors for the statistical analysis, as indicated in Table 10. The selected variables were catalyst concentration and methanol/oil molar ratio in the first and the second step. Four factors in 24 full factorial CCD with five levels culminated in 31 runs of experiments (2k + 2k + 7), where k represents the number of independent variables or factors selected. Seven runs of center point experiments evaluated the pure error increased with 8 axial and 16 factorial experimental runs. The variables were normalized in the range from −2 to +2 to compare between variables according to Equation (4):

$$x_i = \frac{2(X_i - X_{min})}{(X_{max} - X_{min})} - 1 \qquad (4)$$

where x_i is the normalized value of a certain variable (X) at a certain condition i; X_i is the actual value; and X_{min} and X_{max} are the lower and upper limits, respectively. The range of each variable and the decoded values are shown in Table 10. Catalyst concentration in each step was calculated considering the total volume of reaction for each step and the methanol/oil molar ratio for both steps was based on the initial oil amount.

Table 10. Factors and their levels for response surface design.

Variable	Symbol	Coded Factor Levels				
		−2	−1	0	1	2
Catalyst concentration first step (mol·L^{-1})	A	0.02	0.04	0.06	0.08	0.10
Methanol/oil molar ratio first step	B	3.00:1	3.75:1	4.50:1	5.25:1	6.00:1
Catalyst concentration second step (mol·L^{-1})	C	0.01	0.02	0.03	0.04	0.05
Methanol/oil molar ratio second step	D	1:1	2:1	3:1	4:1	5:1

Experimental reactions were performed in a random order to minimize errors due to systematic trends in variables. The results were analyzed through RSM to fit a second-order polynomial model (see Equation (5)):

$$y = \beta_0 + \sum_i \beta_j x_j + \sum_i \beta_{jj} x_j^2 + \sum_{i<jj} \sum \beta_{ij} x_i x_j + \varepsilon \tag{5}$$

where y is the response factor (that is, % methyl ester); x_i is the ith independent factor; β_0 is the intercept; β_i is the first order coefficient of the model; β_{ii} the quadratic coefficient for i factor; β_{ij} the lineal coefficients of the model for the interaction between i and j factors; and ε the experimental error related to y. The quality of the model fit was assessed by ANOVA test and a confidence level of $\alpha = 5\%$ was used to check the statistical significance of the polynomial model. Table 11 shows the experimental conditions, the predicted and experimental values of the response factor and the residual values for each experiment.

Table 11. CCD, predicted and experimental ester content, and residual values of the design.

Runs	A	B	C	D	Predicted Ester Content (wt %)	Experimental Ester Content (wt %)	Residual Value (wt %)
1	−2	0	0	0	84.8	85.9	1.1
2	2	0	0	0	97.4	97.8	0.4
3	0	−2	0	0	84.3	84.2	−0.1
4	0	2	0	0	96.0	97.8	1.8
5	0	0	−2	0	90.3	91.3	1.0
6	0	0	2	0	94.4	95.0	0.6
7	0	0	0	−2	84.4	84.4	0.0
8	0	0	0	2	92.5	94.0	1.5
9	−1	−1	−1	−1	78.3	78.2	−0.1
10	1	−1	−1	−1	86.8	85.7	−1.1
11	−1	1	−1	−1	90.1	89.2	−0.9
12	1	1	−1	−1	95.0	96.0	1.0
13	−1	−1	1	−1	83.4	84.0	0.6
14	1	−1	1	−1	89.4	90.1	0.7
15	−1	1	1	−1	91.2	90.4	−0.8
16	1	1	1	−1	93.6	92.6	−1.0
17	−1	−1	−1	1	83.3	83.0	−0.3
18	1	−1	−1	1	93.6	94.3	0.7
19	−1	1	−1	1	90.9	89.9	−1.0
20	1	1	−1	1	97.6	95.7	−1.9
21	−1	−1	1	1	88.9	87.7	−1.2
22	1	−1	1	1	96.6	96.2	−0.4
23	−1	1	1	1	92.5	92.4	−0.1
24	1	1	1	1	96.7	96.5	−0.2
25	0	0	0	0	93.0	92.5	−0.5
26	0	0	0	0	93.0	93.5	0.5
27	0	0	0	0	93.0	92.3	−0.7

Table 11. *Cont.*

Runs	A	B	C	D	Predicted Ester Content (wt %)	Experimental Ester Content (wt %)	Residual Value (wt %)
28	0	0	0	0	93.0	91.9	−1.1
29	0	0	0	0	93.0	93.5	0.5
30	0	0	0	0	93.0	94.2	1.2
31	0	0	0	0	93.0	92.9	−0.1

3.4. Analytical Procedure

Biodiesel was analyzed by gas chromatography, using a gas chromatograph with an FID detector (VARIAN 3900, Varian, Palo Alto, CA, USA). A polyethylene glycol column (Zebron ZB-WAX PLUS, Phenomenex, CA, USA) was used with the following characteristics: length, 30 m; film thickness, 0.5 μm; and i.d., 0.32 mm. Helium (1.4 mL·min^{-1}) was used as a carrier gas, and the temperature oven for each run was 220 °C for 34 min and at 245 °C for 29 min with a ramp of 20 °C·min^{-1}. The injector and detector temperatures were 270 and 300 °C, respectively. Methyl heptadecanoate and methyl erucate were used as standards for the internal standard method for most methyl esters and methyl ricinoleate, respectively. Ethyl acetate was used as a solvent for the standards, and calibration curves were carried out for each standard.

4. Conclusions

The main findings in this research work are as follows:

Two-step transesterification was an effective and economic method to produce biodiesel from castor oil. An analysis of the main variables of the process—catalyst concentration and methanol/oil molar ratio—in both steps showed that this method was quite robust, since a lot of experimental conditions produced a methyl ester yield in excess of 96.5%. An economic assessment of the main variable costs of biodiesel production showed an important reduction of annual expenses when the optimum conditions were used. This decrease was caused by the use of a two-step process instead of one-step, and the optimization of the conditions allowed for the use of response surface methodology. In the complete economic analysis, raw material costs accounted for a major portion of the total manufacturing costs. For this reason, the total manufacturing costs of biodiesel in the smallest industrial plant were close to the value in the biggest plant. The decrease in castor oil price could considerably improve the profitability of the process.

Supplementary Materials: The following are available online at http://www.mdpi.com/2073-4344/9/10/864/s1, Figure S1: Experimental methyl ester content versus predicted values, Table S1: Experimental conditions which lead to an ester content greater than 96.5 %, according to the SRM.

Author Contributions: Conceptualization, N.S. and J.M.E.; methodology, N.S.; software, N.S.; validation, N.S., J.M.E. and J.F.G.; formal analysis, N.S.; investigation, N.S.; resources, J.M.E and J.F.G.; data curation, N.S. and S.N.; writing—original draft preparation, N.S.; writing—review and editing, S.N.; visualization, S.N. and J.M.E.; supervision, J.M.E. and J.F.G.; project administration, J.M.E. and J.F.G.; funding acquisition, J.M.E. and J.F.G.

Funding: This research was funded by the "Ministerio de Ciencia, Innovación y Universidades", the "Junta de Extremadura" and "FEDER" (ENE2009–13881, PRI09B102, IB18028, GR10159 and GR18150).

Acknowledgments: The authors would like to thank to the "Ministerio de Ciencia, Innovación y Universidades" and the "Gobierno de Extremadura" for the financial support received to perform this study. Nuria Sánchez thanks the Spanish Ministry of Education for the grant received.

Conflicts of Interest: The authors declare no conflict of interest.

References

1. Balat, M.; Balat, H. Progress in biodiesel processing. *Appl. Energy* **2010**, *87*, 1815–1835. [CrossRef]
2. Santori, G.; Di Nicola, G.; Moglie, M.; Polonara, F. A review analyzing the industrial biodiesel production practice starting from vegetable oil refining. *Appl. Energy* **2012**, *92*, 109–132. [CrossRef]

3. Leung, D.Y.C.; Wu, X.; Leung, M.K.H. A review on biodiesel production using catalyzed transesterification. *Appl. Energy* **2010**, *87*, 1083–1095. [CrossRef]
4. Scholz, V.; da Silva, J.N. Prospects and risks of the use of castor oil as a fuel. *Biomass Bioenergy* **2008**, *32*, 95–100. [CrossRef]
5. Al-Esawi, N.; Al Qubeissi, M.; Kolodnytska, R. The impact of biodiesel fuel on ethanol/diesel blends. *Energies* **2019**, *12*, 1804. [CrossRef]
6. Hurtado, B.; Posadillo, A.; Luna, D.; Bautista, F.M.; Hidalgo, J.M.; Luna, C.; Calero, J.; Romero, A.A.; Estevez, R. Synthesis, performance and emission quality assessment of ecodiesel from castor oil in diesel/biofuel/alcohol triple blends in a diesel engine. *Catalysts* **2019**, *9*, 40. [CrossRef]
7. Estevez, R.; Aguado-Deblas, L.; Posadillo, A.; Hurtado, B.; Bautista, F.M.; Hidalgo, J.M.; Luna, C.; Calero, J.; Romero, A.A.; Luna, D. Performance and emission quality assessment in a diesel engine of straight castor and sunflower vegetable oils, in diesel/gasoline/oil triple blends. *Energies* **2019**, *12*, 2181. [CrossRef]
8. Sánchez, N.; Sánchez, R.; Encinar, J.M.; González, J.F.; Martínez, G. Complete analysis of castor oil methanolysis to obtain biodiesel. *Fuel* **2015**, *147*, 95–99. [CrossRef]
9. Meneghetti, S.M.P.; Meneghetti, M.R.; Serra, T.M.; Barbosa, D.C.; Wolf, C.R. Biodiesel production from vegetable oil mixtures: Cottonseed, soybean, and castor oils. *Energy Fuel* **2007**, *21*, 3746–3757. [CrossRef]
10. Hincapié, G.; Mondragón, F.; López, D. Conventional and in situ transesterification of castor seed oil for biodiesel production. *Fuel* **2011**, *90*, 1618–1623. [CrossRef]
11. Encinar, J.M.; González, J.F.; Pardal, A. Transesterification of castor oil under ultrasonic irradiation conditions. Preliminary results. *Fuel Process. Technol.* **2012**, *103*, 9–15. [CrossRef]
12. Zieba, A.; Matachowski, L.; Gurgul, J.; Bielanska, E.; Drelinkiewicz, A. Transesterification reaction of triglycerides in the presence of Ag-doped $H_3PW_{12}O_{40}$. *J. Mol. Catal. A Chem.* **2010**, *316*, 30–44. [CrossRef]
13. Martinez-Guerra, E.; Gnaneswar Gude, V. Assessment of sustainability indicators for biodiesel production. *Appl. Sci.* **2017**, *7*, 869. [CrossRef]
14. Coronado, C.R.; Tuna, C.E.; Zanzi, R.; Vane, L.F.; Silveira, J.L. Development of a thermoeconomic methodology for optimizing biodiesel production. Part II: Manufacture exergetic cost and biodiesel production cost incorporating carbon credits, a Brazilian case study. *Renew. Sustain. Energy Rev.* **2014**, *29*, 565–572. [CrossRef]
15. Olkiewicz, M.; Torres, C.M.; Jiménez, L.; Font, J.; Bengoa, C. Scale-up and economic analysis of biodiesel production from municipal primary sewage sludge. *Bioresour. Technol.* **2016**, *214*, 122–131. [CrossRef]
16. Zhang, Y.; Dubé, M.A.; McLean, D.D.; Kates, M. Biodiesel production from waste cooking oil: 2. Economic assessment and sensitivity analysis. *Bioresour. Technol.* **2003**, *90*, 229–240. [CrossRef]
17. Mohammadshirazi, A.; Akram, A.; Rafiee, S.; Bagheri Kalhor, E. Energy and cost analyses of biodiesel production from waste cooking oil. *Renew. Sustain. Energy Rev.* **2014**, *33*, 44–49. [CrossRef]
18. Tang, Z.-C.; Lu, Z.; Liu, Z.; Xiao, N. Uncertainty analysis and global sensitivity analysis of techno-economic assessments for biodiesel production. *Bioresour. Technol.* **2015**, *175*, 502–508. [CrossRef]
19. Santana, G.C.S.; Martins, P.F.; de Lima da Silva, N.; Batistella, C.B.; Maciel Filho, R.; Wolf Maciel, M.R. Simulation and cost estimate for biodiesel production using castor oil. *Chem. Eng. Res. Des.* **2010**, *88*, 626–632. [CrossRef]
20. Dias, J.M.; Araújo, J.M.; Costa, J.F.; Alvim-Ferraz, M.C.M.; Almeida, M.F. Biodiesel production from raw castor oil. *Energy* **2013**, *53*, 58–66. [CrossRef]
21. Banković-Ilić, I.B.; Stamenković, O.S.; Veljković, V.B. Biodiesel production from non-edible plant oils. *Renew. Sustain. Energy Rev.* **2012**, *16*, 3621–3647. [CrossRef]
22. Karmakar, A.; Karmakar, S.; Mukherjee, S. Properties of various plants and animals feedstocks for biodiesel production. *Bioresour. Technol.* **2010**, *101*, 7201–7210. [CrossRef] [PubMed]
23. Canakci, M.; Van Gerpen, J. Biodiesel production from oils and fats with high free fatty acids. *Trans. Am. Soc. Agric. Eng.* **2001**, *44*, 1429–1436. [CrossRef]
24. Avasthi, K.S.; Reddy, R.N.; Patel, S. Challenges in the production of hydrogen from glycerol—A biodiesel byproduct via steam reforming process. *Procedia Eng.* **2013**, *51*, 423–429. [CrossRef]
25. Martínez, G.; Sánchez, N.; Encinar, J.M.; González, J.F. Fuel properties of biodiesel from vegetable oils and oil mixtures. Influence of methyl esters distribution. *Biomass Bioenergy* **2014**, *63*, 22–32. [CrossRef]
26. Coronado, C.R.; Tuna, C.E.; Zanzi, R.; Vane, L.F.; Silveira, J.L. Development of a thermoeconomic methodology for the optimization of biodiesel production—Part I: Biodiesel plant and thermoeconomic functional diagram. *Renew. Sustain. Energy Rev.* **2013**, *23*, 138–146. [CrossRef]

27. Encinar, J.M.; Sánchez, N.; Martínez, G.; García, L. Study of biodiesel production from animal fats with high free fatty acid content. *Bioresour. Technol.* **2011**, *102*, 10907–10914. [CrossRef]
28. Encinar, J.M.; Pardal, A.; Sánchez, N. An improvement to the transesterification process by the use of co-solvents to produce biodiesel. *Fuel* **2016**, *166*, 51–58. [CrossRef]
29. Canoira, L.; García Galeán, J.; Alcántara, R.; Lapuerta, M.; García-Contreras, R. Fatty acid methyl esters (FAMEs) from castor oil: Production process assessment and synergistic effects in its properties. *Renew. Energy* **2010**, *35*, 208–217. [CrossRef]
30. Meneghetti, S.M.P.; Meneghetti, M.R.; Wolf, C.R.; Silva, E.C.; Lima, G.E.S.; de Lira Silva, L. Biodiesel from castor oil: A comparison of ethanolysis versus methanolysis. *Energy Fuel* **2006**, *20*, 2262–2265. [CrossRef]
31. Barbosa, D.D.C.; Serra, T.M.; Meneghetti, S.M.P.; Meneghetti, M.R. Biodiesel production by ethanolysis of mixed castor and soybean oils. *Fuel* **2010**, *89*, 3791–3794. [CrossRef]
32. Peña, R.; Romero, R.; Martínez, S.L.; Ramos, M.J.; Martínez, A.; Natividad, R. Transesterification of castor oil: Effect of catalyst and co-solvent. *Ind. Eng. Chem. Res.* **2009**, *48*, 1186–1189. [CrossRef]
33. Abuhabaya, A.; Fieldhouse, J.; Brown, D. The optimization of biodiesel production by using response surface methodology and its effect on compression ignition engine. *Fuel Process. Technol.* **2013**, *113*, 57–62. [CrossRef]
34. Jeong, G.-T.; Park, D.-H. Optimization of biodiesel production from castor oil using response surface methodology. *Appl. Biochem. Biotechnol.* **2009**, *156*, 1–11. [CrossRef] [PubMed]
35. Santos, O.O., Jr.; Maruyama, S.A.; Claus, T.; de Souza, N.E.; Matsushita, M.; Visentainer, J.V. A novel response surface methodology optimization of base-catalyzed soybean oil methanolysis. *Fuel* **2013**, *113*, 580–585. [CrossRef]
36. Methanex Methanol Price Sheet. Available online: http://www.methanex.com (accessed on 28 October 2015).
37. Sales-Cruz, M.; Aca-Aca, G.; Sánchez-Daza, O.; López-Arenas, T. Predicting critical properties, density and viscosity of fatty acids, triacylglycerols and methyl esters by group contribution methods. In Proceedings of the 20th European Symposium on Computer Aided Process, Engineering—ESCAPE20, Naples, Italy, 6 June 2010.
38. França, B.B.; Pinto, F.M.; Pessoa, F.L.P.; Uller, A.M.C. Liquid-Liquid Equilibria for Castor Oil Biodiesel + Glycerol + Alcohol. *J. Chem. Eng. Data* **2008**, *54*, 2359–2364. [CrossRef]
39. Machado, A.B.; Ardila, Y.C.; de Oliveira, L.H.; Aznar, M.; Wolf Maciel, M.R. Liquid–Liquid Equilibrium Study in Ternary Castor Oil Biodiesel + Ethanol + Glycerol and Quaternary Castor Oil Biodiesel + Ethanol + Glycerol + NaOH Systems at (298.2 and 333.2) K. *J. Chem. Eng. Data* **2011**, *56*, 2196–2201. [CrossRef]
40. Zhang, Y.; Dubé, M.A.; McLean, D.D.; Kates, M. Biodiesel production from waste cooking oil: 1. Process design and technological assessment. *Bioresour. Technol.* **2003**, *89*, 1–16. [CrossRef]
41. Geankoplis, C.J. *Transport Processes and Unit Operations*, 3rd ed.; Prentice-Hall International, Inc.: Upper Saddle River, NJ, USA, 1993.
42. West, A.H.; Posarac, D.; Ellis, N. Assessment of four biodiesel production processes using HYSYS. Plant. *Bioresour. Technol.* **2008**, *99*, 6587–6601. [CrossRef]
43. Silla, H. *Chemical Process Engineering. Design and Economics*; Marcel Dekker, Inc.: New York, NY, USA, 2003.
44. Treybal, R.E. *Mass-Transfer Operations*, 3rd ed.; McGraw Hill: Singapore, 1981.
45. Loh, H.P.; Lyons, J. *Process Equipment Cost Estimation*; Final Report; DOE/NETL-2002/1169; EG&G Technical Services Inc.: Morgantown, WV, USA, 2002.
46. Institution Chemical Engineering. *A New Guide to Capital Cost Estimating*; I Chem E Services: Rugby, Warkwicksire, UK, 1985.
47. Peters, M.S.; Timmerhaus, K.D. *Plant Design and Economics for Chemical Engineers*, 4th ed.; McGraw Hill: New York, NY, USA, 1991.
48. Castor Oil Industry Reference & Resources. Available online: http://castoroil.in (accessed on 18 December 2015).
49. Panwar, N.L.; Shrirame, H.Y.; Rathore, N.S.; Jindal, S.; Kurchania, A.K. Performance evaluation of a diesel engine fueled with methyl ester of castor seed oil. *Appl. Therm. Eng.* **2010**, *30*, 245–249. [CrossRef]

 © 2019 by the authors. Licensee MDPI, Basel, Switzerland. This article is an open access article distributed under the terms and conditions of the Creative Commons Attribution (CC BY) license (http://creativecommons.org/licenses/by/4.0/).

Article

Catalytic Performance of Bulk and Al$_2$O$_3$-Supported Molybdenum Oxide for the Production of Biodiesel from Oil with High Free Fatty Acids Content

Alberto Navajas [1,2], Inés Reyero [1,2], Elena Jiménez-Barrera [3], Francisca Romero-Sarria [3], Jordi Llorca [4] and Luis M. Gandía [1,2,*]

1. Departamento de Ciencias, Edificio de los Acebos, Universidad Pública de Navarra, Campus de Arrosadía s/n, 31006 Pamplona, Spain; alberto.navajas@unavarra.es (A.N.); ines.reyero@unavarra.es (I.R.)
2. Institute for Advanced Materials (InaMat), Universidad Pública de Navarra, Campus de Arrosadía, 31006 Pamplona, Spain
3. Departamento de Química Inorgánica e Instituto de Ciencia de Materiales de Sevilla, Centro mixto Universidad de Sevilla-CSIC, Av. Américo Vespucio 49, 41092 Seville, Spain; elenamaria_jb@hotmail.com (E.J.-B.); francisca@us.es (F.R.-S.)
4. Institute of Energy Technologies, Department of Chemical Engineering and Barcelona Research Center in Multiscale Science and Engineering. Universitat Politècnica de Catalunya, EEBE, Eduard Maristany 16, 08019 Barcelona, Spain; jordi.llorca@upc.edu
* Correspondence: lgandia@unavarra.es

Received: 16 January 2020; Accepted: 29 January 2020; Published: 1 February 2020

Abstract: Non-edible vegetable oils are characterized by high contents of free fatty acids (FFAs) that prevent from using the conventional basic catalysts for the production of biodiesel. In this work, solid acid catalysts are used for the simultaneous esterification and transesterification with methanol of the FFAs and triglycerides contained in sunflower oil acidified with oleic acid. Molybdenum oxide (MoO$_3$), which has been seldom considered as a catalyst for the production of biodiesel, was used in bulk and alumina-supported forms. Results showed that bulk MoO$_3$ is very active for both transesterification and esterification reactions, but it suffered from severe molybdenum leaching in the reaction medium. When supported on Al$_2$O$_3$, the MoO$_3$ performance improved in terms of active phase utilization and stability though molybdenum leaching remained significant. The improvement of catalytic performance was ascribed to the establishment of MoO$_3$-Al$_2$O$_3$ interactions that favored the anchorage of molybdenum to the support and the formation of new strong acidic centers, although this effect was offset by a decrease of specific surface area. It is concluded that the development of stable catalysts based on MoO$_3$ offers an attractive route for the valorization of oils with high FFAs content.

Keywords: acid catalysis; biodiesel; biofuel; esterification; fatty acid; methanolysis; molybdenum oxide; transesterification; vegetable oil

---

1. Introduction

Biodiesel (a mixture of fatty acid methyl esters, FAMEs) has been the most important alternative fuel for diesel engines for over 25 years [1]. It is typically produced by the catalytic transesterification of refined vegetable oils such as soybean, palm and rapeseed in liquid phase (homogeneous catalysis). To make biodiesel more cost-competitive with petroleum diesel, the use of refined oils, whose cost has been estimated to account for 70–95% of the total costs, should be avoided by replacing them with low-cost feedstocks, such as waste greases, brown grease, non-edible vegetable oils, dark oil generated by the vegetable oil refining industry, or used cooking oils [1–3]. An obvious and important additional benefit of this strategy is obtaining a more sustainable alternative fuel that contributes to reducing the CO$_2$

emissions to the atmosphere and the dependency of the energy system on petroleum. On the other hand, the basic catalysts usually used in the biodiesel industry (KOH, NaOH, and potassium and sodium methoxides) have important drawbacks [4]. Although they could be replaced by heterogeneous solid catalysts to make their re-utilization possible, some issues concerning those materials such as the lack of the required chemical stability are still not well solved [5–7]. Moreover, basic catalysts are not capable of suitably processing virgin, non-edible or waste feedstocks, which are characterized by being relatively rich in free fatty acids (FFAs), because they result deactivated and/or consumed, e.g., by the formation of soaps that also complicate the separation of the products due to their emulsifying properties. In these cases, acid catalysts can be used to perform a pre-esterification of the FFAs followed by base-catalyzed transesterification giving rise to the so-called integrated process for biodiesel production from high FFAs-containing triglyceride feedstocks [1,8]. Nevertheless, this is a multistep process that introduces complexity, thus reducing the competitive advantage of using low-cost feedstocks. It would be more convenient to perform the simultaneous transformation of the triglycerides and FFAs into FAMEs, which can be in fact accomplished using heterogeneous acid catalysts [9–11].

Heterogeneous acid catalysts considered for biodiesel production comprise a large variety of materials [8–18]. The most representative ones consist on sulfated and tungstated zirconia, niobia, silica and alumina, bulk and supported polyoxometalates and heteropoly acids (HPAs) of e.g., Si^{4+} or P^{5+} with W^{6+} or Mo^{6+}, mixed metal oxides such as titania-zirconia and silica-zirconia, immobilized acidic ionic liquids, sulfonic ion-exchange resins, zeolites, mesoporous silicas (SBA, MCM and KIT series), and carbonaceous materials functionalized with sulfonic acid groups. Each of these families of compounds presents advantages and drawbacks whereas their catalytic performance is intimately related to their textural and acidic properties that depend on the exact composition, synthesis method, activation conditions, etc. In this regard, the low activity reported for zeolites has been related to the strong diffusional limitations suffered by the bulky triglyceride and FFA molecules. As for the acidic properties, Brönsted sites are considered significantly more active than the Lewis ones [10], though the necessity of having both strong Brönsted and Lewis sites has also been claimed [1]. Another typical feature of the production of biodiesel through acid catalysis is that temperatures up to around 190 °C are required to obtain reasonable product yields which is in contrast with the temperatures close to the normal boiling point of methanol (65 °C) that are employed with the basic catalysts, even the heterogeneous ones. This is obviously related to the well-known lower activity for this process of the acid catalysts compared to the basic ones.

Among the several materials proposed, sulfated zirconia (SO_4^{2-}/ZrO_2) is considered a super-acid solid exhibiting the strongest Brönsted acidity; therefore, it has been widely investigated as a catalyst for the production of biodiesel. A significant lack of stability has been reported for SO_4^{2-}/ZrO_2 due to sulfate leaching, as well as deactivation by fouling associated with surface deposition of reaction products and reactants causing the blockage of the active sites [1]. On the other hand, some compounds of tungsten and molybdenum, both metals belonging to group 6 of the Periodic Table, have been also identified as potential catalysts for the synthesis of biodiesel. Several works can be found in the literature showing the good catalytic performance of supported WO_3 for the conversion of oils with high contents of FFA [19,20]. As for molybdenum, their compounds have received little attention in this field up to date. Molybdenum compounds are characterized by a remarkable versatility as catalysts, a fact that is related to the ability of this metal to be present on the solid surface in different oxidation states, ranging from Mo^{6+} to metallic Mo (Mo^0) [21]. Anhydrous sodium molybdate [22], bulk MoO_3 [23], and molybdenum supported on alumina [24], silica, silica-alumina, and titania [25,26], as well as carbon [27] have been used as esterification and transesterification catalysts for biodiesel production from several oils, including waste oil.

In this work, both bulk (unsupported) and alumina-supported MoO_3 have been used as catalysts to convert sunflower oil with added FFAs into biodiesel. The main objective is to contribute to a better understanding of the parameters controlling the catalytic performance of these materials in pursuit

2. Results and Discussion

2.1. Catalysts Characterization

Three bulk molybdenum oxide catalysts (Mo (300), Mo (500) and Mo (700)) were prepared through the calcination of ammonium heptamolybdate tetrahydrate (AHM) at 300, 500, and 700 °C. In addition, a series of alumina-supported molybdenum oxide catalysts (s-Mo (x), where x is the MoO_3 content (wt.%)) were obtained from the incipient wetness impregnation of alumina with aqueous AHM solutions and calcination at 800 °C.

The X-ray diffraction (XRD) patterns of the bulk and a selection of the alumina-supported molybdenum catalysts are shown in Figure 1. The thermal decomposition AHM is a complex process in which several molybdenum compounds are involved. Chithambararaj et al. [28] found that ammonium was still present in the crystalline structures at calcination temperatures up to about 300 °C. According to these authors, at relatively low temperatures AHM evolves from $[NH_4]_4(Mo_8O_{24.8}(O_2)_{1.2}(H_2O)_2)(H_2O)_4$ at 50–75 °C to $[NH_4]_8Mo_{10}O_{34}$ between 100 °C and 200 °C; finally, $[NH_4]_4Mo_8O_{26}$ is formed near 300 °C. Then, $MoO_{2.5}(OH)_{0.5}$ is present between 325 and 475 °C that evolves to $MoO_{2.69}(OH)_{0.3}$ at ca. 575 °C, finally resulting MoO_{3-x} at about 700 °C. It should be noted that molybdenum trioxide can exist both as layered orthorhombic α-MoO_3 and monoclinic β-MoO_3 which has monoclinic rutile type structure formed by edge-sharing MoO_6 octahedra [29]. The XRD patterns of Mo (300), Mo (500) and Mo (700) show also an evolution as the calcination temperature increases. In the case of Mo (700), a mixture of crystalline phases is present that apparently contains MoO_3 and $MoO_{2.69}(OH)_{0.3}$. On the other hand, Mo (500) seems to contain both $MoO_{2.5}(OH)_{0.5}$ and $MoO_{2.69}(OH)_{0.3}$. As for the Mo (300) sample, in addition to $MoO_{2.5}(OH)_{0.5}$ and $MoO_{2.69}(OH)_{0.3}$, the presence of compounds containing ammonium cannot be ruled out.

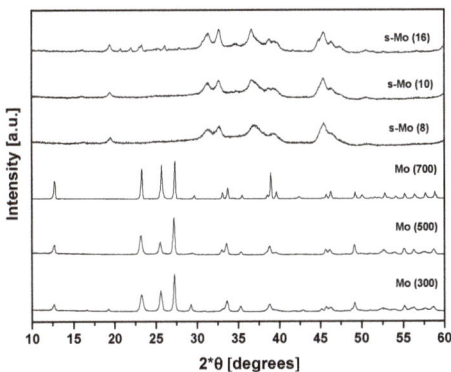

Figure 1. X-ray diffraction (XRD) patterns of the bulk molybdenum catalysts (Mo) calcined at 300, 500, and 700 °C, and of the alumina-supported ones (s-Mo) containing 8, 10 and 16 wt.% MoO_3 and calcined at 800 °C.

In the case of the supported catalysts (calcined at 800 °C), no crystalline molybdenum compound could be identified in the XRD patterns of the samples containing less than 16 wt.% MoO_3 which suggests that the supported species are very well dispersed over the alumina support. However, new diffraction peaks appeared between 20 and 26° (2θ) in the XRD pattern of the s-Mo (16) sample that could correspond to aluminum molybdate ($Al_2(MoO_4)_3$) and to a series of polyoxo Mo species, as reported in the literature for 16 wt.% MoO_3 supported on alumina and calcined at different temperatures between 527 and 827 °C [26]. Kitano et al. [30] investigated a series of alumina-supported molybdenum catalysts calcined at 800 °C containing between 5 and 30 wt.% MoO_3. No XRD peaks corresponding to Mo compounds were found in the diffraction patterns of the catalysts with MoO_3 loadings below 11 wt.% whereas for contents above 13 wt.% $Al_2(MoO_4)_3$ could be detected. The authors estimated at 10 wt.% the MoO_3 loading required to form a surface monolayer. It was proposed that two-dimensional molybdenum oxide domains and some three-dimensional MoO_3 clusters were formed on 11 wt.% MoO_3/Al_2O_3 whereas the MoO_3 clusters were transformed into aluminum molybdate at higher metallic contents. As can be seen, the XRD results reveal a complex nature of the Mo catalysts that is due to the rich chemistry of this metal.

As expected for the bulk catalysts, these solids showed low specific surface areas that decreased from 17.0 m^2/g for Mo (300) to 4.9 m^2/g for Mo (500) and finally 1.6 m^2/g for Mo (700) as the AHM calcination temperature increased. As for the supported catalysts, their specific surface areas (S_{BET}) are compiled in Table 1; all the samples, included the support, were calcined at 800 °C.

Table 1. Physicochemical characterization data of the supported MoO_3 catalysts.

Sample	S_{BET} (m^2/g)	$[NH_3]_{des.}$ ($\mu mol/m^2$) [1]
Al_2O_3 support	112	2.7
s-Mo (6)	99	2.8
s-Mo (8)	72	2.4
s-Mo (10)	63	2.4
s-Mo (13)	47	1.9
s-Mo (16)	44	2.0

[1] Data obtained from the desorption peak centered at 147–150 °C in all cases.

It can be seen that a gradual decrease of the S_{BET} takes place as the MoO_3 loading increases up to 13 wt.%. However, the additional decrease of specific surface area taking place upon an additional MoO_3 loading increase up to 16 wt.% is low resulting in a S_{BET} of 44 m^2/g for s-Mo (16), which represents a decrease of 60% with respect to the alumina support. This indicates a considerable blockage of the porous network, which is not accompanied by the formation of detectable Mo crystalline phases until a 16 wt.% MoO_3 loading is reached, as evidenced by the XRD results. Following the calculations performed by Kitano et al. [30], who assigned a cross-sectional area of 0.22 nm^2 to the octahedral MoO_6 unit, the molybdenum oxide required to form a monolayer over the alumina support used in this work can be estimated at about 11 wt.%. Taking into account these results it could be suggested that in the s-Mo (6) to s-Mo (13) series of samples, Mo oxide seems to be present mainly in the form of a two-dimensional structure well dispersed over the support surface.

The X-ray photoelectron spectroscopy (XPS) results of the bulk molybdenum catalysts are shown in Figure 2.

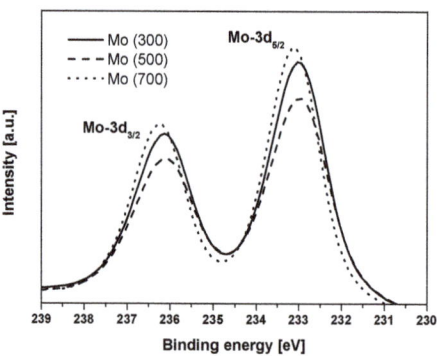

Figure 2. X-ray photoelectron spectra of the Mo 3d doublet corresponding to the bulk MoO_3 catalysts.

In all cases, the XP spectra are characterized by two well-resolved contributions at binding energies of 233.0–233.1 and 236.2–236.3 eV that can be assigned to the Mo $3d_{5/2}$ and Mo $3d_{3/2}$ spin-orbit components of Mo (VI), respectively. The splitting energy of this doublet (3.2 eV) agrees well with the values reported in the literature for Mo (VI) [31]. The presence of molybdenum in oxidation states other than Mo (VI) seems to be not significant though corresponding $3d_{5/2}$ spectral lines have been reported at significantly lower binding energy values (231.6 eV for Mo (V)) [32]. Choi and Thompson indicated values for the Mo $3d_{5/2}$ line of polycrystalline MoO_3 between 231.6 and 232.7 eV [31]. Baltrusaitis et al. [29] compiled a series of values for Mo (VI) taken from the literature and the binding energies reported were 233.0–233.2 eV (Mo $3d_{5/2}$) and 236.1–236.3 eV (Mo $3d_{3/2}$), which coincide with the values found in this work. These values were measured for a layer of oxide thermally developed on molybdenum metal under controlled O_2 pressure [33] as well as for alumina-supported cobalt-molybdena catalysts [34].

Regarding acidity, it was first measured through temperature-programmed desorption of NH_3 (NH_3-TPD), a technique that is not capable of distinguishing between Brönsted and Lewis sites due to the strong basicity of ammonia. As a matter of fact, NH_3 can adsorb on both Lewis and Brönsted sites of MoO_3 as molecular NH_3 and ammonium ion, respectively, though adsorption on Lewis sites is more favorable energetically [35]. The NH_3-TPD patterns corresponding to the alumina support and a selection of the alumina-supported catalysts are included in Figure 3.

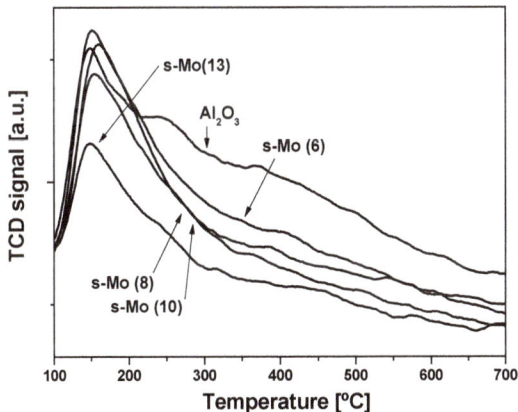

Figure 3. Temperature-programmed desorption of NH_3 (NH_3-TPD) patterns of the samples indicated.

A main peak showing a maximum within the 147–150 °C temperature range characterizes the patterns. The peaks are broad and show a long tail that extends up to 300–600 °C, depending on the case. Very small peaks centered at about 240 °C and 375 °C can be observed in the alumina NH_3-TPD pattern. It is also apparent that baseline drifting complicated the integration of the peaks, which resulted in the specific acidity values compiled in Table 1. It can be seen that the specific acidity tends to decrease as the molybdenum oxide content increases. This result suggests that the acidity of the solids decreases as the alumina surface becomes covered by molybdenum oxide species. Sankaranarayanan et al. [24] found similar NH_3-TPD results for a series of alumina-supported molybdenum oxide catalysts calcined at 527 °C, 677 °C and 827 °C and containing 8wt.%, 12wt.% and 16 wt.% MoO_3. The acidity of the samples calcined at 827 °C decreased from 3.8 µmol/m² for the sample loaded with 8 wt.% MoO_3 to 2.8 µmol/m² for the catalyst containing 16 wt.% MoO_3. This last result can be compared with the value of 2.0 µmol/m² obtained in this work for a sample containing 16 wt.% MoO_3 and calcined at 800 °C (see Table 1). Accurate assessment of the acidity of the unsupported solids through NH_3-TPD was difficult due to the low specific surface area of these materials and the technical limitations of the equipment used (see Section 3) to load high amounts of sample. Their specific acidity was somewhat lower than that of the supported solids, ranging between 0.6 µmol/m² for Mo (300) and 2.0 µmol/m² for Mo (500) and Mo (700).

Representative FTIR spectra of adsorbed pyridine corresponding to bulk (Mo (500)) and Al$_2$O$_3$-supported catalysts (s-Mo (8)) are shown in Figure 4. Pyridinium ions adsorbed over Brönsted sites exhibit characteristic bands at wavenumbers of 1546–1548 cm^{-1} (ν_{19b}), and 1638–1640 cm^{-1} (ν_{8a}) [36,37]. The spectra of Mo (500) and s-Mo (8) show weak and broad bands at 1634 and 1638 cm^{-1} as well as at 1532 and 1542 cm^{-1}, respectively. These bands are ascribed to Brönsted sites that due to their different nature appear at different wavenumbers. As for the Lewis sites on molybdenum oxide, bands of adsorbed pyridine at 1451 (ν_{19b}) and 1611 cm^{-1} (ν_{8a}) have been reported [37]. These values are similar to the ones reported for adsorbed pyridine on Al^{3+} in alumina surface: 1450 and 1615 cm^{-1} [36]. The spectrum of the alumina-supported catalyst shows a weak band at 1616 cm^{-1} and another very intense absorption signal at 1442 cm^{-1} that seems compatible with pyridine adsorption on acidic Lewis sites. In the case of the bulk molybdenum oxide catalyst, the band at 1443 cm^{-1} is much weaker and the broad band observed at a wavelength of 1606 cm^{-1} could correspond to the one reported at 1611 cm^{-1} [37]. This difference in intensity indicates a lower number of Lewis sites on the surface of Mo(500) catalyst. Moreover, the position of the band corresponding to the (ν_{8a}) vibration mode of pyridine adsorbed on Lewis sites is used as an indicator of the Lewis strength. Comparing spectra in Figure 4, we can conclude that acidic Lewis sites are stronger in the supported catalyst (1616 cm^{-1} compared to 1606 cm^{-1} in the Mo(500)) [38]. It is interesting to stress that an intense band at 1594 cm^{-1}, absent in the Mo(500) spectrum, is detected in the supported catalyst after pyridine adsorption. This band may be ascribed to species formed by interaction of the probe molecule via N-atom with a weakly acidic H from the surface (denoted as HPy) [39]. Therefore, the new Brönsted sites (unable to protonate pyridine) appear on the supported catalyst, probably due to the interaction of MoO$_3$ with the alumina. It is concluded that both, bulk and supported molybdenum oxide catalysts, have Brönsted and Lewis acid sites, though they are much more abundant and stronger in the case of the supported solids, which is logical due to the presence of the alumina support. Moreover, new Brönsted sites are detected in the supported solid that are not present in the unsupported one.

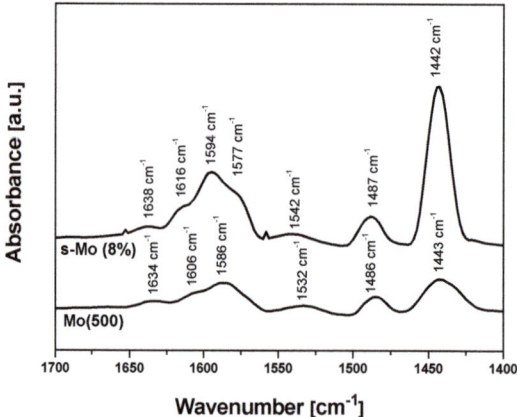

Figure 4. FTIR spectra of adsorbed pyridine on Mo (500) and s-Mo (8) catalysts.

2.2. Catalytic Performance

Figure 5 shows the evolution of the conversion of triglycerides (X_{TG}) and FFAs (X_{FFA}) with reaction time for the unsupported Mo (300), Mo (500), and Mo (700) catalysts.

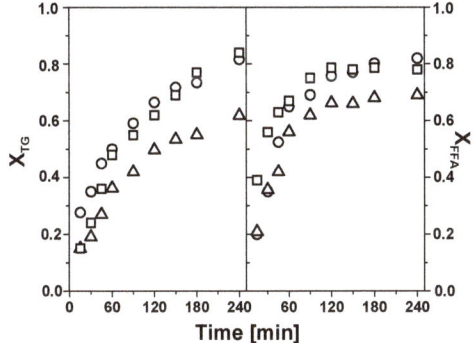

Figure 5. Conversion of the triglycerides (X_{TG}) and FFAs (X_{FFA}) with reaction time for the unsupported Mo (300) (squares), Mo (500) (circles), and Mo (700) (triangles) catalysts. Reaction conditions: 100 °C, 30 atm, methanol/feedstock molar ratio of 12:1, 2 wt.% catalyst referred to the feedstock mass. Feedstock: refined sunflower oil containing 5 wt.% free oleic acid.

It can be seen that both triglycerides and FFAs reach high degrees of conversion (between about 60 and 80%) after 240 min of reaction under the conditions indicated (see Figure 5 caption). The conversion over Mo (300) and Mo (500) are similar, about 80–85% and 78–82% for the triglycerides and FFAs, respectively, after 240 min of reaction. The conversion over Mo (700) is lower, about 60% (triglycerides) and 70% (FFAs) compared to its counterparts calcined at lower temperatures. As the catalyst concentration is fixed on a mass basis (2 wt.%) it is likely that the main cause of these results is the large decrease of the specific surface area experienced by Mo (700) upon calcination. It is also noteworthy that, irrespective of the catalyst used, X_{TG} gradually increases with reaction time, whereas FFAs conversion shows a faster initial increase until about 120 min of reaction, with a much slower increase taking place afterward.

Regarding the yields of the several reaction products, Figure 6a shows the evolution with X_{TG} of the diglycerides (Y_{DG}), monoglycerides (Y_{MG}), and methyl esters produced by transesterification ($Y_{ME,trans}$) (see Equations (3)–(5) in Section 3).

The evolution of the products yields is in accordance with the scheme of the triglycerides (TG) methanolysis reaction consisting of three consecutive steps [4]. In the first step, a molecule of TG is converted into a diglyceride (DG), which then evolves to a monoglyceride (MG) that finally gives glycerol. In each step, a molecule of methanol is consumed and a methyl ester (biodiesel) molecule is formed. This explains why initially, at low X_{TG}, DG were clearly the most abundant products. As the conversion increases, Y_{DG} also increases reaching maximum values of about 35% for X_{TG} within the 60–70% range; then, DG started to decline. The yield of both monoglycerides, and especially methyl esters, increase with the conversion of triglycerides. Maximum biodiesel yields achieved are about 30% and correspond to the highest conversions reached at the end of the catalytic runs (80–85%). It is noteworthy that there are no significant differences among the yields provided by the three unsupported catalysts at similar X_{TG}, meaning that the selectivity is not affected by the calcination temperature of the unsupported catalysts.

Ferreira Pinto et al. [23] reported a maximum FAME yield of 64.2% for an unsupported MoO_3 catalyst calcined at 600 °C after 4 h of reaction during the methanolysis of soybean oil acidified with 10 wt.% of oleic acid. Yields were slightly lower (50–56%) for solids calcined at lower temperatures (300–500 °C), and dropped to 43% for MoO_3 calcined at 700 °C. This trend is similar to the one observed in our work. Ferreira Pinto et al. [23] performed the catalytic tests at considerably higher temperature (150 °C) and methanol to feedstock molar ratio (30:1) than in this work, being the catalyst concentration also higher (5 wt.%), though the vegetable oil contained double FFAs concentration (10 wt.%). These authors found a correlation between the catalytic activity and the acidity of the solids as determined through a titration technique in which aqueous NaOH previously contacted with the catalyst was neutralized with HCl. It was suggested that the changes provoked by the thermal treatment affected the acid properties of the catalysts. The lower biodiesel yields found in this work can be mainly attributed to the milder reaction conditions and lower catalyst concentration employed. On the other hand, the slightly higher specific acidity, developed as the calcination temperature increases, is virtually offset by the accompanying specific surface area decrease.

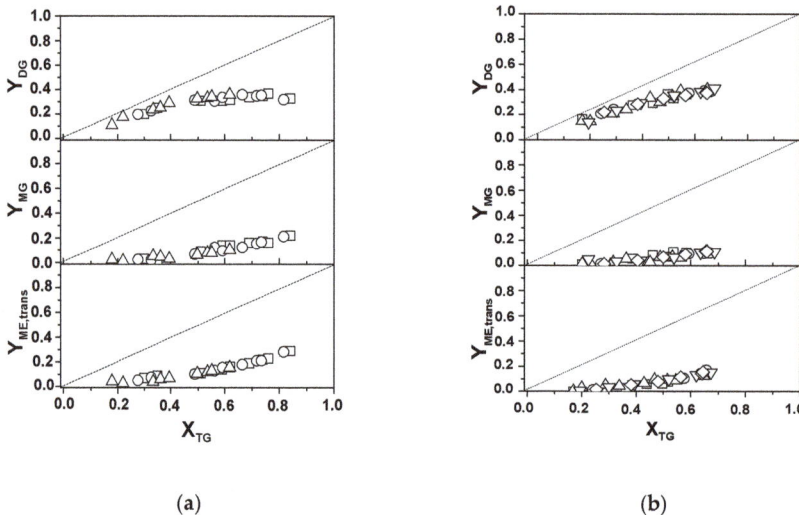

Figure 6. Evolution with the triglycerides conversion of the yields of diglycerides (Y_{DG}), monoglycerides (Y_{MG}), and methyl esters produced by transesterification ($Y_{ME,trans.}$) for: (**a**) unsupported Mo (300) (squares), Mo (500) (circles), and Mo (700) (triangles) catalysts; (**b**) supported s-Mo (6) (squares); s-Mo (8) (circles); s-Mo (10) (triangles); s-Mo (13) (inverted triangles); s-Mo (16) (rhombus) catalysts. Reaction conditions: 100 °C, 30 atm, methanol:feedstock molar ratio of 12:1, 2 wt.% catalyst referred to the feedstock mass. Feedstock: refined sunflower oil containing 5 wt.% free oleic acid.

As for the supported solids, Figure 7 shows the conversions of triglycerides and FFAs reached by the alumina-supported catalysts loaded with 6 to 16 wt.% MoO_3. It should be noted that experiments carried out with the alumina support under the same reaction conditions provided negligible transesterification and esterification conversions. It merits to be highlighted that the feedstock conversions provided by the bulk catalysts are much higher than those of the supported ones. In this regard, after 240 min of reaction, X_{TG} and X_{FFA} reach values between 40–60% and 65–75%, respectively, which are significantly lower than the conversions provided by the unsupported catalysts under similar reaction time (see Figure 5). Final conversions after 480 min range between 55% and 75% for X_{TG} and 70–80% for X_{FFA}. Concerning the effect of the MoO_3 oxide content, it seems that the activity goes through a maximum as the molybdenum oxide content increases. Indeed, both transesterification and esterification conversions increase when passing from s-Mo (6) to s-Mo (8) which is the catalyst providing the highest conversions. Then, the conversions decrease for s-Mo (10), s-Mo (13), and s-Mo (16), which reach relatively similar values.

The findings that the alumina support is not active and that the unsupported catalysts are significantly more active than the alumina-supported ones indicate that the catalytic activity is mainly associated with molybdenum oxide species. Nevertheless, a positive effect of the support in dispersing the active phase is also apparent because the MoO_3 contents were as low as 0.12–0.32 wt.% in the case of the supported catalysts compared with 2 wt.% for the unsupported ones. On the other hand, the acidity of the catalyst as determined by NH_3-TPD does not constitute in this case a suitable measure to predict the activity in the simultaneous transesterification and esterification reactions. This is because alumina showed specific acidities comparable and even higher than those of the supported catalysts (see Table 1). Notwithstanding, additional effects of the MoO$_3$-Al$_2$O$_3$ interactions on the catalytic activity cannot be ruled out; for instance, the presence of Brönsted acid sites of the [=Mo-O(H)-Al=] bridging type, similar to the ones proposed for MoO$_3$-Al$_2$O$_3$ catalytic systems in other works [30,34,37]. These sites could correspond to the adsorbed pyridinium and HPy species detected by FTIR spectroscopy upon pyridine adsorption over the supported catalysts (see previous section). As the specific surface area decreases with the increase of the MoO$_3$ content (see Table 1), the opposite effects of molybdenum oxide on the textural and certain acidic properties could explain that the catalytic activity within the s-Mo (6) to s-Mo (16) series of catalysts reaches a maximum value for an intermediate MoO$_3$ content, in this case, 8 wt.%.

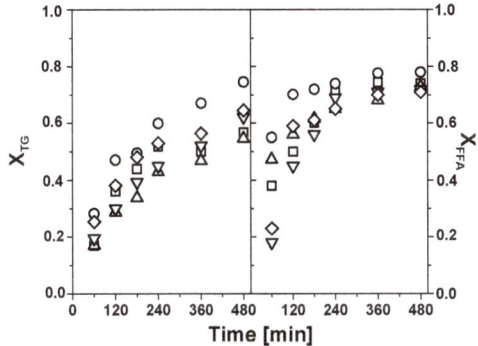

Figure 7. Evolution with reaction time of the conversion of triglycerides (X$_{TG}$) and FFAs (X$_{FFA}$) for supported s-Mo (6) (squares); s-Mo (8) (circles); s-Mo (10) (triangles); s-Mo (13) (inverted triangles); s-Mo (16) (rhombus) catalysts. Reaction conditions: 100 °C, 30 atm, methanol/feedstock molar ratio of 12:1, 2 wt.% catalyst referred to the feedstock mass. Feedstock: refined sunflower oil containing 5 wt.% free oleic acid.

Sankaranarayanan et al. [24] found a triglycerides conversion of about 58% (FAME yield of 47%) after 8 h of reaction at 100 °C with methanol to oil molar ratio of 9 using a 16 wt.% MoO$_3$/Al$_2$O$_3$ catalyst calcined at 677 °C at a concentration of 5 wt.%. The feedstock used consisted of refined sunflower oil (non-acidified). This result can be compared with about 64% triglycerides conversion in this work after 8 h of reaction at the same temperature, slightly higher methanol to oil molar ratio (12), but significantly lower concentration (2 wt.%) of the s-Mo (16) catalyst during the methanolysis of sunflower oil acidified with 5 wt.% of FFAs (see Figure 7). As in our case, Sankaranarayanan et al. [24] also found that the alumina support was not active for the transesterification reaction within the 60–110 °C range of temperature. As for the effect of the MoO$_3$ content of the catalysts, these authors found that it depended on the calcination temperature. In this regard, a large increase of activity was found when passing from 12 to 16 wt.% for catalysts calcined at 527 °C that greatly decreased when the calcination temperature increased to 827 °C; moreover, the catalysts calcined at 677 °C were more active than the ones calcined at 827 °C regardless the MoO$_3$ content [24]. These results were explained in terms of a higher specific activity of the well-dispersed molybdenum oxide species resulting at relatively low calcination temperatures and the presence of poorly active aluminum molybdate formed at high Mo loadings and calcination temperatures.

The yields of the several products given by the supported catalysts are presented in Figure 6b. It can be seen that there are no significant differences among the solids, meaning that the MoO$_3$ content does not affect the catalyst selectivity in the transesterification reaction. The yields of methyl esters were significantly lower than the ones reported by Sankaranarayanan et al. [24], though it should be remembered that these authors used non-acidified refined sunflower oil. The yields of methyl esters and monoglycerides were also lower than the ones obtained with the unsupported catalysts when the results are compared at the same triglycerides conversion, indicating a different performance, which is ascribed to the different nature of the active sites in both types of solids.

2.3. Catalysts Stability

As it is well-known, a key issue in the field of biodiesel production with heterogeneous catalysts is the chemical stability of the solids in the reaction medium, which is directly connected with both the possibility of reutilizing the catalyst and the quality of the produced biodiesel and glycerol, in particular, the fulfillment of the corresponding standards for their typical uses.

In this work, Mo (500) and s-Mo (8) catalysts were recovered after reaction, thoroughly washed with tetrahydrofuran (THF), which is an excellent solvent for the reaction mixture medium [40], calcined at their respective original temperatures (500 °C and 800 °C, respectively), and used again. Mo (500) yielded almost the same catalytic activity results than during its first use. XRD and N_2-adsorption data evidenced that no substantial changes took place during reaction. To check the possible occurrence of molybdenum leaching into the reaction mixture, the Mo (500) catalyst was removed after reaction by centrifugation. The upper fraction was transferred to a rotary evaporator to remove the unreacted methanol, resulting in a liquid that acquired an intense blue color that evidenced the presence of molybdenum. The molybdenum concentration was measured by ICP atomic emission spectroscopy resulting in ca. 1000 ppm of Mo. Additionally, Mo (500) was mixed with methanol for 4 h under the typical conditions used in this work. The solid was removed afterwards by centrifugation and the methanol was used in a reaction run. An X_{TG} value of 75% was obtained after 4 h of reaction, which is only slightly lower than the value of 82% achieved in the presence of the solid catalyst (see Figure 5). Clearly, leached molybdenum species were capable of homogeneously catalyzing the transesterification and esterification reactions. The fact that the recovered solid gives the same feedstock conversion than the fresh catalyst is not strange provided that the solid represents a sufficiently excess amount with respect to its solubility in the reaction medium, thus guaranteeing the activity during several re-utilization cycles. As a matter of fact, it is likely that the concentration of ca. 1000 ppm of Mo measured is close to the solubility limit of the molybdenum species in the polar phase. Ferreira Pinto et al. [23] reutilized unsupported MoO_3 in eight consecutive cycles of acidified soybean oil methanolysis. The catalyst was filtered from run to run and was used without washing. A very low loss of activity was observed only during the two last cycles; however, information about Mo leaching was not provided.

Regarding s-Mo (8), catalyst recovery, washing with THF, calcination, and reuse was repeated four times. The X_{TG} values recorded after 8 h of reaction are shown in Figure 8. A gradual decrease of the triglycerides conversion takes place during the first three reaction cycles. The conversion stabilizes at values about 40% after a fourth reaction cycle; a conversion 47% lower than the original value.

As described before for Mo (500), s-Mo (8) was mixed also with methanol for 4 h under the typical conditions used in this work. When used in a reaction run, the recovered methanol yielded X_{TG} and X_{FFA} values of 41% and 46%, respectively, after 4 h of reaction. These conversions were substantially lower than the 60% and 73% values obtained in the presence of fresh s-Mo (8). Mo leaching after the first reaction cycle was investigated following the procedure described above for the unsupported catalyst. Although the Mo concentration could not be measured, the color of the liquid phase resulting after methanol removal evidenced the presence of dissolved Mo. On the whole, the results point to a somewhat improved stability of the supported catalysts compared to the unsupported ones that could be attributed to the interaction established between MoO_3 and Al_2O_3. This interaction has been evidenced by the XRD results that showed the formation of aluminum molybdate, which seems to contribute to an improved dispersion and anchorage of the molybdenum species. This leads to a comparatively high activity for low MoO_3 loadings and reduced leaching in comparison with the unsupported catalysts.

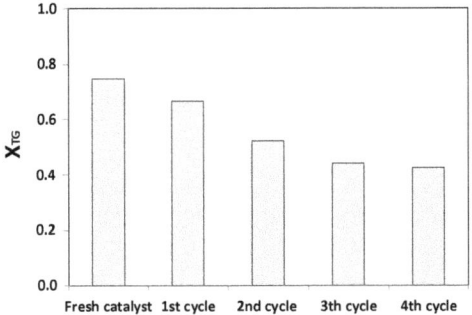

Figure 8. Triglycerides conversion (X_{TG}) obtained after 8 h of reaction during the number of reuse cycles indicated for the catalyst s-Mo (8). Reaction conditions: 100 °C, 30 atm, methanol/feedstock molar ratio of 12:1, 2 wt.% catalyst referred to the feedstock mass. Feedstock: refined sunflower oil containing 5 wt.% free oleic acid.

3. Materials and Methods

MoO_3 was synthesized by thermal decomposition of ammonium heptamolybdate tetrahydrate (AHM, from Merck, Darmstadt, Germany) in air (6 h) using a muffle furnace at different temperatures (300, 500 and 700 °C). The resulting solids were labeled as Mo (300), Mo (500) and Mo (700), respectively. Supported molybdenum catalysts were prepared by incipient wetness impregnation of γ–Al_2O_3 (Spheralite 505, Procatalyse) in powder form (particle size between 100 and 200 µm) previously calcined at 800 °C (6 h) in order to remove adsorbed impurities. Different impregnating AHM aqueous solutions were prepared to adjust the Mo salt concentration in order to obtain several MoO_3 contents in the final catalyst, namely 6, 8, 10, 13 and 16 wt.%. After impregnation, the solids were dried for 8 h at 80 °C and calcined for 6 h at 800 °C in a muffle furnace. The calcination temperature was selected in order to promote the interaction of molybdenum species with the alumina support with the aim of improving their resistance to leaching during the reaction. The supported catalysts were referred to as s-Mo followed by the corresponding nominal MoO_3 content between parentheses.

As for the catalysts characterization, XRD analyses were carried out in a D-Max Rigaku diffractometer (Akishima-shi, Tokyo, Japan). Specific surface areas were measured through N_2 adsorption at −196 °C using a Micromeritics Gemini V 2380 apparatus (Norcross (Atlanta), GA, USA). Acidity was characterized by NH_3-TPD in a Micromeritics Autochem 2920 equipment (Norcross (Atlanta), GA, USA). Pyridine adsorption was carried out using a purpose-made quartz IR cell connected to a vacuum adsorption device with a residual pressure lower than 10^{-4} Pa. The samples, in the form of wafers, were activated and pyridine introduced into the cell at room temperature. Spectra were recorded in a Thermo Nicolet 380 spectrophotometer (Waltham, MA, USA) with a DTGS/KBr detector (Waltham, MA, USA) and accumulating 128 scans at a spectral resolution of 4 cm^{-1}. X-ray photoelectron spectroscopy (XPS) analyses were carried out on a SPECS system equipped with an Al anode XR50 source operating at 150 W and a Phoibos 150 MCD-9 detector (Berlin, Germany). The pass energy of the hemispherical analyzer was set at 25 eV and the energy step was set at 0.1 eV. Charge stabilization was achieved by using a SPECS Flood Gun FG 15/40. The sample powders were pressed to self-consistent disks. Data processing was performed with the CasaXPS program (Casa Software Ltd., Teignmouth, UK).

Catalytic tests were carried out in batch mode in a Parr 4843 stainless steel autoclave reactor with mechanical stirring under controlled temperature and pressure. The feedstock consisted of refined sunflower oil (Urzante, Navarra, Spain; Acid Value of 0.07 mg KOH/g) to which pure oleic acid (Sigma Aldrich, San Luis, MO, USA) was added until it reached an FFAs content of 5 wt.% (Acid Value of 10.0 mg KOH/g). Catalyst concentration was set at a relatively low value of 2 wt.% referred to the feedstock (oil and FFAs mixture) mass. Simultaneous esterification of FFAs and transesterification of the triglycerides (TG) with methanol (Scharlau, HPLC grade, Barcelona, Spain) was carried out at 100 °C and methanol/feedstock molar ratio of 12:1. The amounts of the several substances used were: sunflower oil 60 g, FFAs 3 g, catalyst 1.26 g, and methanol 27.5 g. After reaching the reaction temperature, the reactor was pressurized with nitrogen until reaching an absolute pressure of 30 atm. Samples withdrawn from the reactor were analyzed by size exclusion chromatography as described elsewhere [40]. Conversion of triglycerides (X_{TG}) and free fatty acids (X_{FFA}), and yields of diglycerides (Y_{DG}), monoglycerides (Y_{MG}), and methyl esters (FAMEs) produced by transesterification ($Y_{ME,trans.}$) were calculated through Equations (1)–(5), respectively:

$$X_{TG} = \frac{N_{TG,0} - N_{TG}}{N_{TG,0}} \quad (1)$$

$$X_{FFA} = \frac{N_{FFA,0} - N_{FFA}}{N_{FFA,0}} \quad (2)$$

$$Y_{DG} = \frac{N_{DG}}{N_{TG,0}} \quad (3)$$

$$Y_{MG} = \frac{N_{MG}}{N_{TG,0}} \quad (4)$$

$$Y_{ME,trans.} = \frac{N_{ME} - (N_{FFA,0} \cdot X_{FFA})}{N_{TG,0}}, \quad (5)$$

where $N_{TG,0}$ and $N_{FFA,0}$ stand for the initial number of moles of triglycerides and FFAs, respectively, and N_{TG}, N_{DG}, N_{MG}, N_{ME}, and N_{FFA} stand for the number of moles of triglycerides, diglycerides, monoglycerides, FAMEs and FFAs, respectively, present in a sample taken at a given reaction time. Note that in Equation (5), in order to calculate the yield of methyl esters produced by transesterification, the number of moles of methyl esters produced by the esterification of FFAs ($N_{FFA,0} \cdot X_{FFA}$) is subtracted from the total number of methyl esters present in the sample.

4. Conclusions

The MoO_3-Al_2O_3 catalytic system has provided positive results consisting of high triglycerides conversions into biodiesel under conditions of high FFAs content that would have made the conventional transesterification reaction impossible in a single step with basic catalysts. Moreover, the FFAs are efficiently converted into biodiesel as well. Compared to bulk MoO_3, alumina-supported MoO_3 leads to a more efficient utilization of the active phase and enhanced stability towards molybdenum leaching by the reaction medium. These effects likely arise from the interaction established between MoO_3 and alumina that affect the textural and acidic properties of the catalysts. However, much effort is still required to develop a sufficiently stable catalyst based on molybdenum oxide for transforming triglycerides with high FFAs content into biodiesel.

Author Contributions: All the authors participated actively in the writing and editing of the manuscript. Furthermore: conceptualization, A.N. and L.M.G.; methodology, A.N.; catalysts preparation, textural and structural characterization, and catalytic testing, A.N. and I.R.; acid properties characterization, E.J.-B. and F.R.-S.; XPS characterization J.L.; results interpretation and discussion, and literature review, all the authors; project administration and funding acquisition, L.M.G. All authors have read and agreed to the published version of the manuscript.

Funding: Financial support from Spanish Ministerio de Ciencia, Innovación y Universidades, and the European Regional Development Fund (ERDF/FEDER) (grant RTI2018-096294-B-C31) is gratefully acknowledged. L.M.G. thanks Banco de Santander and Universidad Pública de Navarra for their financial support under "Programa de Intensificación de la Investigación 2018" initiative. JL is a Serra Húnter Fellow and is grateful to ICREA Academia program and 2017 SGR 128.

Conflicts of Interest: The authors declare no conflict of interest. The funders had no role in the design of the study; in the collection, analyses, or interpretation of data; in the writing of the manuscript, or in the decision to publish the results.

References

1. Mittelbach, M. Fuels from oils and Fats: Recent developments and perspectives. *Eur. J. Lipid Sci. Technol.* **2015**, *117*, 1832–1846. [CrossRef]
2. Chhetri, A.B.; Watts, K.C.; Islam, M.R. Waste Cooking Oil as an Alternate Feedstock for Biodiesel Production. *Energies* **2008**, *1*, 3–18. [CrossRef]
3. Navajas, A.; Issariyakul, T.; Arzamendi, G.; Gandía, L.M.; Dalai, A.K. Development of eggshell derived catalyst for transesterification of used cooking oil for biodiesel production. *Asia-Pac. J. Chem. Eng.* **2013**, *8*, 742–748. [CrossRef]
4. Arzamendi, G.; Campo, I.; Arguiñarena, E.; Sánchez, M.; Montes, M.; Gandía, L.M. Synthesis of biodiesel with heterogeneous NaOH/alumina catalysts: Comparison with homogeneous NaOH. *Chem. Eng. J.* **2007**, *134*, 123–130. [CrossRef]
5. Navajas, A.; Campo, I.; Arzamendi, G.; Hernández, W.Y.; Bobadilla, L.F.; Centeno, M.A.; Odriozola, J.A.; Gandía, L.M. Synthesis of biodiesel from the methanolysis of sunflower oil using PURAL® Mg–Al hydrotalcites as catalyst precursors. *Appl. Catal. B Environ.* **2010**, *100*, 299–309. [CrossRef]
6. Reyero, I.; Bimbela, F.; Navajas, A.; Arzamendi, G.; Gandía, L.M. Issues concerning the use of renewable Ca-based solids as transesterification catalysts. *Fuel* **2015**, *158*, 558–564. [CrossRef]
7. Reyero, I.; Moral, A.; Bimbela, F.; Radosevic, J.; Sanz, O.; Montes, M.; Gandía, L.M. Metallic monolithic catalysts based on calcium and cerium for the production of biodiesel. *Fuel* **2016**, *182*, 668–676. [CrossRef]
8. Lotero, E.; Liu, Y.; Lopez, D.E.; Suwannakarn, K.; Bruce, D.A.; Goodwin, J.G., Jr. Synthesis of Biodiesel via Acid Catalysis. *Ind. Eng. Chem. Res.* **2005**, *44*, 5353–5363. [CrossRef]
9. Tesser, R.; Di Serio, M.; Guida, M.; Nastasi, M.; Santacesaria, E. Kinetics of oleic acid esterification with methanol in the presence of triglycerides. *Ind. Eng. Chem. Res.* **2005**, *44*, 7978–7982. [CrossRef]
10. Di Serio, M.; Tesser, R.; Pengmei, L.; Santacesaria, E. Heterogeneous Catalysts for Biodiesel Production. *Energy Fuels* **2008**, *22*, 207–217. [CrossRef]
11. Melero, J.A.; Iglesias, J.; Morales, G. Heterogeneous acid catalysts for biodiesel production: Current status and future challenges. *Green Chem.* **2009**, *11*, 1285–1308. [CrossRef]
12. Sreeprasanth, P.S.; Srivastava, R.; Srinivas, D.; Ratnasamy, P. Hydrophobic, solid acid catalysts for production of biofuels and lubricants. *Appl. Catal. A: Gen.* **2006**, *314*, 148–159. [CrossRef]
13. Thanh, L.T.; Okitsu, K.; Boi, L.V.; Maeda, Y. Catalytic Technologies for Biodiesel Fuel Production and Utilization of Glycerol: A Review. *Catalysts* **2012**, *2*, 191–222. [CrossRef]
14. Su, F.; Guo, Y. Advancements in solid acid catalysts for biodiesel production. *Green Chem.* **2014**, *16*, 2934–2957. [CrossRef]
15. Sani, Y.M.; Daud, W.M.A.W.; Aziz, A.R.A. Activity of solid acid catalysts for biodiesel production: A critical review. *Appl. Catal. A: Gen.* **2014**, *470*, 140–161. [CrossRef]
16. Gupta, P.; Paul, S. Solid acids: Green alternatives for acid catalysis. *Catal. Today* **2014**, *236*, 153–170. [CrossRef]
17. Mansir, N.; Taufiq-Yap, Y.H.; Rashid, U.; Lokman, I.M. Investigation of heterogeneous solid acid catalyst performance on low grade feedstocks for biodiesel production: A review. *Energ. Convers. Manag.* **2017**, *141*, 171–182. [CrossRef]
18. Hanif, M.A.; Nisar, S.; Rashid, U. Supported solid and heteropoly acid catalyst for production of biodiesel. *Catal. Rev.* **2017**, *59*, 165–188. [CrossRef]
19. Park, Y.-M.; Chung, S.-H.; Eom, H.J.; Lee, J.-S.; Lee, K.Y. Tungsten oxide zirconia as solid superacid catalyst for esterification of waste acid oil (dark oil). *Bioresour. Technol.* **2010**, *101*, 6589–6593. [CrossRef]
20. Kaur, M.; Malhotra, R.; Ali, A. Tungsten supported Ti/SiO_2 nanoflowers as reusable heterogeneous catalyst for biodiesel production. *Renew. Energ.* **2018**, *116*, 109–119. [CrossRef]
21. Haber, J. Molybdenum Compounds in Heterogeneous Catalysis. In *Molybdenum: An Outline of its Chemistry and Uses*; Braithwaite, E.R., Haber, J., Eds.; Elsevier: Amsterdam, The Netherlands, 1994; Chapter 10; pp. 477–617.
22. Nakagaki, S.; Bail, A.; dos Santos, V.C.; Rodrigues de Souza, V.H.; Vrubel, H.; Souza Nunes, F.; Pereira Ramos, L. Use of anhydrous sodium molybdate as an efficient heterogeneous catalyst for soybean oil methanolysis. *Appl. Catal. A: Gen.* **2008**, *351*, 267–274. [CrossRef]

23. Ferreira Pinto, B.; Suller Garcia, M.S.; Santos Costa, J.C.; Rodarte de Moura, C.V.; Chaves de Abreu, W.; Miranda de Moura, E. Effect of calcination temperature on the application of molybdenum trioxide acid catalyst: Screening of substrates for biodiesel production. *Fuel* **2019**, *239*, 290–296. [CrossRef]
24. Sankaranarayanan, T.M.; Pandurangan, A.; Banu, M.; Sivansanker, S. Transesterification of sunflower oil over MoO_3 supported on alumina. *Appl. Catal. A: Gen.* **2011**, *409–410*, 239–247. [CrossRef]
25. Bail, A.; dos Santos, V.C.; Roque de Freitas, M.; Pereira Ramos, L.; Schreiner, W.H.; Ricci, G.P.; Ciuffi, K.J.; Nakagaki, S. Investigation of a molybdenum-containing silica catalyst synthesized by the sol-gel process in heterogeneous catalytic esterification reactions using methanol and ethanol. *Appl. Catal. B: Environ.* **2013**, *130–131*, 314–324. [CrossRef]
26. Sankaranarayanan, T.M.; Thirunavukkarasu, K.; Banu, M.; Pandurangan, K.; Sivansanker, S. Activity of supported MoO_3 catalysts for the transesterification of sunflower oil. *Int. J. Adv. Eng. Sci. Appl. Math.* **2013**, *5*, 197–209. [CrossRef]
27. Mouat, A.R.; Lohr, T.L.; Wegener, E.C.; Miller, J.T.; Delferro, M.; Stair, P.C.; Marks, T.J. Reactivity of a carbon-supported single-site molybdenum dioxo catalyst for biodiesel synthesis. *ACS Catal.* **2016**, *6*, 6762–6769. [CrossRef]
28. Chithambararaj, A.; Bhagya Mathi, D.; Rajeswari Yogamalar, N.; Chandra Bose, A. Structural evolution and phase transition of $[NH_4]_6Mo_7O_{24}.4H_2O$ to 2D layered MoO_{3-x}. *Mater. Res. Express* **2015**, *2*, 055004. [CrossRef]
29. Baltrusaitis, J.; Mendoza-Sanchez, B.; Fernandez, V.; Veenstra, R.; Dukstiene, N.; Roberts, A.; Fairley, N. Generalized molybdenum oxide surface chemical state XPS determination via informed amorphous sample model. *Appl. Surf. Sci.* **2015**, *326*, 151–161. [CrossRef]
30. Kitano, T.; Okazaki, S.; Shishido, T.; Teramura, K.; Tanaka, T. Brønsted acid generation of alumina-supported molybdenum oxide calcined at high temperatures: Characterization by acid-catalyzed reactions and spectroscopic methods. *J. Mol. Catal. A: Chem.* **2013**, *371*, 21–28. [CrossRef]
31. Choi, J.-G.; Thompson, L.T. XPS study of as-prepared and reduced molybdenum oxides. *Appl. Surf. Sci.* **1996**, *93*, 143–149. [CrossRef]
32. Deng, X.; Ying Quek, S.; Biener, M.M.; Biener, J.; Hyuk Kang, D.; Schalek, R.; Kaxiras, E.; Friend, C.M. Selective thermal reduction of single-layer MoO_3 nanostructures on Au(111). *Surf. Sci.* **2008**, *602*, 1166–1174. [CrossRef]
33. Światowska-Mrowiecka, J.; de Diesbach, S.; Maurice, V.; Zanna, S.; Klein, L.; Briand, E.; Vickridge, I.; Marcus, P. Li-Ion Intercalation in Thermal Oxide Thin Films of MoO_3 as Studied by XPS, RBS, and NRA. *J. Phys. Chem. C* **2008**, *112*, 11050–11058. [CrossRef]
34. Patterson, T.A.; Carver, J.C.; Leyden, D.E.; Hercules, D.M. A surface study of cobalt-molybdena-alumina catalyst using X-Ray Photoelectron Spectroscopy. *J. Phys. Chem.* **1976**, *80*, 1700–1708. [CrossRef]
35. Yan, Z.; Fan, J.; Zuo, Z.; Li, Z.; Zhang, J. NH_3 adsorption on the Lewis and Brønsted acid sites of MoO_3 (010) surface: A cluster DFT study. *Appl. Surf. Sci.* **2014**, *288*, 690–694. [CrossRef]
36. Morterra, C.; Magnacca, G. A case study: Surface chemistry and surface structure of catalytic aluminas, as studied by vibrational spectroscopy of adsorbed species. *Catal. Today* **1996**, *27*, 497–532. [CrossRef]
37. Skara, G.; Baran, R.; Onfroy, T.; De Proft, F.; Dzwigaj, S.; Tielens, F. Characterization of zeolitic intraframework molybdenum sites. *Micropor. Mesopor. Mater.* **2016**, *225*, 355–364. [CrossRef]
38. Bagshaw, S.A.; Cooney, R.P. FTIR Surface Site Analysis of Pillared Clays Using Pyridine Probe Species. *Chem. Mater.* **1993**, *5*, 1101–1109. [CrossRef]
39. Penkova, A.; Bobadilla, L.F.; Romero-Sarria, F.; Centeno, M.A.; Odriozola, J.A. Pyridine adsorption on $NiSn/MgO–Al_2O_3$: An FTIR spectroscopic study of surface acidity. *Appl. Surf. Sci.* **2014**, *317*, 241–251. [CrossRef]
40. Arzamendi, G.; Arguiñarena, E.; Campo, I.; Gandía, L.M. Monitoring of biodiesel production: Simultaneous analysis of the transesterification products using size-exclusion chromatography. *Chem. Eng. J.* **2006**, *122*, 31–40. [CrossRef]

© 2020 by the authors. Licensee MDPI, Basel, Switzerland. This article is an open access article distributed under the terms and conditions of the Creative Commons Attribution (CC BY) license (http://creativecommons.org/licenses/by/4.0/).

Article

Irrigation Combined with Aeration Promoted Soil Respiration through Increasing Soil Microbes, Enzymes, and Crop Growth in Tomato Fields

Hui Chen [1,2], Zihui Shang [2,3], Huanjie Cai [2,3,*] and Yan Zhu [2,3]

1. College of Engineering, Huazhong Agricultural University, Wuhan 430070, China; chenhui2014@nwafu.edu.cn
2. Key Laboratory of Agricultural Soil and Water Engineering in Arid and Semiarid Areas of Ministry of Education, Northwest A & F University, Yangling 712100, China; shangzh@nwafu.edu.cn (Z.S.); feifeixu2016@163.com (Y.Z.)
3. College of Water Resources and Architectural Engineering, Northwest A & F University, Yangling 712100, China
* Correspondence: huanjiec@yahoo.com; Tel.: +86-029-8708-2133

Received: 26 September 2019; Accepted: 6 November 2019; Published: 11 November 2019

Abstract: Soil respiration (Rs) is one of the major components controlling the carbon budget of terrestrial ecosystems. Aerated irrigation has been proven to increase Rs compared with the control, but the mechanisms of CO_2 release remain poorly understood. The objective of this study was (1) to test the effects of irrigation, aeration, and their interaction on Rs, soil physical and biotic properties (soil water-filled pore space, temperature, bacteria, fungi, actinomycetes, microbial biomass carbon, cellulose activity, dehydrogenase activity, root morphology, and dry biomass of tomato), and (2) to assess how soil physical and biotic variables control Rs. Therefore, three irrigation levels were included (60%, 80%, and 100% of full irrigation). Each irrigation level contained aeration and control. A total of six treatments were included. The results showed that aeration significantly increased total root length, dry biomass of leaf, stem, and fruit compared with the control ($p < 0.05$). The positive effect of irrigation on dry biomass of leaf, fruit, and root was significant ($p < 0.05$). With respect to the control, greater Rs under aeration (averaging 6.2% increase) was mainly driven by soil water-filled pore space, soil bacteria, and soil fungi. The results of this study are helpful for understanding the mechanisms of soil CO_2 release under aerated subsurface drip irrigation.

Keywords: aerated irrigation; soil enzyme activity; soil microbial biomass; soil respiration

1. Introduction

Subsurface drip irrigation (SDI) has been largely applied in arid and semi-arid regions to supply water due to greater yield production and water-saving characteristics [1,2]. Nevertheless, a large number of wetting fronts are generated near emitters, producing ethylene and CO_2, which are harmful for crop growth [3]. Aerated irrigation (AI), a modified irrigation technique that involves injecting air into soils based on SDI, has been extensively proven to improve soil aeration, thus increasing crop yields and fruit quality [4–6]. Even so, the effect of AI on soil environmental pollution is relatively sparse.

Soil respiration, originating primarily from heterotrophic respiration and autotrophic respiration [7,8], is a principal component in the global carbon cycle. A few studies have reported an increase of soil respiration under AI [6,9,10], while the cause of CO_2 release needs to be further explored. Previously, studies on drivers of soil respiration have been largely conducted on soil water content, temperature, and the interaction of these two parameters [9,11–14]. For AI treatment, a close correlation between soil CO_2 fluxes with soil water content and temperature has been confirmed [9,10]. Soil microbes and enzymes as biocatalysts for all biochemical reactions in the soil would decompose and oxidize soil organic matter [15] and intrinsically affect heterotrophic respiration, while the effect of AI on soil microbes and enzymes has been less tested [16]. Additionally, the properties of root morphology (total length, surface area, and volume) not only determine the ability of water and nutrient absorption but also determine the intensity of autotrophic respiration. Studies of root morphology under AI have been conducted in multiple crops [1,17–19], but the effect of AI on roots of greenhouse vegetables is still scarce. In recent years, researchers began to focus on the effects of soil microorganisms and plant growth on soil respiration [14,20,21]. However, the relationship between soil respiration and biotic components (microbes and plants) under AI remains unknown. Hence, studies of soil physical and biotic properties under AI are of critical significance to improve our mechanistic understanding of processes that release CO_2 to the atmosphere.

To better understand the mechanism of soil respiration change under different irrigation levels with and without aeration, soil respiration from greenhouse tomato fields, as well as soil physical and biotic components (soil water-filled pore space, temperature, abundance of soil bacteria, fungi, and actinomycetes, soil microbial biomass carbon, soil cellulase and dehydrogenase activity, tomato root morphology, and plant dry biomass) were investigated in the present study. We hypothesized that irrigation in combination with AI would increase soil respiration, soil microbes, soil enzyme activity, and plant growth. We also hypothesized that soil respiration would be closely related to soil physical and biotic components. Our results were used to manage irrigation measures under AI for CO_2 mitigation and to reveal the mechanism of soil respiration.

2. Results and Discussion

2.1. Environmental Variables

2.1.1. Soil Water-Filled Pore Space and Temperature

Soil water availability influences organic carbon decomposition, and soil temperature affects microbial growth and activity. They are considered as two major factors driving the variation of soil respiration [14].

A distinct seasonal difference of soil water-filled pore space (WFPS) and soil temperature can be observed (Figure 1). A sharp increase of WFPS occurred before 35 days after transplanting (DAT), and a decrease pattern was shown between 35 to 53 DAT. WFPS presented a total increase then decrease trend from 53 to 98 DAT. There was an upward trend of WFPS since 98 DAT (Figure 1a–c). As for soil temperature, a total decreasing trend was found throughout the whole tomato growing period except for a general increase between 35 to 49 DAT, between 70 to 83 DAT, as well as between 133 to 141 DAT (Figure 1d–f), which coincided with the seasonal patterns of air temperature. WFPS and soil temperature under aeration and high irrigation level were higher than the control and low irrigation level most of the time, which were in accordance with the findings of a previous study [9]. However, analysis of variance indicated that the effects of irrigation, aeration, and their interaction on mean WFPS and soil temperature were not significant (Table 1, $p > 0.05$).

Table 1. The effects of irrigation, aeration, and their interaction on mean values of soil water-filled pore space (WFPS), temperature, the abundance of soil bacteria (*cfu*b), fungi (*cfu*f) and actinomycetes (*cfu*a), soil microbial biomass carbon (MBC), soil cellulase activity (CA), dehydrogenase activity (DHA), and soil respiration (Rs) using a two-way ANOVA.

Factor	Analysis of Variance (*p*-Value)								
	WFPS	Temperature	*cfu*b	*cfu*f	*cfu*a	MBC	CA	DHA	Rs
Irrigation	ns	ns	ns	ns	ns	ns	ns	ns	ns
Aeration	ns	ns	ns	ns	ns	ns	ns	ns	ns
Irrigation × Aeration	ns	ns	ns	ns	ns	ns	ns	ns	ns

Note: ns—significance at $p > 0.05$.

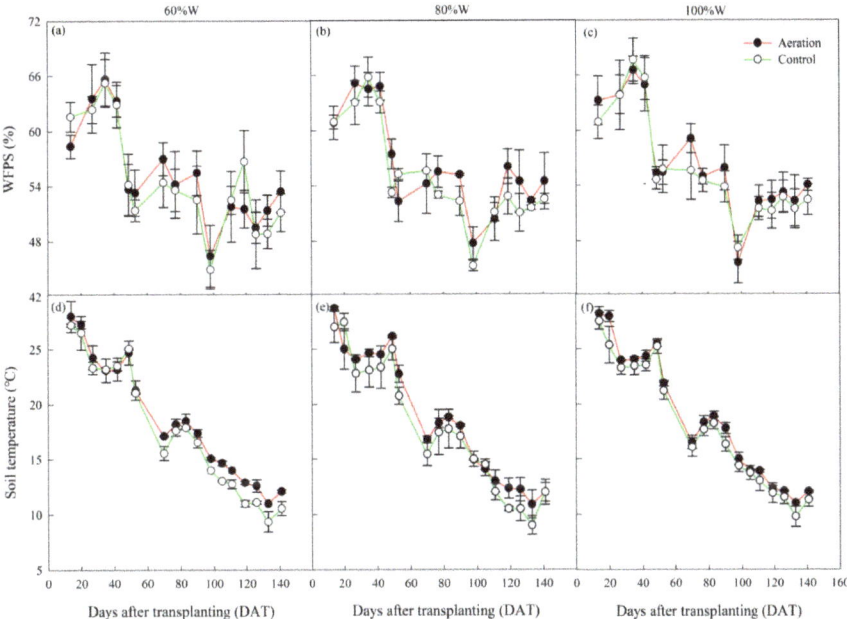

Figure 1. Soil water-filled pore space (WFPS) (**a,b,c**) and soil temperature (**d,e,f**) with and without aeration under the irrigation level of 60%W (**a,d**), 80%W (**b,e**), and 100%W (**c,f**) (mean ± SD, n = 3). W refers to full irrigation.

2.1.2. Soil Microbe and Enzyme Activity

Heterotrophic respiration, as a primary contributor to the soil respiration, is impacted by soil carbon-use efficiency which varies based on soil microbial abundance and richness [22]. A previous study demonstrated that the abundance of bacteria (*cfu*b), fungi (*cfu*f), and actinomycetes (*cfu*a) are involved in the soil carbon cycle by decomposing organic matter, degrading cellulose, and forming antibiotic substances [16]. Soil microbial biomass carbon (MBC), which affects the transformation of all organic matter entering the soil, is the key and driving force of the nutrient and energy cycle in the whole ecosystem and is also an important source and reservoir of soil nutrient transformation. Soil cellulose activity (CA), which participates in the decomposition and release of CO_2 from soil organic substances, is the main enzyme activity in soil carbon cycle. Soil dehydrogenase activity (DHA), which catalyzes dehydrogenation of organic substances and plays an intermediate role in hydrogen transformation and transfer, can be used as an indicator of the microbial redox system and is considered to be a global indicator of microbial metabolic activity in soil. However, soil biological activity can be limited by many factors, i.e., soil water and soil aeration conditions [23,24]. Very few pieces of literature were concerned with the soil microbes under AI [23,25], and the effect of AI on MBC, CA, and DHA have rarely been reported. Hence, study of the soil microbe and enzyme activity (*cfu*b, *cfu*f, *cfu*a, MBC, CA, and DHA) has great significance to reveal the mechanism of CO_2 release under AI.

As seen in Figure 2, *cfu*b made up the majority of soil microbes, followed by *cfu*a and *cfu*f, which was generally supported by the results of Li et al. [16] and Zhu et al. [25]. Nevertheless, the microorganism abundance in the study of Zhu et al. [25] were greater than the values of the current research (Figure 2), which was influenced by higher soil temperature in their study (their study vs. our study = 18–32 °C vs. 9–29 °C). Furthermore, there were different results about the changing trends of soil microbes in the tomato growing period. Zhu et al. [25] pointed out that *cfu*b, *cfu*f, and *cfu*a integrally presented an increase pattern. Chen et al. [26] concluded that *cfu*f and *cfu*a showed an initial increase then decrease trend, and peaks were observed on 50 d. In our study (Figure 2), the number of *cfu*b as a function of the days after transplanting was normally distributed, with the highest values observed on 98 DAT (Figure 2a,b). The number of *cfu*f peaked on 98 DAT, and the values during other periods were relatively stable (Figure 2c,d). The number of *cfu*a peaked on 35 DAT, but remained at a relatively constant level during other periods (Figure 2e,f). The differences of changing patterns could have resulted from the combined effects of the availability of different rhizosphere secretions and substrates, changes of soil moisture, temperature, and fertility, as well as plant growth. Soil microbial abundance (especially for *cfu*b and *cfu*f, Figure 2) peaked when soil hydrothermal conditions were good (Figure 1) and crops were growing vigorously on 98 DAT. Peaks of *cfu*a during the early tomato growing period (on 35 DAT) were probably ascribed to the highest WFPS (64.5%–67.7%) and greater soil temperature (23.1–24.7 °C), as well as greater soil substrates resulted from base fertilizer application [9]. Compared with the control, aeration under each irrigation level slightly increased mean values of *cfu*b, *cfu*f, and *cfu*a (Table 1, $p > 0.05$), with average increases of 4.6%, 5.5%, and 3.4%, respectively. Similar results were also reported by Li et al. [16], Du et al. [23], and Zhu et al. [25]. The increases of soil microbes under the aeration were likely due to the frequent alternation of soil dry and wet zones, thereby enhancing soil nutrient mineralization to improve microbial growth. Additionally, in line with previous researches [24,25], *cfu*b, *cfu*f, and *cfu*a in this study increased as irrigation amount increased (Figure 2), which was in order of 60% full irrigation (W) level without aeration (S) ($W_{0.6}S$) < 80%W irrigation level without aeration ($W_{0.8}S$) < 100%W irrigation level without aeration ($W_{1.0}S$). The enhancement of soil microbes under aeration or high irrigation level was also probably ascribed to greater temperature (Figure 1d–f), which stimulated more microbial growth and activity [14].

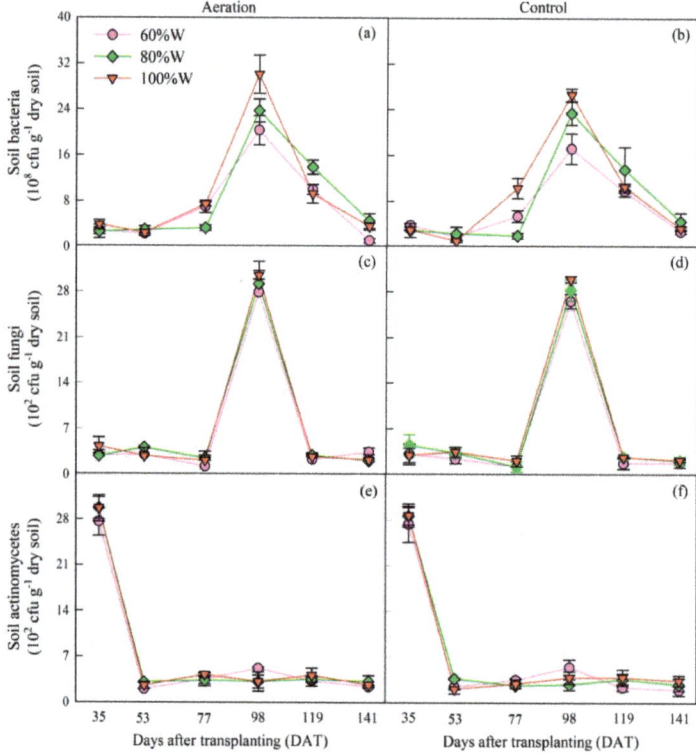

Figure 2. The abundance of soil bacteria (**a,b**), fungi (**c,d**), and actinomycetes (**e,f**) with the irrigation level of 60%W, 80%W, and 100% W under the aeration (**a,c,e**) and control (**b,d,f**) (mean ± SD, n = 3).

MBC generally exhibited an initial increase followed by a volatility within the range of 210.43 to 289.75 mg·kg^{-1} throughout the whole tomato growing period (Figure 3a,b). Across all sampling periods, CA among treatments varied from 0.63 to 1.00 mg·kg^{-1} and peaked on 35 DAT except for W$_{0.8}$S treatment on 119 DAT (Figure 3c,d). Contrary to the changing rule of CA, DHA generally increased throughout the tomato growing period (Figure 3e,f). The changing patterns of soil enzyme activities were primarily because soil enzymes were correlated with the growth stages, soil texture, soil water content, soil temperature, air availability, and other factors [16]. Compared with the control, mean MBC, CA, and DHA under the aeration were slightly greater (Table 1, $p > 0.05$). As noted by Li et al. [16], soil enzymes are secreted by crop roots and rhizosphere microorganisms, as well as the decomposition of plant residues and microbial cells. Under the aeration, enhanced tomato root (Figure 4) and increased soil microbes (Figure 2) could immobilize and release nutrients into the soil and ameliorate soil fertility [23,27], which ultimately improved the CA and DHA (Figure 3). Additionally, soil water availability affects substrate availability, O$_2$ concentrations, osmotic potential, gas diffusion, and cellular metabolism [24,28], thus impacting soil microbes. Difference in mean MBC, CA, and DHA values among treatments in this experiment was not significant (Table 1, $p > 0.05$).

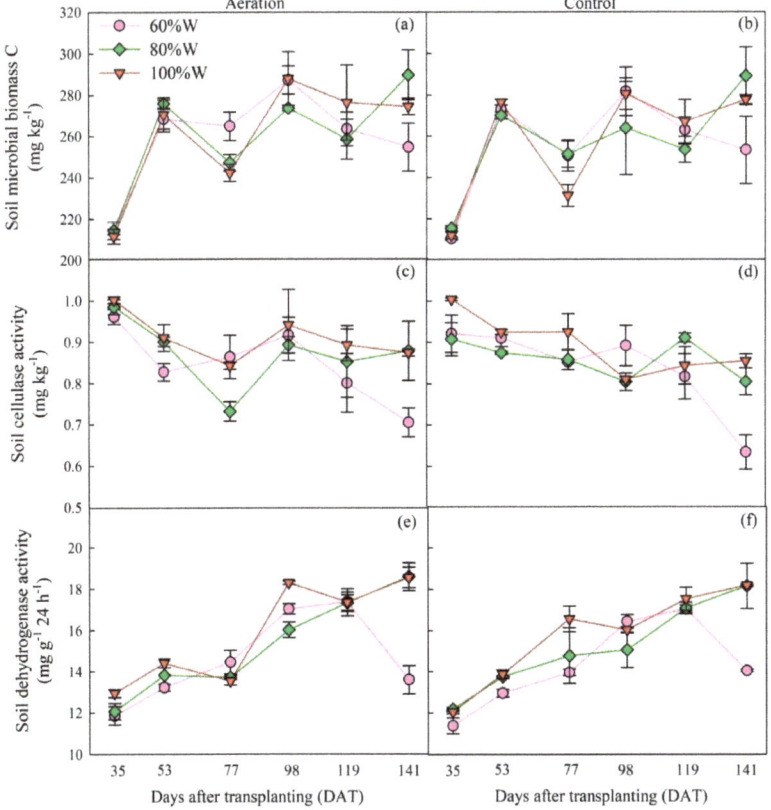

Figure 3. Soil microbial biomass carbon (**a,b**), soil cellulase activity (**c,d**), and soil dehydrogenase activity (**e,f**) with the irrigation level of 60%W, 80%W, and 100%W under the aeration (**a,c,e**) and control (**b,d,f**) (mean ± SD, n = 3).

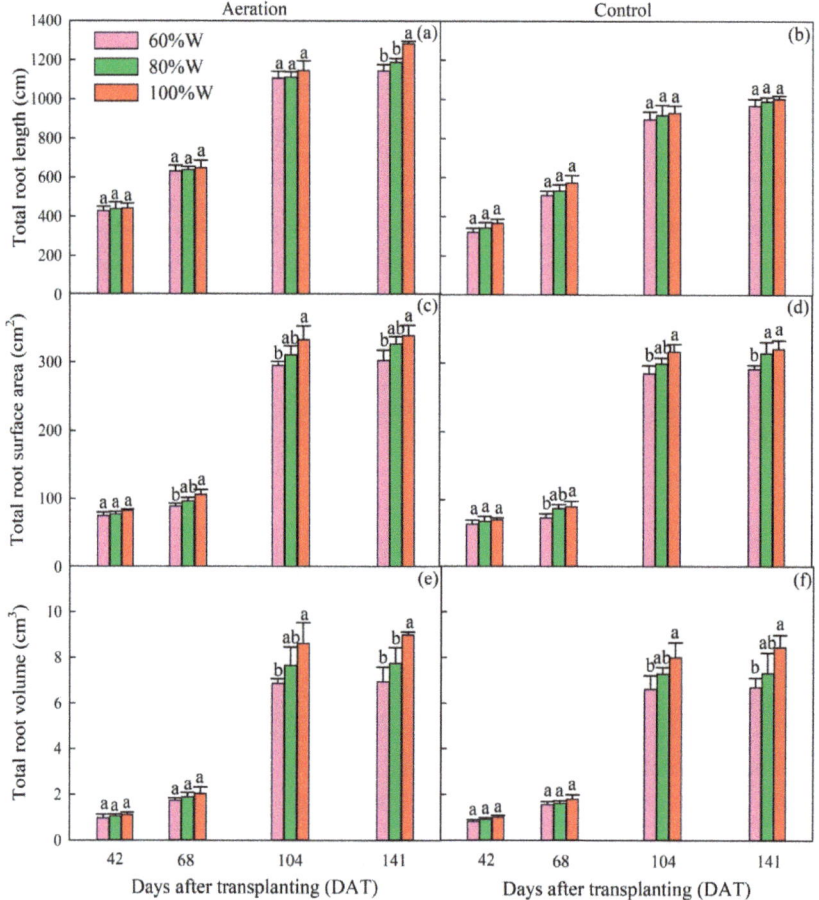

Figure 4. Total root length (**a,b**), surface area (**c,d**), and volume (**e,f**) with the irrigation level of 60%W, 80%W, and 100%W under the aeration (**a,c,e**) and control (**b,d,f**) (mean ± SD, n = 3).

In our study, the highest mean values of soil microbe and enzyme activity were obtained when 100%W was applied coupled with AI. This indicated that in a way the effect of irrigation on soil microbe and enzyme activity was enhanced under AI and that the soil biological environment was improved.

2.1.3. Root Morphology

The root system plays a decisive role in water and nutrient absorption. The size of crop roots also determines autotrophic respiration. Hence, studies on root morphology are of great practical significance to the study of plant growth and root respiration.

Aeration has been determined to increase root dry biomass and root morphology in cucumber [17,18], soybean [1], and even in the conventional staple grain crop [19]. However, there have been few studies regarding tomato root morphology under AI. Our results showed that total root length, surface area, and volume on 104 and 141 DAT were significantly greater than those on 42 and 68 DAT (Figure 4, $p < 0.05$). Compared with the control, total root length was significantly increased by 22.2% on average under aeration (Table 2, $p < 0.05$). Meanwhile, total root surface area and volume under the aeration was 6.6% and 6.7% higher than that of the control, respectively ($p > 0.05$). Li et al. [18] also showed that root morphology (root length, surface area, and volume) increased with increasing frequency of aeration. Root length of greenhouse muskmelon was 7076, 5839, 5207, and 3864 cm, and root surface area was 1217, 1023, 998, and 746 cm^2, while root volume was 31.0, 26.1, 25.7, and 20.1 cm^3 for daily, 2-day, 4-day, and no aeration, respectively ($p < 0.05$) [18]. These increases of root morphology under the aeration were attributed to elongation, branching, and curving, influenced by the shape and dimensions of the wetted soil volume [18]. The injected air changed the soil structure owing to the shrinking and movement of soil particles, and it also pushed the water downwards [29]. All these characters in conjunction with higher soil moisture under aeration (Figure 1) were conducive to elongation of roots due to hydro-tropism. With respect to $W_{1.0}S$, $W_{0.6}S$ significantly decreased the total root volume by 18.6% ($p < 0.05$), while the effects of other irrigation levels on root morphology were not significant ($p > 0.05$). Contrary to the results of the current study, Li et al. [18] stated that high irrigation levels has a negative effect on total root length, surface area, and volume with root length of 5981, 5364, and 5145 cm, surface area of 1114, 947, and 927 cm^2, and volume of 30.8, 22.7, and 23.6 cm^3 for the 70%, 80%, and 90% of field capacity level, respectively. Xu et al. [30] demonstrated that root length and surface area presented an increasing then decreasing trend as soil changes from dry to moist. Differences among literature were likely due to different hydrophily of crops controlled by the genes and tropic response to stimuli [18].

Table 2. The effects of irrigation, aeration, and their interaction on mean root morphology and dry biomass using a two-way ANOVA.

Factor	Analysis of Variance (p-Value)						
	Root Morphology			Dry Biomass			
	Length	Surface Area	Volume	Leaf	Stem	Fruit	Root
Irrigation	ns	ns	ns	*	ns	**	**
Aeration	*	ns	ns	*	*	*	ns
Irrigation × Aeration	ns	ns	ns	ns	ns	ns	ns

Note: ns, *, and **—significance at $p > 0.05$, $p < 0.05$, and $p < 0.01$, respectively.

2.1.4. Dry Biomass

Soil respiration was influenced by not only soil physical environment (i.e., soil temperature and moisture), but also plant growth [14]. Study of dry biomass throughout the whole tomato growing period was an effective way to analyze the changes of soil respiration, especially for the autotrophic component.

An increasing trend of dry biomass was observed throughout the whole tomato growing period, and dry biomass on 42, 68, 104, and 142 DAT showed a similar changing pattern among treatments (Figure 5). Taking dry biomass at harvest (142 DAT) as an example, dry biomass of tomato leaf, stem, fruit, and root under aeration were higher than the control (Figure 5). As reported previously [31], the average increases of each part were 17.8%, 17.7%, 17.8%, and 8.4%, respectively, and the effect of aeration on leaf, stem, and fruit was significant (Table 2, $p < 0.05$). These improvements of dry biomass were in agreement with the results of former research [4,32], which were beneficial from increased soil aeration and reduced phytohormones under AI [31]. Dry biomass of tomato leaf, stem, fruit, and root increased as irrigation amount increased, and the effect was significant on leaf, fruit, and root (Table 2, $p < 0.05$). As noted previously [31], dry weight of root, stem, leaf, and fruit under 100%W was increased by 22.2%, 19.3%, 22.5%, and 19.0%, and by 20.1%, 5.4%, 7.0%, and 12.1% than that under 60%W and 80%W treatment, respectively. Zhu et al. [4] demonstrated that with crop-pan coefficient increasing from 0.6 to 1.0, dry biomass of root, stem, and leaf was increased by 24.0%, 17.2%, and 22.8%, respectively. The enhancement of dry biomass as irrigation amount increased was primarily ascribed to the greater canopy and leaf area index [4], as well as increased assimilation rate under high irrigation level [33].

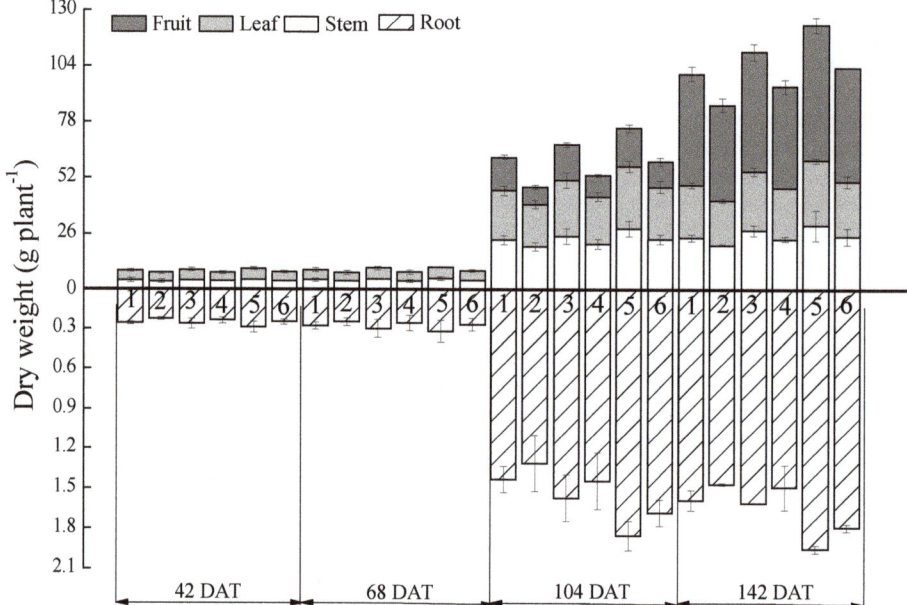

Figure 5. Dry biomass of tomato fruit, leaf, stem, and root among treatments on 42, 68, 104, and 142 days after transplanting (DAT). The number 1, 2, 3, 4, 5, and 6 represented treatment of 60%W with aeration, 60%W without aeration, 80%W with aeration, 80%W without aeration, 100%W with aeration, and 100%W without aeration, respectively.

2.2. Soil Respiration

As presented in Figure 6, soil respiration showed fluctuated patterns during the whole tomato growing period, which varied from 139.19 to 748.64 mg·m^{-2}·h^{-1} among treatments. Ranges of soil respiration in the present study was similar to the results of Hou et al. [10] but was higher than the research of the same tomato cultivations [9]. Differences might be the results of different irrigation amount and weather condition based on the year of cultivation. The changing patterns of soil respiration could be explained mostly by the abiotic and biotic factors (Figures 1–5). The lowest values on 9 DAT were mainly due to lower soil microbes (especially for *cfub* and *cfuf*, Figure 2) and undeveloped tomato roots (Figures 4 and 5) at the onset of transplantation. As days after transplanting increased, *cfub* and DHA increased gradually (Figures 2 and 3), and the root growth enhanced slightly (Figures 4 and 5), inducing larger emissions on 83 DAT. Relatively lower WFPS and obvious increases of soil temperature on 49 DAT (Figure 1) resulted in the peaks of soil respiration under $W_{0.6}O$, $W_{0.8}S$, and $W_{1.0}S$ treatment. Higher soil respiration on 62 DAT was attributed to increased WFPS, resulting in peaks under $W_{0.8}O$ and $W_{1.0}O$ treatment. Lower soil respiration on 98 and 133 DAT was primarily ascribed to a sharp decline of WFPS (Figure 1). An increasing trend of soil respiration was detected since 133 DAT, which was probably due to the increase of WFPS and soil temperature.

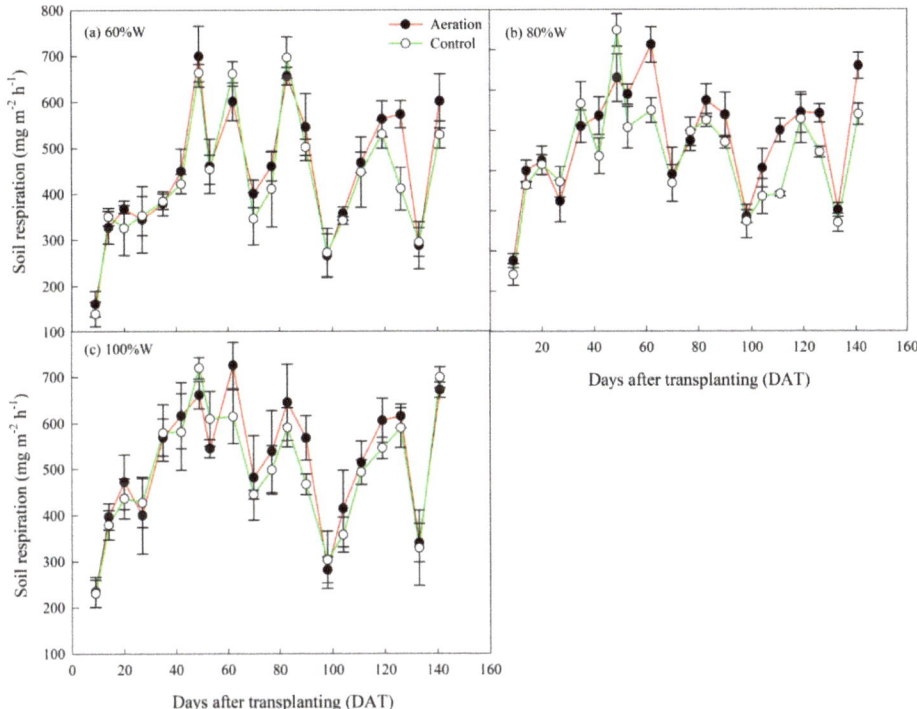

Figure 6. Soil respiration with and without aeration under the irrigation level of 60%W (**a**), 80%W (**b**), and 100%W (**c**) (mean ± SD, n = 3).

Previous research has shown a good correlation between soil respiration and soil temperature, oxygen concentration, and air-filled porosity [34]. Nevertheless, the correlation between soil respiration and soil microbe and enzyme activity, as well as plant growth under the aeration and irrigation treatments has not yet been well studied. In our study, regression analysis (linear, polynomial, and exponential) between soil respiration and WFPS was conducted, and a significant polynomial function was observed (Figure 7a,b, $p < 0.05$), similar to previous studies [11,12]. Further analysis found that a polynomial correlation was detected between soil respiration and WFPS when WFPS was below 60% ($p < 0.01$), while a linear positive correlation was observed when WFPS was above 60% (Figure 7c,d, $p = 0.245$ and 0.001 for the aeration and control, respectively). Moreover, there were significant negative correlations between soil respiration and *cfu*b, as well as between soil respiration and *cfu*f (Table 3, $p < 0.01$), which was different from the result of Zhu et al. [25] where soil respiration showed strong positive correlations with *cfu*b, *cfu*f, and *cfu*a. The reason for the inconsistent conclusions was probably due to the different growing seasons. Zhu et al. [25] conducted the experiment in the spring–summer period where the weather was gradually raised, while the present experiment was finished in the autumn–winter period where the weather was gradually reduced. Different variation of soil temperature would lead to different changing rules of soil respiration, microbial activity, and water content. In the present study, the interactive effect of WFPS, *cfu*b, and *cfu*f on soil respiration was extremely significant (Table 3, $p < 0.01$), which collectively accounted for 70.2% and 61.6% of changes in soil respiration under aeration and control, respectively. Unfortunately, correlations between soil respiration and other soil physical and biotic components (soil temperature, *cfu*a, MBC, CA, DHA, tomato root morphology, and plant dry biomass) were not significant ($p > 0.05$, data not shown), which required further study.

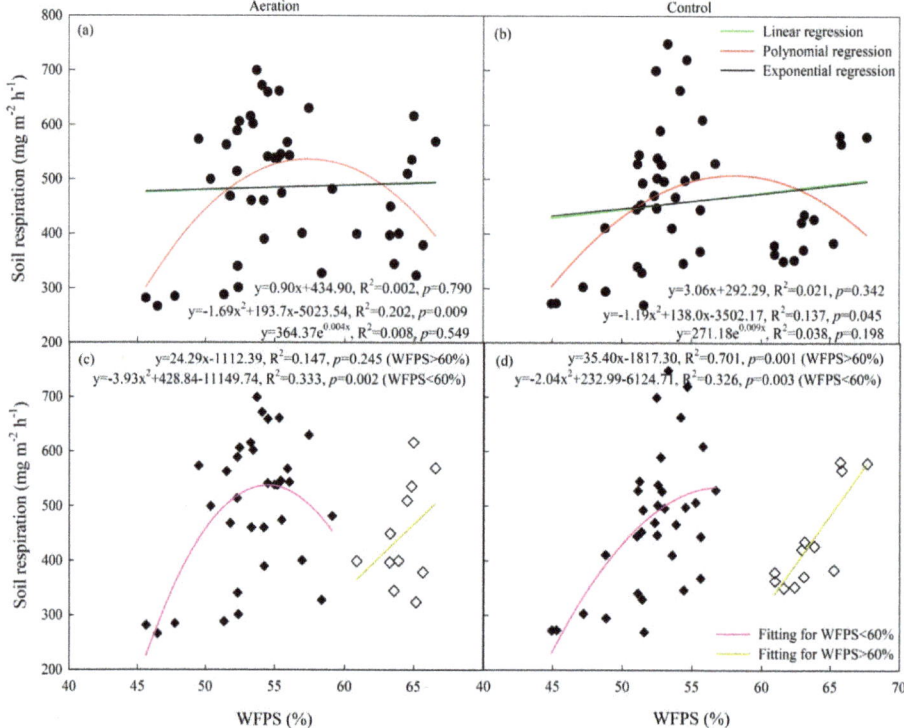

Figure 7. Correlation between soil respiration with soil water-filled pore space (WFPS) for the all WFPS data (**a,b**) and data with WFPS piecewise analysis (**c,d**). (**a**)(**c**) and (**b**)(**d**) represented correlation under the aeration and control, respectively.

Table 3. Relationships of soil respiration (Rs) with soil water-filled pore space (WFPS), the abundance of soil bacteria (*cfu*b) and fungi (*cfu*f) under the aeration and control treatments.

Treatment	Model	R^2	p
Aeration	Rs = −10.579*cfu*b + 588.685	0.503	0.001
	Rs = −9.997*cfu*f + 571.845	0.660	0.000
	Rs = 0.257WFPS^2 − 36.187WFPS − 0.943*cfu*b − 12.586*cfu*f + 1799	0.702	0.001
Control	Rs = −10.576*cfu*b + 569.316	0.491	0.001
	Rs = −9.142*cfu*f + 546.848	0.608	0.000
	Rs = −0.351WFPS^2 + 41.536WFPS − 2.142*cfu*b − 5.858*cfu*f − 664.275	0.616	0.003

Similar to previous results [6,9,10,34], soil respiration under the aeration in our study was typically and on average 6.2% greater but no significant different to that under the control according to ANOVA (Figure 6, Table 1, $p > 0.05$). Chen et al. [6] found that soil respiration increased by 42%–100% for oxygenation compared to control. Hou et al. [10] stated that aeration increased soil CO_2 emissions by 11.8% ($p = 0.394$) compared to the control. Zhu et al. [34] revealed that mean soil respiration under the aeration was 22.5% higher than the control. Potential reasons explaining the enhancement of soil respiration under aeration include: (1) aeration increased soil microbes (Figure 2,3) and root growth (Figures 4 and 5), which essentially controlled heterotrophic respiration and autotrophic respiration [22,35]; (2) greater CA and DHA under AI (Figure 3), which were involved in the decomposition and release of CO_2 from soil organic substances, in turn promoted soil respiration; and (3) as a result of the enhanced aboveground dry biomass under AI, the increased demand for nutrients stimulated belowground C allocation and root growth (Figures 4 and 5), which increased substrates to soil organisms and stimulated organic matter turnover [36,37], leading to higher biomass and/or activity that might stimulate the decomposition of soil organic matter [38,39]. All of these factors increased soil respiration conclusively. Consistent with previous research [9,40], soil respiration increased in the order of $W_{0.6}S < W_{0.8}S < W_{1.0}S$ (Figure 6), which resulted from increased soil microbial biomass (Figure 2), soil enzyme activity (Figure 3), root biomass (Figure 5), mineralization and decomposition rate of soil organic matter, as well as the diffusion rate of gases in soil pores [41]. Soil respiration under $W_{1.0}S$ was 16.0% and 13.9% higher than that under $W_{0.6}S$ and $W_{0.8}S$, respectively. Nevertheless, the effect of irrigation on soil respiration was not significant (Table 1, $p > 0.05$).

Although this paper analyzed the response of soil respiration to soil physical and biotic variables, we do not know the proportion of root or microbial respiration to soil respiration as no measurement was made in this study, which was the deficiency of this study, and needs to be further carried out in future experiments.

3. Materials and Methods

3.1. Experimental Site

The experiment was conducted from August to December 2017 in a solar greenhouse located at 34°20′ N, 108°04′ E, at the Key Laboratory of Agricultural Soil and Water Engineering in Arid and Semi-Arid Areas of the Ministry of Education, Northwest A and F University, Yangling, Shaanxi Province, China. The site is in a semi-arid climate zone with an annual mean sunshine duration of 2163.8 h and frost-free period of 210 days. Lou soil was used in the experimental site. The texture was a silt clay loam (sand 26.0%; silt 33.0%; clay 41.0%). Soil properties of the top 20 cm were: field capacity 23.8% by weight; dry bulk density 1.35 g·cm^{-3}; organic matter 14.62 g·kg^{-1}; total N 1.88 g·kg^{-1}; total P 1.37 g·kg^{-1}; total K 20.21 g·kg^{-1}; and pH 7.82.

Daily maximum and minimum temperatures inside the greenhouse during the experimental period, collected by an Automatic Meteorological Observing Station (Hobo event logger, Onset Computer Corporation, Bourne, MA, USA), are shown in Figure 8. The weather station, which was placed 2 m above the ground, recorded meteorological data at an interval of 15 min [31]. Higher temperatures were observed in August, while lower temperatures were recorded in December (Figure 8).

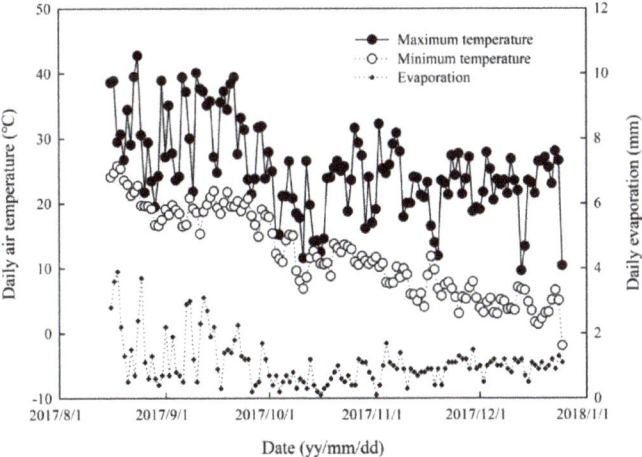

Figure 8. Daily evaporation, maximum and minimum temperature during the growing season at the experimental site.

3.2. Experimental Design

Based on the irrigation amount of full irrigation (W) calculated as Equation (1), three irrigation levels were set: 60%, 80%, and 100% of W. Non-aerated SDI (S) was used as a control for aeration (O). Therefore, six treatments were designed ($W_{0.6}O$, $W_{0.6}S$, $W_{0.8}O$, $W_{0.8}S$, $W_{1.0}O$, and $W_{1.0}S$). Three replicates of each treatment were used (18 total plots), and the experiment was arranged using a randomized block [31]. Each plot with one row was 4 × 0.8 m in size, with eleven tomato plants of cultivar "JINGPENG SEED" planted on 6 August 2017. The plants were spaced 35 cm apart. All plots were mulched with a layer of low-density polyethylene to minimize surface evaporation [42]. SDI was applied in the experiment, which was buried at a depth of 15 cm below the soil surface with a dripper interval of 35 cm [9,10]. Additionally, a Mazzei air injector Model 287 (Mazzei Injector Company, LLC, Bakersfield, CA, USA) was installed at the head of each irrigation line for AI (inlet pressure: 0.1 MPa; outlet pressure: 0.02 MPa) [42]. Definitively, the air injectors were set to inject 17% volumetric air concentration in the water [25].

Daily evaporation measured by an E601 evaporation pan is shown in Figure 8. In all growth stages, 20 irrigation events were applied every seven days, with a total irrigation amount for W of 19.80 L per plot [31]. Irrigation amount was determined following the Equation (1) [9,10,42]:

$$W = k_{cp} \times E_{pan} \times A \qquad (1)$$

where k_{cp} is the crop-pan coefficient, being 1.0; E_{pan} is the total evaporation quantity following the last irrigation event (mm); and A is the area controlled by one irrigation dripper in this experiment, being 0.14 m^2 (0.35 m × 0.4 m).

Only basal fertilizer, including organic fertilizer (N–P_2O_5–K_2O ≥ 10%, organic matter ≥ 45%) and compound fertilizer (total nutrients ≥ 45%, N, P_2O_5 and K_2O each at 15%), was applied for all plots. The application was achieved at a rate of 1875 and 1250 kg·ha^{-1} on 3 August 2017 for organic and compound fertilizer, respectively. Other agronomic managements were consistent with local production practices [42]. The experiment ended on 25 December 2017 with a total growth period of 142 days.

3.3. Measurement Index and Methods

Soil samples from 0 to 10 cm depth were collected when gas samples were collected except on 9, 20, 62, 83, and 104 DAT. Soil samples were taken through a diameter gauge with the three-point sampling method to measure soil water content via oven drying at 105 °C for 12 h, and then converted to WFPS by the following equation:

$$WFPS(\%) = \frac{gravimetric\,water\,content}{total\,soil\,porosity} \times soil\,bulk\,density \times 100 \qquad (2)$$

where total soil porosity = 1 − soil bulk density/2.65, with 1.35 g·cm^{-3} as the assumed particle density of the soil.

Soil temperature at a depth of 10 cm was recorded using a geothermometer (RM-004, Hengshui, China) when gas samples were collected, excluding on 9 and 62 DAT.

Soil samples of top-soil (0–20 cm) were collected to measure soil microbe and enzyme activity on 35, 53, 77, 98, 119, and 141 DAT. The *cfub*, *cfuf*, and *cfua* were estimated using the plate dilution counting method in beef extract and peptone medium, Martin's medium, and the improved Gao's No. 1 medium, respectively. Media plates were incubated at 37 and 25 °C, and the number of colonies after approximately 3 to 5 d was counted [23]. MBC was measured by the chloroform fumigation–K_2SO_4 extraction method. MBC in the extracts was determined by the $K_2Cr_2O_7$–$FeSO_4$ additional heating method. Detailed measurement steps regarding CA and DHA are described by Хазиев[43].

On 42, 68, 104, and 142 DAT, one plant from per plot was sampled to measure dry biomass and root morphology (total root length, surface area, and volume). All plant samples were first separated into leaves, stems, fruits, and roots. The roots collected from soil by digging were gently washed, scanned (Epson Perfection V700 photo, Seiko Epson Crop., Nagano-ken, Japan) to obtain a gray-scale JPG image, and then analyzed with the WinRHIZO Pro image processing system (Regent Instrument Inc., 2672 Chemin Sainte-Foy, Quebec City, Quebec G1V 1V4, Canada) to obtain root morphology [18]. After that, every part of the tomato plant including root was put into a 105 °C oven for 1 h to deactivate enzymes and then dried at 75 °C until the parts reached a constant weight. The dry biomass of each part was weighed on an electronic scale [31].

Gas samples of soil respiration was measured using the static closed chamber method described by Hou et al. [10]. All chambers, which were made of polyvinyl chloride (PVC) materials and wrapped with sponge and aluminum foil, were 25 × 25 × 25 cm in dimension. The bases of the chambers were installed between two plants in the middle of each plot on the day of transplanting and remained there until the end of the experiment. There was a 3-cm-deep groove on the top edge of the bottom layer and on the base of the chamber that was filled with water to seal the rim of the chamber. A mercury thermometer (WNG-01, Hengshui, China) at the top of each chamber was equipped to measure air temperature when gas sampling for calculating gas emission flux. Gas samples at an average interval of seven days were collected at 10:00, 10:10, 10:20 and 10:30 a.m. of each sampling time. A 30-mL air sample was drawn each time with a syringe. Gas samples in the syringes were analyzed for CO_2 concentrations using a gas chromatograph (7890A GC System, Agilent Technologies, Santa Clara, USA) within a few hours. Sample sets were discarded unless they yielded an R^2 linear regression value higher than 0.90. Then, soil CO_2 fluxes (soil respiration), which were the sum of autotrophic and heterotrophic respiration, were calculated following the equation given by Hou et al. [10]:

$$F = \rho \cdot h \cdot \frac{273}{273+T} \cdot \frac{dc}{dt} \qquad (3)$$

where F is the soil respiration (mg·m^{-2}·h^{-1}); ρ is the gas density at standard state (1.964 kg·m^{-3}); h is the height of chamber above the water surface (m); $\frac{dc}{dt}$ is the gas mixing ratio concentration (µL·L^{-1}·h^{-1}); and T is the mean air temperature inside the chamber during sampling (°C).

3.4. Statistical Analysis

A two-way analysis of variance (ANOVA) followed by an LSD test (95% confidence level, $p < 0.05$) was used to test for the effects of irrigation, aeration, and their interaction on soil respiration, soil physical and biotic properties (WFPS, temperature, cfub, cfuf, cfua, MBC, CA, DHA, root morphology, dry biomass). Regression analysis of soil respiration with soil physical and biotic variables was conducted. All statistical and regression analysis were performed using the software SPSS Statistics 22.0 (SPSS Inc., Chicago, IL, USA), and figures were generated using SigmaPlot 12.5 (Systat Software, Inc., Chicago, IL, USA).

4. Conclusions

This study investigated the variation of soil respiration and its influencing factors under different irrigation levels with and without aeration in a greenhouse tomato system. Aeration had a significant effect on tomato root length, as well as dry biomass of leaf, stem, and fruit, while no significant differences on other parameters were observed. As irrigation amount applied into soils increased, soil respiration increased in conjunction with its influencing factors, and the effect was significant on the dry biomass of leaf, fruit, and root. Soil respiration was significantly correlated with soil water-filled pore space, the abundance of soil bacteria and fungi. These results indicate that irrigation combined with aeration would increase soil physical and biotic variables, which stimulate more CO_2 release. The application of reduced irrigation and aeration has potential for alleviating CO_2 emissions.

Author Contributions: Conceptualization, H.C. (Hui Chen) and H.C. (Huanjie Cai); Data curation, H.C. (Hui Chen) and Z.S.; Formal analysis, Z.S.; Funding acquisition, H.C. (Huanjie Cai); Investigation, H.C. (Hui Chen), Z.S. and Y.Z.; Methodology, H.C. (Hui Chen), H.C. (Huanjie Cai) and Y.Z.; Software, H.C. (Hui Chen); Validation, H.C. (Huanjie Cai); Writing—original draft, H.C. (Hui Chen); Writing—review and editing, Z.S., H.C. (Huanjie Cai) and Y.Z.

Funding: This work was supported by the National Key Research and Development Program of China (2016YFC0400201) and the National Natural Science Foundation of China (51309192).

Conflicts of Interest: The authors declare no conflict of interest.

References

1. Bhattarai, S.P.; Midmore, D.J.; Pendergast, L. Yield, water-use efficiencies and root distribution of soybean, chickpea and pumpkin under different subsurface drip irrigation depths and oxygation treatments in vertisols. *Irrig. Sci.* **2008**, *26*, 439–450. [CrossRef]
2. Mo, Y.; Li, G.; Wang, D. A sowing method for subsurface drip irrigation that increases the emergence rate, yield, and water use efficiency in spring corn. *Agric. Water Manag.* **2017**, *179*, 288–295. [CrossRef]
3. Růžička, K.; Ljung, K.; Vanneste, S.; Podhorská, R.; Beeckman, T.; Friml, J.; Benková, E. Ethylene Regulates Root Growth through Effects on Auxin Biosynthesis and Transport-Dependent Auxin Distribution. *Plant Cell* **2007**, *19*, 2197–2212. [CrossRef] [PubMed]
4. Zhu, Y.; Cai, H.J.; Song, L.B.; Chen, H. Impacts of oxygation on plant growth, yield and fruit quality of tomato. *Trans. Chin. Soc. Agric. Mach.* **2017**, *48*, 199–211. (In Chinese)
5. Bhattarai, S.P.; Pendergast, L.; Midmore, D.J. Root aeration improves yield and water use efficiency of tomato in heavy clay and saline soils. *Sci. Hortic.* **2006**, *108*, 278–288. [CrossRef]
6. Dhungel, J.; Torabi, M.; Pendergast, L.; Chen, X.; Bhattarai, S.P.; Midmore, D.J. Impact of oxygation on soil respiration, yield and water use efficiency of three crop species. *J. Plant Ecol.* **2010**, *4*, 236–248.
7. Hanson, P.J.; Edwards, N.; Garten, C.; Andrews, J. Separating root and soil microbial contributions to soil respiration: A review of methods and observations. *Biogeochemistry* **2000**, *48*, 115–146. [CrossRef]
8. Kuzyakov, Y. Sources of CO_2 efflux from soil and review of partitioning methods. *Soil Boil. Biochem.* **2006**, *38*, 425–448. [CrossRef]
9. Chen, H.; Hou, H.-J.; Wang, X.-Y.; Zhu, Y.; Saddique, Q.; Wang, Y.-F.; Cai, H. The effects of aeration and irrigation regimes on soil CO_2 and N_2O emissions in a greenhouse tomato production system. *J. Integr. Agric.* **2018**, *17*, 449–460. [CrossRef]
10. Hou, H.; Chen, H.; Cai, H.; Yang, F.; Li, D.; Wang, F. CO_2 and N_2O emissions from Lou soils of greenhouse tomato fields under aerated irrigation. *Atmos. Environ.* **2016**, *132*, 69–76. [CrossRef]
11. Qi, Y.; Xu, M. Separating the effects of moisture and temperature on soil CO_2 efflux in a coniferous forest in the Sierra Nevada mountains. *Plant Soil* **2001**, *237*, 15–23. [CrossRef]
12. Shi, W.-Y.; Du, S.; Morina, J.C.; Guan, J.-H.; Wang, K.-B.; Ma, M.-G.; Yamanaka, N.; Tateno, R. Physical and biogeochemical controls on soil respiration along a topographical gradient in a semiarid forest. *Agric. For. Meteorol.* **2017**, *247*, 1–11. [CrossRef]
13. Jia, S.; Wang, Z.; Li, X.; Sun, Y.; Zhang, X.; Liang, A. N fertilization affects on soil respiration, microbial biomass and root respiration in Larix gmelinii and Fraxinus mandshurica plantations in China. *Plant Soil* **2010**, *333*, 325–336. [CrossRef]
14. Yu, C.-L.; Hui, D.; Deng, Q.; Dzantor, E.K.; Fay, P.A.; Shen, W.; Luo, Y. Responses of switchgrass soil respiration and its components to precipitation gradient in a mesocosm study. *Plant Soil* **2017**, *420*, 105–117. [CrossRef]
15. Makoi, J.; Ndakidemi, P. Selected soil enzymes: Examples of their potential roles in the ecosystem. *Afr. J. Biotechnol.* **2008**, *7*, 181–191.
16. Li, Y.; Niu, W.; Wang, J.; Liu, L.; Zhang, M.; Xu, J. Effects of Artificial Soil Aeration Volume and Frequency on Soil Enzyme Activity and Microbial Abundance when Cultivating Greenhouse Tomato. *Soil Sci. Soc. Am. J.* **2016**, *80*, 1208. [CrossRef]
17. Li, J.; Sun, J.; Yang, Y.; Guo, S.; Glick, B.R. Identification of hypoxic-responsive proteins in cucumber roots using a proteomic approach. *Plant Physiol. Biochem.* **2012**, *51*, 74–80. [CrossRef]
18. Li, Y.; Niu, W.; Xu, J.; Wang, J.; Zhang, M.; Lv, W. Root morphology of greenhouse produced muskmelon under sub-surface drip irrigation with supplemental soil aeration. *Sci. Hortic.* **2016**, *201*, 287–294. [CrossRef]

19. Xu, C.-M.; Wang, D.-Y.; Chen, S.; Chen, L.-P.; Zhang, X.-F. Effects of Aeration on Root Physiology and Nitrogen Metabolism in Rice. *Rice Sci.* **2013**, *20*, 148–153. [CrossRef]
20. Jia, B.; Zhou, G.; Wang, F.; Wang, Y.; Yuan, W.; Zhou, L. Partitioning root and microbial contributions to soil respiration in Leymus chinensis populations. *Soil Boil. Biochem.* **2006**, *38*, 653–660. [CrossRef]
21. Iqbal, J.; Hu, R.; Feng, M.; Lin, S.; Malghani, S.; Ali, I.M. Microbial biomass, and dissolved organic carbon and nitrogen strongly affect soil respiration in different land uses: A case study at Three Gorges Reservoir Area, South China. *Agric. Ecosyst. Environ.* **2010**, *137*, 294–307. [CrossRef]
22. Lange, M.; Eisenhauer, N.; Sierra, C.A.; Bessler, H.; Engels, C.; Griffiths, R.I.; Mellado-Vázquez, P.G.; Malik, A.A.; Roy, J.; Scheu, S.; et al. Plant diversity increases soil microbial activity and soil carbon storage. *Nat. Commun.* **2015**, *6*, 6707. [CrossRef] [PubMed]
23. Du, Y.; Niu, W.; Zhang, Q.; Cui, B.; Gu, X.; Guo, L.; Liang, B. Effects of Nitrogen on Soil Microbial Abundance, Enzyme Activity, and Nitrogen Use Efficiency in Greenhouse Celery under Aerated Irrigation. *Soil Sci. Soc. Am. J.* **2018**, *82*, 606. [CrossRef]
24. Manzoni, S.; Schimel, J.P.; Porporato, A. Responses of soil microbial communities to water stress: Results from a meta-analysis. *Ecology* **2012**, *93*, 930–938. [CrossRef] [PubMed]
25. Zhu, Y.; Cai, H.; Song, L.; Chen, H. Aerated Irrigation Promotes Soil Respiration and Microorganism Abundance around Tomato Rhizosphere. *Soil Sci. Soc. Am. J.* **2019**, *83*, 1343. [CrossRef]
26. Chen, S.; Liu, A.; He, C.; Zou, Z. Microbial and enzyme activities in tomato rhizosphere with organic soil cultivation in solar greenhouse. *Chin. J. Soil Sci.* **2010**, *41*, 815–818. (In Chinese)
27. Chaparro, J.M.; Sheflin, A.M.; Manter, D.K.; Vivanco, J.M. Manipulating the soil microbiome to increase soil health and plant fertility. *Boil. Fertil. Soils* **2012**, *48*, 489–499. [CrossRef]
28. Drenovsky, R.; Vo, D.; Graham, K.; Scow, K. Soil Water Content and Organic Carbon Availability Are Major Determinants of Soil Microbial Community Composition. *Microb. Ecol.* **2004**, *48*, 424–430. [CrossRef]
29. Ben-Noah, I.; Friedman, S. Aeration of clayey soils by injecting air through subsurface drippers: Lysimetric and field experiments. *Agric. Water Manag.* **2016**, *176*, 222–233. [CrossRef]
30. Xu, G.-W.; Lu, D.-K.; Wang, H.-Z.; Li, Y. Morphological and physiological traits of rice roots and their relationships to yield and nitrogen utilization as influenced by irrigation regime and nitrogen rate. *Agric. Water Manag.* **2018**, *203*, 385–394. [CrossRef]
31. Chen, H.; Shang, Z.-H.; Cai, H.-J.; Zhu, Y. An Optimum Irrigation Schedule with Aeration for Greenhouse Tomato Cultivations Based on Entropy Evaluation Method. *Sustainability* **2019**, *11*, 4490. [CrossRef]
32. Bai, T.; Li, C.; Li, C.; Liang, D.; Ma, F. Contrasting hypoxia tolerance and adaptation in Malus species is linked to differences in stomatal behavior and photosynthesis. *Physiol. Plantarum.* **2013**, *147*, 514–523. [CrossRef] [PubMed]
33. Intrigliolo, D.S.; Bonet, L.; Nortes, P.A.; Puerto, H.; Nicolas, E.; Bartual, J. Pomegranate trees performance under sustained and regulated deficit irrigation. *Irrig. Sci.* **2013**, *31*, 959–970. [CrossRef]
34. Zhu, Y.; Cai, H.; Song, L.; Chen, H. Oxygation improving soil aeration around tomato root zone in greenhouse. *Trans. Chin. Soc. Agric. Eng.* **2017**, *33*, 163–172. (In Chinese)
35. Borkhuu, B.; Peckham, S.; Ewers, B.; Norton, U.; Pendall, E.; Ewers, B. Does soil respiration decline following bark beetle induced forest mortality? Evidence from a lodgepole pine forest. *Agric. For. Meteorol.* **2015**, *214*, 201–207. [CrossRef]
36. Dieleman, W.I.J.; Luyssaert, S.; Rey, A.; de Angelis, P.; Barton, C.V.M.; Broadmeadow, M.S.J.; Broadmeadow, S.B.; Chigwerewe, K.S.; Crookshanks, M.; Dufrêne, E.; et al. Soil [N] modulates soil C cycling in CO_2-fumigated tree stands: A meta-analysis. *Plant Cell Environ.* **2010**, *33*, 2001–2011. [CrossRef]
37. Cheng, W. Rhizosphere priming effect: Its functional relationships with microbial turnover, evapotranspiration, and C–N budgets. *Soil Boil. Biochem.* **2009**, *41*, 1795–1801. [CrossRef]
38. Dijkstra, F.A.; Cheng, W. Interactions between soil and tree roots accelerate long-term soil carbon decomposition. *Ecol. Lett.* **2007**, *10*, 1046–1053. [CrossRef]
39. Zak, D.R.; Pregitzer, K.S.; King, J.S.; Holmes, W.E. Elevated atmospheric CO_2, fine roots and the response of soil microorganisms: A review and hypothesis. *New Phytol.* **2000**, *147*, 201–222. [CrossRef]
40. Scheer, C.; Grace, P.R.; Rowlings, D.W.; Payero, J. Soil N_2O and CO_2 emissions from cotton in Australia under varying irrigation management. *Nutr. Cycl. Agroecosys.* **2013**, *95*, 43–56. [CrossRef]
41. Reth, S.; Göckede, M.; Falge, E. CO_2 efflux from agricultural soils in Eastern Germany-comparison of a closed chamber system with eddy covariance measurements. *Theor. Appl. Climatol.* **2005**, *80*, 105–120. [CrossRef]

42. Chen, H.; Hou, H.; Hu, H.; Shang, Z.; Zhu, Y.; Cai, H.; Qaisar, S. Aeration of different irrigation levels affects net global warming potential and carbon footprint for greenhouse tomato systems. *Sci. Hortic.* **2018**, *242*, 10–19. [CrossRef]
43. Хазиев, ф.Х. *Soil Enzymes Activities*; Science Press: Beijing, China, 1980. (In Chinese)

 © 2019 by the authors. Licensee MDPI, Basel, Switzerland. This article is an open access article distributed under the terms and conditions of the Creative Commons Attribution (CC BY) license (http://creativecommons.org/licenses/by/4.0/).

Review

Recent Developments in Metal-Based Catalysts for the Catalytic Aerobic Oxidation of 5-Hydroxymethyl-Furfural to 2,5-Furandicarboxylic Acid

Sohaib Hameed [1,2], Lu Lin [1], Aiqin Wang [1] and Wenhao Luo [1,*]

[1] CAS Key Laboratory of Science and Technology on Applied Catalysis, Dalian Institute of Chemical Physics, Chinese Academy of Sciences, 457 Zhongshan Road, Dalian 116023, China; sohaibhameed@dicp.ac.cn (S.H.); linlu@dicp.ac.cn (L.L.); aqwang@dicp.ac.cn (A.W.)
[2] University of Chinese Academy of Sciences, Beijing 100049, China
* Correspondence: w.luo@dicp.ac.cn; Tel.: +86-411-8437-9738

Received: 23 December 2019; Accepted: 9 January 2020; Published: 15 January 2020

Abstract: Biomass can be used as an alternative feedstock for the production of fuels and valuable chemicals, which can alleviate the current global dependence on fossil resources. One of the biomass-derived molecules, 2,5-furandicarboxylic acid (FDCA), has attracted great interest due to its broad applications in various fields. In particular, it is considered a potential substitute of petrochemical-derived terephthalic acid (PTA), and can be used for the preparation of valuable bio-based polyesters such as polyethylene furanoate (PEF). Therefore, significant attempts have been made for efficient production of FDCA and the catalytic chemical approach for FDCA production, typically from a biomass-derived platform molecule, 5-hydroxymethylfurfural (HMF), over metal catalysts is the focus of great research attention. In this review, we provide a systematic critical overview of recent progress in the use of different metal-based catalysts for the catalytic aerobic oxidation of HMF to FDCA. Catalytic performance and reaction mechanisms are described and discussed to understand the details of this reaction. Special emphasis is also placed on the base-free system, which is a more green process considering the environmental aspect. Finally, conclusions are given and perspectives related to further development of the catalysts are also provided, for the potential production of FDCA on a large scale in an economical and environmentally friendly manner.

Keywords: biomass; 5-hydroxymethylfurfural; 2,5-furandicrboxylic acid; aerobic oxidation; metal catalysts

---

1. Introduction

The significantly growing demand for energy driven by an ever increasing global population and the strong economic growth in developing countries, the potential threats to the environment associated with the utilization of non-renewable fossil resources (oil, coal and natural gas) and difficulties with exploitation of the dwindling reserves have stimulated the research for alternative feedstocks that can be used for production of fuels and chemicals [1–5]. Biomass is one of the more promising and attractive alternative feedstocks, given its general abundance and wide-ranging availability in Nature and human society [1]. In addition, biomass is the only renewable carbon resource which can serve as a feedstock for the various carbon-containing chemicals and fuels that our society relies on [6].

5-Hydroxymethylfurfural (HMF), being a furan derivative, has been recognized as an important compound in our foods and as a versatile platform molecule derived from biomass for bulk chemicals and fuels production [7]. HMF can be used as biomass substrate compound for the production of a plethora of end products such as biofuels, biodegradable plastics, additives, macromolecules and functional polymers via different reactions such as oxidation, hydration, hydrogenation, etherification and decarbonylation (Scheme 1a) [8,9]. In this review, we focus on one particular route, namely, the oxidation of HMF to a value-added chemical, 2,5-furandicarboxylic acid (FDCA). HMF can be produced from the dehydration of C_6 carbohydrates or direct transformation of cellulose, involving several steps including catalytic hydrolysis of cellulose to glucose, followed by isomerization of glucose to form fructose, and finally dehydration of fructose to produce HMF. Acid sites are normally required for the production of HMF, and the formation of a multitude of other components, i.e., formic acid, levulinic acid and humin, is also observed during the production of HMF, which hampers the purity and yield of HMF [10].

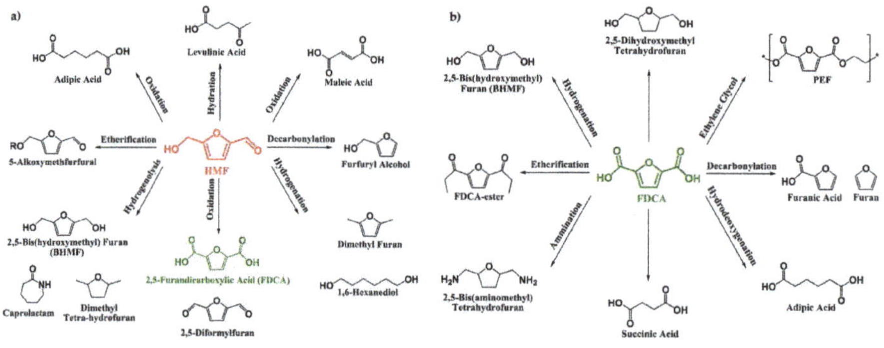

Scheme 1. Possible value-added platform chemicals from (**a**) HMF and (**b**) FDCA (reproduced from [11] with permission from Wiley-VCH, copyright 2019).

FDCA, being a promising molecule in the furan family, contains a multifunctional cyclic structure with two carboxylic acid groups attached at the *para* positions of the furan ring. It is listed on the top in all biomass-derived value-added chemicals by the US Department of Energy [12]. FDCA is considered stable, even at high temperatures, because of its high melting point (342 °C) due to which it is not easily soluble in common organic and inorganic solvents [13]. Scheme 1b summarizes the possible value-added chemicals which can be derived from FDCA in different reaction systems. FDCA is actually used as a starting material for the synthesis of a new class of bioderived polymers such as polyethylene 2,5-furandicaboxylate (PEF) [14,15]. PEF displays a series of excellent properties (thermal, chemical and mechanical resistance as well as easy depolymerization in Nature), compared to its petrochemical counterpart polyethylene terephthalate (PET) [16,17]. PET is the most common thermoplastic polymer material of the polyester family, and widely used in containers for foods and liquids, thermoforming for manufacturing, fibers for clothing, etc. [18]. The potential of substituting PET by this new bio-polyester PEF has stimulated great research efforts on this topic [19–22].

The production of FDCA has been studied since the 19th century. Firstly in 1876, Fittig et al successfully prepared FDCA by conducting the dehydration reaction of mucic acid over an acidic catalyst [23]. Concisely, aqueous hydrobromic acid (HBr) which acts both as catalyst and solvent was reacted with mucic acid to produce FDCA. Later, different dehydrating agents were also applied, along with some modifications to get higher efficiency in the dehydration process but this process is limited because of the severe reaction conditions i.e., high temperature (>120 °C), >20 h reaction time, the use of highly concentrated acids and less FDCA selectivity as well as moderate yield (<50%) [24]. Later HMF emerged as a promising biomass feedstock to produce FDCA. Direct oxidation of HMF is a simple method for the efficient and economical production of bio-based FDCA. Scheme 2 presents the general reaction scheme for the production of FDCA by catalytic oxidation of HMF monomer. In the typical reaction pathway, the synthesis of FDCA through catalytic oxidation of HMF proceeds initially through the selective oxidation of the hydroxyl group of HMF to produce 2,5-diformylfuran (DFF) (Scheme 2, Path 1), or via oxidation of the aldehyde group to form 5-hydroxymethyl-2-furan carboxylic acid (HMFCA) (Scheme 2, Path 2). Both intermediates are then further oxidized to 5-formyl-2-furancarboxylic acid (FFCA), which is finally transformed to FDCA [25]. Several reaction systems with different oxidants are used to activate the oxygen species for catalytic oxidation of HMF such as oxygen, air, H_2O_2, and $KMnO_4$ [26]. Oxygen or air are normally preferred for HMF oxidation due to the broad availability, low price, and benignity to the environment.

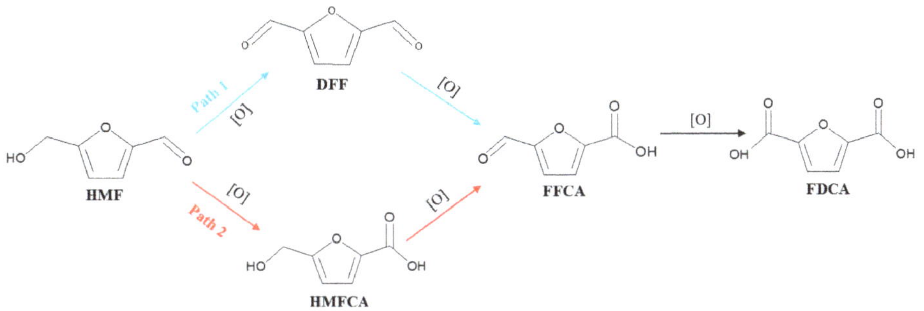

![Scheme 2]

Scheme 2. Reaction scheme for catalytic oxidation of HMF into FDCA.

Figure 1 depicts the total number of research publications on HMF and FDCA per year from 2000 to 2019. The past few years in particular have seen a significant increase in the number of publications on HMF and FDCA chemistry. Due to maximum product separation capability, oxidation of HMF is considered as advantageous process to produce FDCA in an economical way [27].

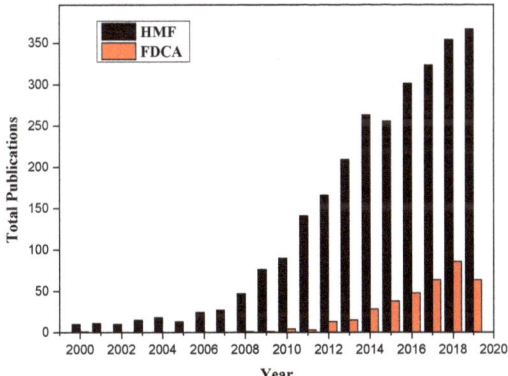

Figure 1. Total number of reported publications on HMF and FDCA chemistry per year from 2000 to 2019 (topic keywords searched in *Web of Science*: "5-hydroxymethylfurfural HMF" and "2,5-furan-dicarboxylic acid FDCA").

Various catalytic systems involving heterogeneous catalysts [28], homogeneous catalysts [29], and bio-catalysts [30] have been reported for this process in aqueous media as well as organic and biphasic systems [31]. Electrochemical [32] and photocatalytic [33] processes have also been studied for HMF oxidation to FDCA. Owing to the good recycling capability and stability, heterogeneous catalysts are the most studied for this process. In this review, only metal-based solid catalysts will be discussed, which thus excludes the other catalytic systems, i.e., homogeneous catalysts and bio-catalysts. We first highlight the recent developments and current state of art of the application of metal-based heterogeneous catalysts for the catalytic aerobic oxidation of HMF to FDCA. The emphasis of this review is thus put on comparing the catalytic performance of different catalysts categorized by metal, with specific examples using molecular oxygen (O_2) as oxidant and without using base additives. Reaction mechanisms of the aerobic oxidation of HMF to FDCA are also demonstrated in detail. Finally, conclusions are provided and perspectives referring to further development of the catalysts for the practical production are also highlighted, especially in designing efficient catalysts for green and cost-effective catalytic oxidation of HMF to FDCA.

2. Noble Metal Catalysts for FDCA Production

Oxidation can be catalyzed by metal sites in heterogeneous catalysis. Various noble metals such as gold (Au), platinum (Pt), palladium (Pd), ruthenium (Ru), and rhodium (Rh) catalysts have been reported for the catalytic oxidation of HMF to FDCA. The choice of molecular oxygen (O_2) as oxidant offers advantages of availability and benignity to the environment, in accordance with the concept of "green chemistry". As oxygen is not easy to activate, supported noble metal catalysts are the main heterogeneous catalysts used for the catalytic aerobic oxidation of HMF to FDCA. We will first summarize the recent advances in applying noble metal catalysts for the oxidation of HMF to FDCA under relatively mild conditions.

2.1. Au-Based Catalysts

Gold (Au) was generally considered as inactive in the field of catalysis, but in the 1980s, it was a great achievement in research to discover the catalytic ability of Au, which opened a new gateway for researchers to develop highly active catalysts for many processes [34,35]. Recently, many catalysts are developed using Au that present promising catalytic performance for the aerobic oxidation of HMF to FDCA in aqueous solutions. The performance of different Au-based catalysts for catalytic aerobic oxidation of HMF to FDCA is summarized, and details are also given, in Table 1.

Table 1. Summary of reported results for the aerobic oxidation of HMF to FDCA over Au catalysts.

Catalysts	Base	Reaction Conditions			HMF Conv. (%)	FDCA Yield (%)	Ref.
		T (°C)	Oxidant P (bar)	Time (h)			
Au/CeO$_2$	NaOH	130	Air, 10	8	100	>99	[36]
Au/TiO$_2$	NaOH	130	Air, 10	8	100	>99	[36]
Au/Fe$_2$O$_3$	NaOH	130	Air, 10	8	100	15	[36]
Au/C	NaOH	130	Air, 10	8	100	44	[36]
Au/m-CeO$_2$	NaOH	70	O$_2$, 10	4	100	92	[37]
Au/CeO$_2$	Na$_2$CO$_3$	140	O$_2$, 5.0	15	>99	91	[38]
Au/TiO$_2$	NaOH	30	O$_2$, 20	18	100	71	[39]
Au/HY	NaOH	60	O$_2$, 0.3	6	>99	>99	[40]
Au/TiO$_2$	NaOH	60	O$_2$, 0.3	6	>99	85	[40]
Au/Mg(OH)$_2$	NaOH	60	O$_2$, 0.3	6	>99	76	[40]
Au/CeO$_2$	NaOH	60	O$_2$, 0.3	6	>99	73	[40]
Au/H-MOR	NaOH	60	O$_2$, 0.3	6	96	15	[40]
Au/Na-ZSM5-25	NaOH	60	O$_2$, 0.3	6	92	1	[40]
Au-Cu/TiO$_2$	NaOH	95	O$_2$, 10	4	100	99	[41]
Au$_8$-Pd$_2$/C	NaOH	60	O$_2$, 30	4	>99	>99	[42]
Au/HT	Base free	95	O$_2$, 1	7	>99	>99	[43]
Au/HT-AC	Base free	100	O$_2$, 5	12	100	>99	[44]
Au-Pd/CNT	Base free	100	O$_2$, 5	12	100	94	[45]
Au-Pd/CNT	Base free	100	Air, 10	12	100	96	[45]
Au-Pd/CNT	Base free	100	O$_2$, 5	18	100	91	[45]
AuPd-nNiO [a]	Base free	90	O$_2$, 10	6	95	70	[46]
AuPd-La-CaMgAl-LDH [b]	Base free	100	O$_2$, 5	6	96.1	89.4	[47]

[a] nNiO = nanosized NiO [b] La-CaMgAl-LDH = La doped Ca-Mg-Al layered double hydroxide.

The choice of support for Au-based catalysts can also have a great impact on the catalytic performance in HMF oxidation. When using TiO$_2$ and CeO$_2$ as supports, Au-based catalysts showed nearly quantitative FDCA yields of >99% at 65 °C under 10 bar of air after a reaction time of 8 h; in contrast, Au catalysts supported on carbon and Fe_2O_3 only afforded FDCA yields of 44% and 15% under the same conditions, separately [36]. According to the reaction mechanism discussed in this study, HMFCA was observed as the only intermediate. As shown in Scheme 3, HMF was first oxidized to HMFCA very fast via the formation of a hemiacetal-1 intermediate. Owing to the fact no FFCA was directly observed, the authors proposed that FFCA was transformed via the oxidation of HMFCA was quickly transformed into FDCA through the production of a second intermediate product, hemiacetal-2. Compared to the one-pot reaction, substrate degradation was strongly diminished and the catalysts life increased by performing the reaction in two steps: first the oxidation of HMF into HMFCA at a low reaction temperature of 25 °C and, second, the subsequent oxidation of HMFCA in FDCA at 130 °C. Reductive pretreatment of the Au/CeO_2 was shown to efficiently increase the catalytic activity due to increased amount of Ce^{3+} and oxygen vacancies. The increased Ce^{3+} species and oxygen vacancies on the support were shown to have a great effect on transferring hydride and activating O_2 during the oxidation of the alcohol group. The Lewis acid sites of Ce^{3+} centers and Au^+ species of Au/CeO_2 could easily accept a hydride from the C–H bond in alcohol or in the corresponding alkoxide to form Ce–H and Au–H, with the simultaneous formation of a carbonyl species. The oxygen vacancies of CeO_2 could activate O_2 and form cerium-coordinated superoxide (Ce–OO) species, which subsequently evolved into cerium hydroperoxide by hydrogen abstraction from Au–H. The cerium hydroperoxide then interacted with Ce–H, producing H_2O and recovering the Ce^{3+} centers. Au–H donated H and changed back to the initial Au^+ species. Further improvement of activity of Au/CeO_2 was reported by Lolli et al [37]. An ordered mesoporous CeO_2 (m-CeO_2) supported Au catalyst was synthesized by nanocasting technique using meso-structured silica SBA-15 as hard template. Au nano-particles immobilized on this high surface area mesoporous CeO_2 showed a FDCA yield of 92% with 100% HMF conversion under relatively mild reaction conditions (T = 70 °C, P_{O2} = 10 bar, and t = 4 h).

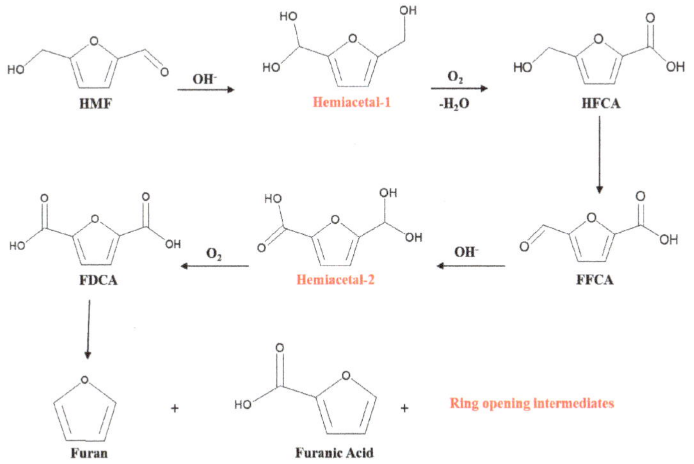

Scheme 3. Reaction mechanisms for aerial oxidation of aqueous HMF over CeO_2 supported Au catalysts (reproduced from [36] with permission from Wiley-VCH, copyright 2009).

Formation of undesired humin is a great issue for HMF transformation, especially in concentrated HMF solutions. Kim et al. reported recently progress in utilizing Au/CeO_2 catalyst for achieving 90–95% yield of FDCA via aerobic oxidation of acetal derivatives of HMF [38]. In this approach, protection of aldehyde group of HMF with 1,3-propanediol was proposed to prevent the formation of undesired humin via decomposition and self-polymerization, and to achieve efficient FDCA yield from the resultant HMF acetal derivative. Even in concentrated solutions of 20% PD-HMF, FDCA could still be obtained in a high yield of 91% at 140 °C and 5 bar O_2, for 15 h reaction. This example presents a significant advance over the conventional oxidation of HMF that gives only reasonable FDCA yields in dilute solutions.

Zeolite-supported Au catalysts have also been investigated for the catalytic oxidation of HMF by Xu et al. [40]. The Au/H-Y catalyst showed high yield of FDCA (>99%) with a quantitative HMF conversion under mild reaction conditions (T = 60 °C, P_{O2} = 0.3 bar and t = 6 h), which was much higher than that of Au supported on Mg(OH)$_2$, TiO$_2$, CeO$_2$, H-MOR, and ZSM-5. Further characterization indicated that Au-nanoclusters (approx. 1 nm) are encapsulated inside the supercages of the H-Y-zeolite, and the confinement of the supercage prevented the further agglomeration of Au nanoclusters into large particles (Figure 2). The interaction between the acidic hydroxyl groups in the zeolite supercage and Au clusters has been shown to be responsible for stabilization of the Au species, to which the high catalytic efficiency for the oxidation of HMF to FDCA was ascribed [40].

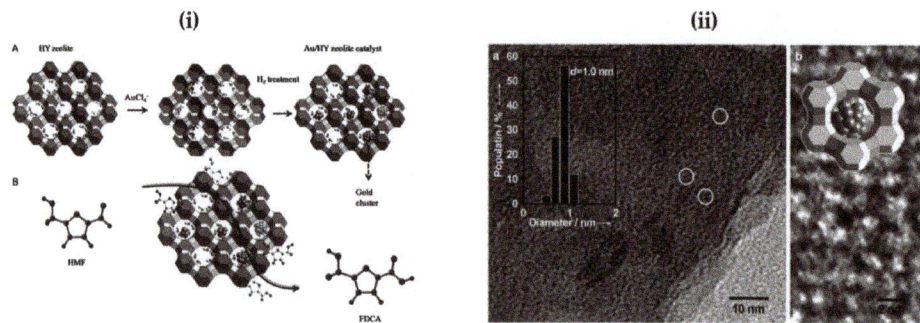

Figure 2. (i)-**A** Schematic illustration of synthesis for Au-nanoclusters in HY zeolite supercages (i)-**B** Catalytic oxidation process (ii)-**a** TEM image of Au/HY (ii)-**b** HR-TEM image of Au/HY (reproduced from [40] with permission from Wiley-VCH, copyright 2013).

Various Au-based bimetallic catalysts were also studied for the oxidation of HMF to FDCA. The physical and chemical characteristics of the prepared bimetallic catalysts can be simply tuned by altering catalytic composition, particle size and mixing equality. Pasini et al. reported that the Au-Cu/TiO$_2$ bimetallic catalyst afforded higher catalytic activity and stability over its corresponding mono-metallic Au/TiO$_2$ for HMF oxidation into FDCA [41]. All the bimetallic Au-Cu/TiO$_2$ catalysts with different Au/Cu mole ratio prepared via a colloidal route showed an improved activity, by at least factor of two compared to their corresponding monometallic Au catalysts. Under the optimal reaction conditions (10 bar O$_2$, 4 equiv of NaOH, 95 °C), HMF conversion of 100%, and FDCA yield of 99% were attained after 4 h (Table 1). The isolation of Au sites caused by AuCu alloying was the main reason for the excellent catalytic activity of the Au-Cu/TiO$_2$ catalysts for HMF oxidation into FDCA. The Au-Cu/TiO$_2$ catalyst could be easily recovered and reused without significant leaching and agglomeration of the metal nanoparticles. Thus, a strong synergistic effect was also evident in term of the catalyst stability and resistance to poisoning. Similar results in catalytic performance were also demonstrated for the Au-Pd/AC catalyst [42]. Alloying Au with Pd at molar ratio of 8:2 on carbon support significantly increased the catalyst stability and activity than the monometallic counterparts for the aerobic oxidation of HMF to FDCA. The Au/AC catalyst showed good product selectivity, but suffered with catalyst deactivation, with a drop in 20% of HMF conversion after the fifth run. No Au leaching from the catalyst was detected, and the deactivation was mainly attributed to irreversible adsorption of the byproducts or intermediates and the agglomeration of Au particles. In contrast, Au-Pd/AC delivered an excellent stability with a FDCA yield of 99% even after the fifth run. The alloying of a second metal (e.g., Pd or Cu) with Au to form bimetallic alloy catalysts can indeed combine the advantages of different components and thus improve the catalyst performance.

To understand the strategic reaction mechanism and role of base, molecular oxygen and Au catalysts, Davis and co-workers studied the reaction route for oxidation of HMF to FDCA over Au/TiO$_2$ catalyst in alkaline medium (NaOH) using an isotope labeling approach [48,49]. By control experiments, base and a metal catalyst were shown to be important to produce FDCA at 22 °C. The aldehyde side-chain of HMF undergoes a rapid reversible hydration to a geminal diol through nucleophilic addition of a hydroxide ion to the carbonyl and subsequent proton transfer from water to the alkoxy ion intermediate (Scheme 4, step 1). This step is due to the incorporation of two ^{18}O atoms in HMFCA when the reaction was performed in H$_2$^{18}O. The second step is the dehydrogenation of the geminal diol intermediate, facilitated by the hydroxide ions adsorbed on the metal surface, to form the carboxylic acid HMFCA (Scheme 4, step 2). Further oxidation of the alcoholic group of HMFCA is required to produce FDCA. Base deprotonates the alcoholic group to form an alkoxy intermediate in solution [50]. Hydroxide ions on the catalyst surface then facilitate the activation of the C–H bond in the alcoholic group to form the aldehyde intermediate, 5-formyl-2-furancarboxylic acid (FFCA) (Scheme 4, step 3). The next two steps (Scheme 4, steps 4 and 5) oxidize the aldehyde function of FFCA to form FDCA. These two steps are expected to proceed similarly to steps 1 and 2 for oxidation of HMF to HFCA. The reversible hydration of the aldehyde group in step 4 to a geminal diol accounts for two more ^{18}O atoms incorporated in FDCA when the oxidation is performed in H$_2$^{18}O. Overall, complete HMF oxidation to FDCA illustrates that water molecules incorporate all four oxygen atoms in FDCA instead of readily available oxidant (O$_2$). The isotope labeling experiments of ^{18}O$_2$ and H$_2$^{18}O revealed that water was the source of oxygen atoms during the oxidation of HMF to HMFCA and FDCA, probably through direct participation of hydroxide in the catalytic cycle. Molecular oxygen was essential for the production of FDCA and played an indirect role during oxidation by removing electrons deposited into the supported metal particles. Those results provided a fundamental understanding of the roles of added base and molecular oxygen for FDCA production from HMF [49].

Most aerobic oxidations of HMF over Au catalysts are conducted in the presence of excess base, however, considering environmental and economic concerns, base-free HMF to FDCA oxidation systems are more desirable. Thus, some reports on the oxidation of HMF to FDCA over Au-based catalysts without using base were published recently. Gupta et al. reported a hydrotalcite-supported Au catalyst (Au/HT) for the oxidation of HMF into FDCA without using base [43]. An excellent yield of 99% FDCA was demonstrated at 95 °C under 1 bar O$_2$ in water after 7 h. Compared to Au deposited on neutral support or acidic SiO$_2$, limited activity was shown, indicating the essential need for basic sites on the catalyst. Although Au/MgO gave a FDCA yield of 21%, it was much lower than that of Au/HT. TEM revealed a much larger size of Au nanoparticles on MgO (>10 nm) than that of Au/HT (3.2 nm), which should be the main reason for the lower catalytic activity of Au/MgO. Although the authors claimed that Au/HT catalyst could be reused, Zope et al. observed a severe leaching of Mg^{2+} from HT over Au/TiO$_2$ catalyst and HT as solid base during the oxidation of HMF, owing to the chemical interaction between the basic HT and the formed FDCA [51]. Further improvement in catalyst stability by modified robust hydrotalcite and activated carbon supported Au catalyst (Au/HT-AC) was demonstrated under base-free conditions [44]. Physical milling of homemade hydrotalcite and commercial activated carbon was applied for catalyst preparation. The Au/HT-AC catalyst showed superior catalytic activity (FDCA Yield = 99.5%) at 100 °C, 5 bar O$_2$ pressure after 12 h of oxidation reaction (Table 1) with excellent catalytic stability (6 times). Availability of enough basic sites, large surface area of catalyst, and presence of hydroxyl and carbonyl groups are the reasons for enhanced catalytic performance and improved reusability.

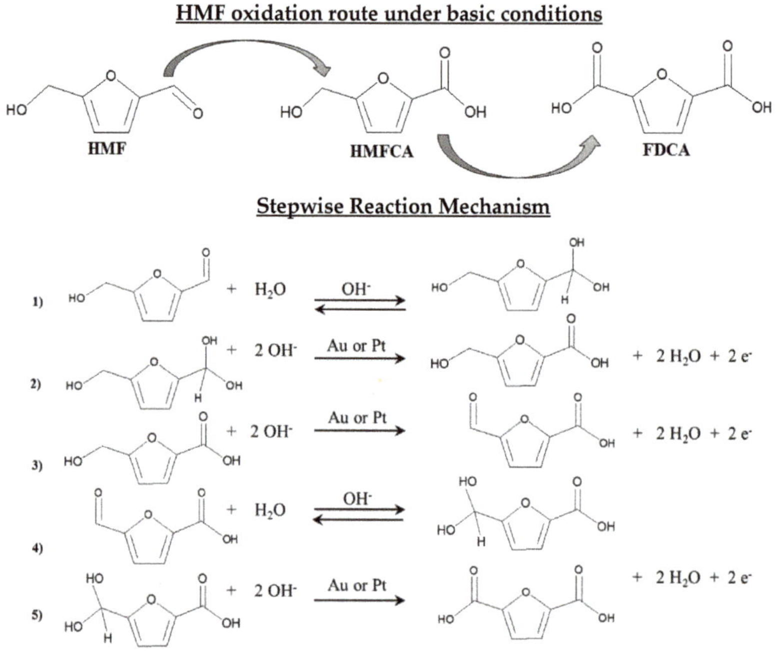

Scheme 4. Expanded reaction pathway of HMF oxidation in basic (OH⁻) media over Au or Pt catalysts (reproduced from [48,49] with permission from Royal Society of Chemistry & Elsevier, copyright 2012 and 2014).

Development of an active and stable bimetallic Au-Pd catalyst also reported by Wan et al [45]. A FDCA yield of 94% could be achieved at 100 °C and 0.5 MPa O_2 for 10 h, and a FDCA yield of 96% was obtained at 100 °C and 1 MPa air for 12 h. The surface carbonyl/quinone and phenol species on CNT was found to facilitate the adsorption of HMF and DFF, rather than FDCA, contributing to the high activity of the Au-Pd/CNT catalyst. In addition, the incorporation of Pd to Au/CNT changed the reaction pathway from HMFCA to DFF route by facilitate the oxidation of the hydroxyl species of HMF, and further enhanced the oxidation of FFCA to FDCA, which is a difficult step for Au catalysts under base-free conditions. Notably, an improved stability with Au-Pd/CNT was also depicted, with marginal loss in activity during the consecutive six runs. Bonincontro et al. further reported an efficient and stable nNiO-supported Au-Pd alloy, with an optimal Au/Pd atomic ratio of 6:4, for base-free oxidation of HMF to FDCA [46]. A nearly quatitative yield of FDCA could be obtained at 90 °C after 14 h. NiO was shown to provide basic sites that can promote the reaction and the suitable choice of Au-Pd chemical composition favors the formation of FDCA. Gao et al. reported a highly efficient and stable bimetallic AuPd nanocatalyst over the La-doped Ca-Mg-Al layered double hydroxide (La-CaMgAl-LDH) support for base-free aerobic oxidation of HMF to FDCA in water [47]. A nearly full yield of FDCA could be achieved at 120 °C and 0.5 MPa O_2 after 6 h. No catalyst deactivation was observed at 100 °C and 0.5 MPa O_2 after four consecutive runs. The high dispersion of a small amount of La_2O_3 on the surface of LDH support were attributed to stabilize the support via preventing the deterioration of LDH support by formed carboxylic acid products during reaction, thus resulting in excellent stability and recyclability of AuPd/La-CaMgAl-LDH catalyst.

2.2. Pt-Based Catalysts

Pt-based catalysts were initially reported to be efficient for the aerobic oxidation of HMF to produce FDCA. Recent results of the aerobic oxidation of HMF to FDCA over Pt-based catalysts are summarized in Table 2. Verdeguer et al. studied the carbon supported Pt catalysts for the oxidation of HMF [52]. The addition of Pb into the Pt/C catalyst could improve the catalytic performance with a high FDCA yield of 99%, than that with the Pd/C catalyst (81%) at ambient conditions for 2 h (25 °C, P = 1 bar, a O_2 flow rate of 2.5 mL/s and 1.25 M NaOH solution). HMFCA was detected as the reaction intermediate, pointing to the preferred oxidation of aldehyde group of HMF with Pt-Pb catalyst. Moreover, the addition of Bi to the Pt/C catalyst also showed a beneficial effect on the FDCA yield (Table 2) [53]. An optimal Pt/Bi molar ratio of about 0.2 showed a high FDCA yield of 98% for the Pt-Bi/C at 100 °C, 40 bar air and the use of four equivalents of $NaHCO_3$ after 6 h, compared to 69% for Pt/C catalyst. Both HMFCA and DFF were detected as intermediates, and the oxidation of FFCA was figured out to be the rate-determining step for this Pt-Bi/C catalyst. Introducing Bi into the Pt/C catalyst improved the catalyst stability, owing to depressing the oxygen poisoning and leaching of Pt. Similar effect was also observed in another study over TiO_2 supported Pt-Bi catalyst [54].

Table 2. Results of the aerobic oxidation of HMF to FDCA over Pt-based catalysts.

Catalysts	Base	Reaction Conditions			HMF Conv. (%)	FDCA Yield (%)	Ref.
		T (°C)	Oxidant P (bar)	Time (h)			
Pt/C	NaOH	25	O_2, 1	2	100	81	[52]
Pt-Pb/C	NaOH	25	O_2, 1	2	100	99	[52]
Pt/C	Na_2CO_3	100	Air, 40	6	99	69	[53]
Pt-Bi/C	Na_2CO_3	100	Air, 40	6	100	>99	[53]
Pt/TiO_2	Na_2CO_3	100	Air, 40	6	90	84	[54]
Pt-Bi/TiO_2	Na_2CO_3	100	Air, 40	6	>99	99	[54]
Pt/Al_2O_3	Na_2CO_3	75	O_2, 1	12	96	96	[55]
Pt/ZrO_2	Na_2CO_3	75	O_2, 1	12	100	94	[55]
Pt/C	Na_2CO_3	75	O_2, 1	12	100	89	[55]
Pt/CeO_2	Na_2CO_3	75	O_2, 1	12	100	8	[55]
Pt/TiO_2	Na_2CO_3	75	O_2, 1	12	96	2	[55]
Pt/$Ce_{0.8}Bi_{0.2}O_{2-\delta}$	NaOH	23	O_2, 10	0.5	100	98	[56]
Pt/CeO_2	NaOH	23	O_2, 10	0.5	100	20	[56]
Pt/RGO [a]	NaOH	25	O_2, 1	24	100	84	[57]
Fe_3O_4@C@Pt	Na_2CO_3	90	O_2, 1	4	100	100	[58]
Pt/Al_2O_3	pH = 9	60	O_2, 0.2	6	100	99	[59]
Pt/ZrO_2	Base free	100	O_2, 4	12	100	97.3	[60]
Pt/C-O-Mg	Base free	110	O_2, 10	12	>99	97	[61]
Pt/C-EDA-x [b]	Base free	110	O_2, 10	12	100	96	[62]
Pt-Ni/AC	Base free	100	O_2, 4	15	100	97.5	[63]
Pt-PVP-GLY [c]	Base free	80	O_2, 1	24	100	94	[64]
Pt-PVP-$NaBH_4$	Base free	80	O_2, 1	24	100	80	[64]
Pt-PVP-EtOH	Base free	80	O_2, 1	24	100	75	[64]
Pt-PVP-H_2	Base free	80	O_2, 1	24	100	19	[64]
Pt-NP-Cl [d]	Base free	80	O_2, 1	6	100	65	[65]
Pt-NP5 [e]	Base free	80	O_2, 1	6	100	60	[65]

[a] RGO = Reduced Graphene Oxide [b] EDA = ethylenediamine [c] PVP = polyvinylpyrrolidone. [d] NP-Cl = Nanoparticle with Cl ionic polymer [e] NP5 = Nanoparticle with $C_{12}H_{23}O_2$ ionic polymer.

The catalytic performance of different Pt-based catalysts is highly dependent on the choice of support. Ramakanta et al. compared the performance of different metal oxide-supported Pt catalysts under 75 °C, 1 bar O_2 and the use of Na_2CO_3 after 12 h. Pt catalysts supported on the non-reducible oxides (ZrO_2, Al_2O_3 and C) showed high FDCA yields of above 90%, while Pt catalysts supported on reducible oxides (TiO_2 and CeO_2) separately gave poor FDCA yields of 2% and 8% [55]. The authors attributed the higher catalytic performance on Pt catalysts to the lower oxygen storage ability of the non-reducible oxides, which could efficiently prevent the oxidation of active metal sites. Addition of Bi into the CeO_2 support could also enhance the catalytic performance of Pt/CeO_2 [56]. Pt-catalysts with bismuth (Bi) modified ceria support, the Pt/Ce$_{0.8}$Bi$_{0.2}$O$_{2-\delta}$ afforded a FDCA yield of 98% while the Pt/CeO$_2$ only showed a FDCA yield of 20 % at 23 °C, 10 bar O$_2$. Moreover, the Pt/Ce$_{0.8}$Bi$_{0.2}$O$_{2-\delta}$ catalyst showed a good catalyst stability, and could be reused for five runs with a marginal loss of its FDCA yield (from 98% in the first run to 97% in the fifth run). In this work, Pt nanoparticles were believed to first react with the hydroxyl group of HMF and form the Pt-alkoxide intermediate, which was further converted to aldehyde by β-H elimination. Bi-containing ceria contained a large amount of oxygen vacancies, which could further promote the oxygen reduction process and the cleavage of the peroxide intermediate by bismuth. Therefore, the surface electrons were used to the reduction of oxygen, and a new catalytic cycle could be smoothly continued.

Carbon materials are widely used as catalyst supports owing to their good properties and easy availability. Apart from the most used active carbon, reduced graphene oxide (RGO) can be also used as support owing to its abundant surface functional groups, which can be used to anchor the metal nanoparticles. Niu et al. studied a series of RGO supported metal nanoparticles for the oxidation of HMF at 25 °C and 1 bar O$_2$, with the presence of NaOH and an O$_2$ flow rate of 50 mL/min [57]. The Pt/RGO catalyst showed a higher FDCA yield than Pd/RGO, Ru/RGO and Rh/RGO, and a FDCA yield of 84% could be obtained after 24 h. During the recycling experiments, an increase in HMFCA and a decrease in FDCA yield could be observed. Similar results that the Pt/C outperforms in catalytic activity than the Pd/C was also reported by Davis et al. [66]. Notably, Zhang et al. prepared a series of novel Fe$_3$O$_4$@C@Pt catalysts containing super-paramagnetic Pt-nanoparticles with a core-shell structure, and used for the oxidation of HMF (Figure 3) [58]. These novel Fe$_3$O$_4$@C@Pt catalysts have a spherical shape with a Fe$_3$O$_4$ core, a protective amorphous carbon shell and Pt nanoparticle clusters decorated on the surface. The preparation temperature showed an impact on the morphology of active Pt on the carbon shell (Figure 3i). The 110-Fe$_3$O$_4$@C@Pt prepared at reflux of 110 °C afforded a nearly full FDCA yield at 90 °C after 4 h during the oxidation of HMF (Figure 3ii). In addition, this catalyst can be reused up to three times without significant loss in performance. Platinum on alumina is also studied in basic conditions (pH = 9) and high FDCA yield (99%) is reported due to strong metal-substrate interaction through π-electrons of furan nucleus [59].

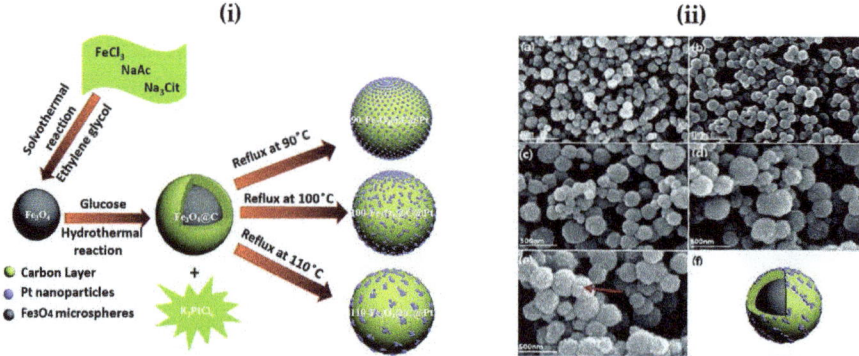

Figure 3. (i) Schematic illustration of synthesis of core-shell structure X-Fe$_3$O$_4$@C@Pt super-paramagnetic microspheres (X = Refluxing Temperature) (ii) (a) SEM images and model of 110-Fe$_3$O$_4$@C@Pt microsphere, (b) core/shell microspheres after coating with an amorphous carbon layer, and Pt-decorated magnetic core/shell microspheres: (c) 90-Fe$_3$O$_4$@C@Pt (d) 100-Fe$_3$O$_4$@C@Pt and (e) 110-Fe$_3$O$_4$@C@Pt synthesized at different temperatures. (f) A model image of a 110-Fe$_3$O$_4$@C@Pt microsphere (reproduced from [58] with permission from the Royal Society of Chemistry, copyright 2016).

Most of the above examples of the aerobic oxidation of HMF over Pt catalysts are generally performed in the presence of excess base. The disadvantages of basic feeds are that product solutions require neutralization of the base, and separation of the resulting formed inorganic salts. Base-free catalytic oxidation of HMF to FDCA has also been reported over Pt-based catalysts. Recently, Chen et al. prepared a Pt/ZrO_2 by the atomic layer deposition (ALD) method and conducted the aerobic oxidation of HMF under base-free conditions [60]. A complete HMF conversion and a FDCA yield of 97.3% under mild reaction conditions (100 °C, O_2, 4 bar, 12 h) were reported. The highly dispersed and uniform particle size of Pt particles was visualized by TEM. An improved C=O adsorption on the catalyst surface was also indicated by temperature programmed desorption of CO (CO-TPD), pointing to a strong interaction between reactants/intermediates. Both factors were attributed to the good catalytic activity for the Pt/ZrO_2. Han et al. designed a novel Pt catalyst with modified C-O-Mg support for base-free aerobic oxidation of HMF to FDCA [61]. A high FDCA yield of 97% could be obtained at 110 °C and 10 bar O_2. In addition, this catalyst could be used for ten times with little loss of activity. Even scaling up the reaction by 20 times at a large scale, a decent yield of isolated FDCA could achieve 74.9% with a high purity of 99.5%. Han et al. further developed a N-doped carbon- supported Pt for base-free aerobic oxidation of HMF to FDCA [62]. The synthesized Pt/C-EDA-x catalyst (where EDA = ethylenediamine and x = nitrogen dose) prepared by using EDA as nitrogen source showed higher catalytic activity than the counterparts using N,N-dimethylaniline (DMA), ammonia (NH_3), and acetonitrile (ACH) as nitrogen source. A FDCA yield of 96% was achieved with the Pt/C-EDA-4.1 catalysts at optimal conditions (T = 110 °C, P_{O2} = 1 bar, and t =12 h). This high catalytic performance is attributed to the formation of a new kind of medium basic site due to the formation of pyridine-type nitrogen in the catalyst.

Limited examples of Pt-based bimetallic catalysts are reported for base-free aerobic oxidation of HMF to FDCA. Shen et al. prepared a Pt-Ni/AC bimetallic catalyst by atomic layer deposition of Pt nanoparticles on the surface of Ni/AC for base-free oxidation of HMF to FDCA [63]. The bimetallic Pt-Ni/AC catalyst showed a higher activity (a FDCA yield of 97.5% yield with 100% HMF conversion) even with only a 0.4 wt % Pt loading at 100 °C and 4 bar O_2 after 15 h, compared to the monometallic counterparts with a higher Pt loading (the 5.6 wt% Ni/AC and 1.6 wt% Pt/AC). In addition, the bimetallic catalyst afforded good reusability for at least four 15 h runs without any obvious loss of catalytic activity. The authors proposed that the addition of Ni to Pt improved the ability of Pt for C=O adsorption and oxidation, thus increasing the activity of the Pt catalyst.

Pt nanoparticles stabilized by polymers were also shown to be effective for base-free aerobic oxidation of HMF to FDCA. Siankevich et al. reported that polyvinylpyrrolidone (PVP) stabilized Pt nanoparticles could promote the base-free aerobic oxidation of HMF into FDCA in water [64]. A high FDCA of 95% were achieved for the Pt-PVP-GLY catalyst at mild reaction conditions (T = 80 °C, P_{O2} = 1 bar, and t = 24 h). Notably, a slight decrease of its catalytic activity was observed during five consecutive runs. The reaction mechanism was investigated for the Pt-PVP-GLY catalyst. Isotope ($H_2^{18}O$) labeling technology was used to elaborate the reaction path, and mass spectrometric analysis of solutions after reaction verified the existence of ^{18}O atomic levels. DFF and FFCA were detected as the reaction intermediates during this oxidation reaction, while HMFCA was not detected during base-free oxidation of HMF.

As shown in Scheme 5, the aldehyde group was proposed to undergo a rapid reversible hydration to a geminal diol by nucleophilic addition of water, and followed by proton transfer to form DFF. Similar results were also reported by other researchers under base-free conditions at relatively low pH values [45,48]. The release of two protons from the hydroxyl group in HMF could form DFF. Mass spectrometric analysis of the reaction mixture confirmed that ^{18}O was incorporated in the oxidation products (FDCA and FFCA). Peaks with m/z 161 and 163 were attributed to three and four ^{18}O atoms incorporated in FDCA, and peaks with m/z 143 and 145 attributed to two and three ^{18}O atoms incorporated in FFCA. Finally, a transfer of two H to the surface of the metal occurred to form the carboxylic acid groups, and molecular oxygen reacted with the surface hydride to release H_2O. Furthermore, Siankevich et al. reported Pt nanoparticles stabilized by an imidazolium-based cross-linked polymer (with chloride as the counter-anion), which could efficiently catalyzed the oxidation of HMF to FDCA in water with oxygen as the oxidant under mild conditions (T = 80 °C, P_{O2} = 1 bar, and t = 6 h) [65]. Various counter-anion, that is, replacing chloride by BF_4^-, PF_6^-, bis(trifluoromethylsulfonyl)imide, hexanoate, or laurate anions, in the cationic polymer has been explored. The counter-anion indeed showed an impact on the structure of the obtained platinum nanoparticles, the surface electronic properties, and their catalytic activity. The highest reaction rates were obtained with the weakly nucleophilic bis(trifluoromethylsulfonyl)imide anion, which also favored Pt in the metallic state, leading to complete conversion of the substrate and a high yield of FDCA (65%).

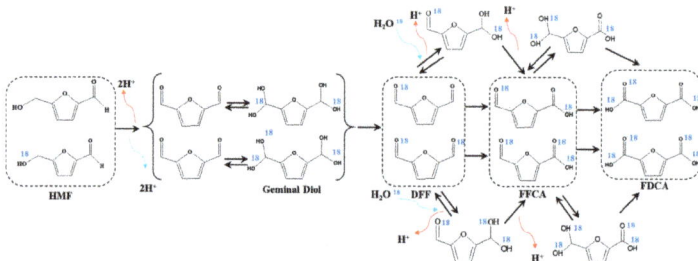

Scheme 5. Reaction mechanism with inclusion of ^{18}O (blue). Main focusing units are boxed in dashed lines (reproduced from [64] with permission from Elsevier, copyright [2014]).

2.3. Pd-Based Catalysts

Pd-based catalysts also showed good catalytic performance for the aerobic oxidation of HMF into FDCA. According to the reported Pd-based catalysts for the aerobic oxidation of HMF, results along with their reaction conditions are summarized in Table 3.

Table 3. Summary of reported results for the aerobic oxidation of HMF to FDCA over Pd catalysts.

Catalysts	Base	Reaction Conditions			HMF Conv. (%)	FDCA Yield (%)	Ref.
		T (°C)	Oxidant, P (bar)	Time (h)			
Pd/C	NaOH	23	O_2, 6.9	6	100	71	[66]
Pd/ZrO$_2$/La$_2$O$_3$	NaOH	90	O_2, 1	6	>99	90	[67]
Pd/Al$_2$O$_3$	NaOH	90	O_2, 1	6	>99	78	[67]
Pd/Ti$_2$O$_3$	NaOH	90	O_2, 1	6	>99	53	[67]
Pd/PVP [a]	NaOH	90	O_2, 1.01	6	>99	90	[68]
Pd/CC [b]	K_2CO_3	100	O_2, 20 mL/min	30	100	85	[69]
γ-Fe$_2$O$_3$@HAP-Pd [c]	K_2CO_3	100	O_2, 1	6	97	92.9	[70]
C-Fe$_2$O$_3$-Pd	K_2CO_3	80	O_2, 1	4	98.2	91.8	[71]
Pd/C@Fe$_2$O$_3$	K_2CO_3	80	O_2, 1	6	98.4	86.7	[72]
Pd-Au/TiO$_2$	NaOH	70	O_2, 10	4	100	85	[73]
Pd/TiO$_2$	NaOH	70	O_2, 10	4	100	9	[73]
Pd-Au/HT [d]	NaOH	60	O_2, 1	6	100	90	[74]
Pd/HT	NaOH	60	O_2, 1	6	85	6	[74]
Pd-Ni/Mg(OH)$_2$	Base free	100	Air	16	>99	89	[75]
Pd/HT	Base free	100	O_2, 1	8	>99	>99	[76]

[a] PVP = Polyvinylpyrrolidone [b] CC = Carbonaceous Catalyst [c] HAP = Hydroxyapatite [d] HT= Hydrotalcite.

The performance and particle size of different Pd-based catalysts are found to be dependent on the choice of the support. Siyo et al. studied the effect of catalytic support by depositing PVP stabilized Pd-nanoparticles on different metal oxide (Al_2O_3, TiO_2, KF/Al_2O_3, and ZrO_2/La_2O_3) supports in ethylene glycol with average particle size of 1.8 nm [67]. $Pd/ZrO_2/La_2O_3$ catalyst showed to be a promising catalyst for HMF oxidation reaction, with a FDCA yield of 90% at 90 °C and 1 bar O_2 pressure after 8 h of reaction time (Table 3). TEM indicated that there was no obvious sintering of Pd nanoparticles in the spent $Pd/ZrO_2/La_2O_3$ catalyst, whereas other catalysts showed serious aggregation. XPS confirmed that the majority of Pd remained as the metallic state and that the electronic structure of the Pd nanoparticles was unchanged in the spent $Pd/ZrO_2/La_2O_3$ catalyst. The strong interaction between Pd nanoparticles and the metal oxide supports makes the catalysts more stable, to which its efficient catalytic performance is attributed. Siyo et al. further studied PVP stabilized Pd NP catalyst using ethylene glycol at three different Pd/NaOH ratios [68]. Differently sized Pd nanoparticles are obtained, with mean diameters between 1.8 nm and 4.4 nm, by changing the Pd/NaOH ratios. It was concluded that smaller Pd-nanoparticles showed higher FDCA yield. The highest FDCA yield obtained is 90% with greater than 99% HMF conversion over 1.8 nm Pd-nanoparticle catalyst synthesized with 1:4 molar ratio of Pd/NaOH [68]. These newly designed Pd-catalysts were able to store for long time (up to one month) in alkaline medium under ambient conditions without any change in their activity.

Owing to the ease in catalyst separation after its use, Zhang et al. prepared a Pd catalyst with magnetic properties [70]. In this catalyst hydroxyapatite (HAP) was layered on Fe_2O_3 using coating technique and Pd^{2+} ions are exchanged with HAP's Ca^{2+} ions with further reduction to produce Pd^0 nanoparticles. This novel catalyst showed excellent catalytic activity with almost 93% of FDCA yield and 97% HMF conversion under optimal reaction conditions (T = 100 °C, P_{O2} = 1 bar, t = 6 h). This catalyst, due to its magnetic properties, is very easy to recycle as it can be easily separated using an external magnet without any change in catalytic activity. In another study, similar super paramagnetic catalyst was prepared by Mei and coworkers, in which graphene-based $C-Fe_2O_3-Pd$ catalysts were synthesized using reliable graphene oxide [71]. Pd nanoparticles and Fe_2O_3 particles are simultaneously deposited on surface of graphene oxide by one pot solvothermal technique. The reaction temperature and the base concentration notably affected the HMF conversion and FDCA selectivity. This carbon catalyst demonstrated high catalytic activity for oxidation reactions of HMF with FDCA with a yield of 91.8% and HMF conversion of 98.2% under mild reaction parameters (80 °C, O_2, 1 bar, 4 h) with a base/substrate ratio of 0.5. The same research group further studied the Pd-based magnetic catalysts and effectively immobilized the Pd-nanoparticles on core-shell structure of $C@Fe_3O_4$ (C acted as shell and Fe_3O_4 as core) with further reduction to generate $Pd/C@Fe_3O_4$ catalysts [72]. In this technique, no excess reducing agents and capping reagents were used which makes it a clean and environmentally benign technique. Under optimal conditions, $Pd/C@Fe_3O_4$ catalyst showed high catalytic activity in the aerobic oxidation of HMF to FDCA with 98.4% HMF conversion and 87.8% FDCA yield at 80 °C for 6 h reaction.

Bimetallic catalysts has also been studied and found to be an efficient approach for the aerobic oxidation of HMF to FDCA. Lolli et al. studied bimetallic Pd-catalysts for HMF oxidation to FDCA by preparing $Pd-Au/TiO_2$ nanoparticle catalysts along with their monometallic counterpart Pd/TiO_2 and revealed that minor alloying of Pd with gold (Pd:Au = 1:6) can enhance the FDCA yield from 9% to 85% with complete HMF conversion at 70 °C and 10 bar O_2 pressure for 4 h reaction [73]. Major reaction routes involved in FDCA production on pyrolized $Pd-Au/TiO_2$ in the presence of base are summarized in Scheme 6. According to this study, once HMFCA is formed, the bimetallic catalyst showed a reaction route with no further oxidation due to the substantial inability of Pd and its alloy to oxidize the hydroxymethyl group of HMFCA [73]. FDCA molecule is produced via oxidation of alcohol part of the HMF molecule to produce DFF, which upon further oxidation gives FFCA. This produced FFCA has the options of producing FDCA or being oxidized to HMFCA which will again show the same behavior of no further oxidation. Recently, similar Pd-Au bimetallic catalysts were also synthesized on hydrotalcite (HT) support by Xia et al and their catalytic performance compared with that of their monometallic counterparts Pd/HT and Au/HT catalysts [74]. High FDCA yield (up to 90%) was achieved over bimetallic Pd-Au/Ht catalyst with Au:Pd = 4:1 at 60 °C and 1 bar O_2 pressure after 6 h of reaction. Similar to the previous report, the enhanced catalytic performance over a bimetallic catalyst is attributed to the synergistic effect between two metals and the formation of smaller particle size of the catalyst that facilities the HMF oxidation reaction to produce FDCA. In addition, Gupta et al. further studied the bimetallic $M_{0.9}$-$Pd_{0.1}$ (M = Ni, Co or Cu) alloy nanoparticles supported on in situ prepared $Mg(OH)_2$ for the aerobic oxidation of HMF to FDCA without using an external base additive [75]. $Ni_{0.9}$-$Pd_{0.1}$/$Mg(OH)_2$ was found to be the most efficient, and afforded superior catalytic activity with a FDCA yield of 89% compared to Co- or Cu-based bimetallic nanoparticles at 100 °C. The basicity of support could facilitate the activation of the hydroxyl group of HMF, leading to the enhanced FDCA production. Furthermore, Wang et al. reported another HT supported Pd nanoparticle catalyst in base-free environment, in which Mg-Al-CO_3 HT supported Pd catalysts are synthesized with different Mg/Al ratios [76]. The Mg/Al molar ratio 5:1 with 2% Pd metal loading (2%Pd/HT-5) appeared to be best in terms of catalytic activity due to formation of weak basic sites (OH^- groups). FDCA yields greater than 99% are achieved with >99% HMF conversion for 8 h under mild reaction conditions (T = 100 °C and P = 1 bar).

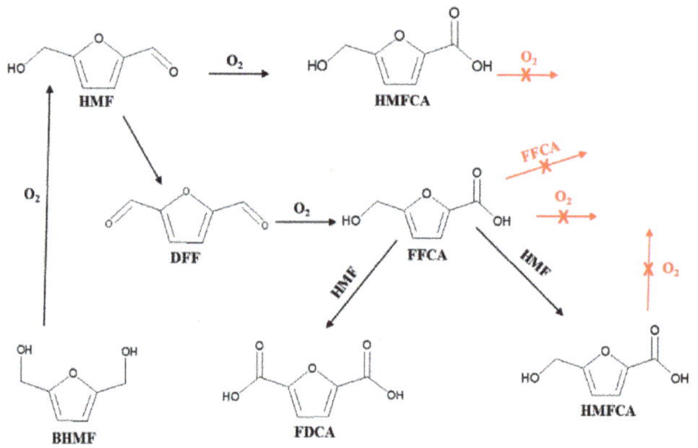

Scheme 6. Reaction pathways for HMF oxidation over Pd-Au/TiO_2 catalyst (reproduced from [73] with permission from Elsevier, copyright 2015).

2.4. Other Noble Metal Catalysts

The aerobic oxidation of HMF to FDCA can also be normally catalyzed with different Ru-based catalysts at moderate reaction temperatures ranging from 100–150 °C, with O_2/air pressure of 1 to 40 bar and base additives [77–83]. Majorly, Ru-based catalysts were reported for the oxidation of HMF to DFF in organic solvents [84]. While only a few examples were extended to the oxidation of HMF to FDCA in water [85]. The performance of different Ru-based catalysts, for catalytic aerobic oxidation of HMF to FDCA is summarized, and details are also given in Table 4.

Yi et al. compared the effect of weak bases (K_2CO_3, Na_2CO_3, HT, and $CaCO_3$) with strong base (NaOH) over commercial Ru/C catalyst and concluded that the stronger base lead towards lower FDCA yield (69%) (Table 1) due to degradation of HMF at higher pH [78]. The weaker the base, the higher the FDCA yield. The maximum FDCA yield (95%) was attained by the use of $CaCO_3$ and FDCA obtained is in the form of its calcium salt because of its lower solubility.

Table 4. Summary of reported literature on HMF oxidation to FDCA over Ru catalysts.

Catalysts	Base	Reaction Conditions			HMF Conv. (%)	FDCA Yield (%)	Ref.
		T (°C)	Oxidant, P (bar)	Time (h)			
Ru/C	CaCO$_3$	120	O$_2$, 2	5	100	95	[78]
Ru/C	Na$_2$CO$_3$	120	O$_2$, 2	5	100	93	[78]
Ru/C	K$_2$CO$_3$	120	O$_2$, 2	5	100	80	[78]
Ru/C	NaOH	120	O$_2$, 2	5	100	69	[78]
Ru/C	HT	120	O$_2$, 2	5	100	60	[78]
Ru/C	Base free	120	O$_2$, 2	10	100	88	[78]
Ru/C	NaHCO$_3$	100	Air, 40	2	100	75	[79]
Ru/AC$_{NaOCl}$ [a]	NaHCO$_3$	100	Air, 40	4	100	55	[79]
Ru(OH)$_x$/La$_2$O$_3$	Base free	100	O$_2$, 30	5	98	48	[80]
Ru(OH)$_x$/HT [b]	Base free	140	Air, 1	24	99	19	[80]
Ru/MnCo$_2$O$_4$	Base free	120	Air, 24	10	100	99.1	[81]
Ru/CoMn$_2$O$_4$	Base free	120	Air, 24	10	100	82.2	[81]
Ru/MnCo$_2$CO$_3$	Base free	120	Air, 24	10	100	69.9	[81]
Ru/HAP [c]	Base free	120	O$_2$, 10	24	100	99.6	[82]
Ru/HAP [c]	Base free	140	O$_2$, 10	24	100	99.9	[82]
Ru/ZrO$_2$	Base free	120	O$_2$, 10	16	100	97	[83]

[a] AC$_{NaOCl}$ = Activated Carbon oxidized with sodium hypochlorite [b] HT= Hydrotalcite [c] HAP = Hydroxyapatite.

In this reaction system, oxidation of HMF occurred through hydroxyl part rather than aldehyde part. Ru-catalyst favors the production of FDCA through DFF route instead of HMFCA route due to oxidation of –OH group of HMF to –CHO group (Scheme 7). It is concluded that the conversion of –OH to –CHO is very fast step as compared to subsequent conversion of –CHO to –COOH, which is the rate determining step [78]. Over Ru/C catalyst, the oxidation of –CHO to –COOH is in less time (5 h) due to presence of base, which is a challenge in base-free conditions. This is why base-free reaction was carried out for relatively longer reaction time (10 h) to get 88% FDCA yield at similar reaction conditions (120 °C, O$_2$, 2 bar).

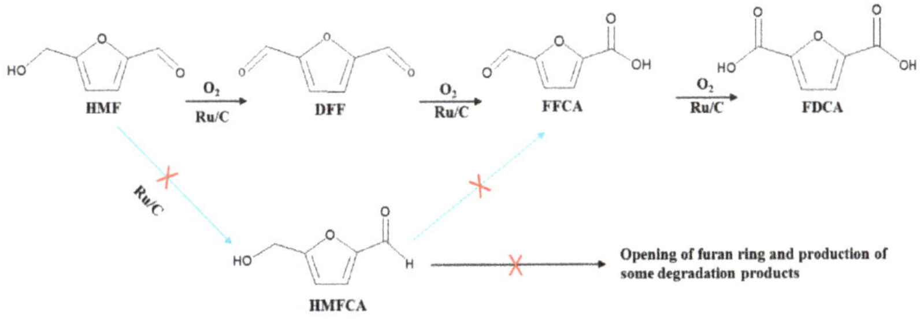

Scheme 7. Reaction route for oxidation of HMF to produce FDCA over Ru/C catalysts (reproduced from [78] with permission from the Royal Society of Chemistry, copyright 2016).

Kerdi and coworkers studied the modification of catalyst support using various doped carbons as supporting materials for Ru-based catalysts to investigate the consequence of surface properties as well as pore structure of carbon on oxidation rates [79]. Activated carbon was oxidized using sodium hypochlorite (NaOCl) and the Ru metal was impregnated on this oxidized AC$_{NaOCl}$ support to prepare Ru/AC$_{NaOCl}$ catalyst with particle size of 2 nm. Moderate results with FDCA yield of 55% with complete HMF conversion are obtained after 4 h reaction at 100 °C over a modified supported Ru catalysts [79].

Normally, the HMF oxidation reaction takes place in organic solvents or water, but this reaction is also investigated in ionic liquids (ILs) over Ru(OH)$_x$ catalysts in base-free environment due to redox stability, negligible vapor pressure, non-flammability and unique dissolving abilities of ILs [80]. In this study, several different supports (TiO$_2$, Fe$_2$O$_3$, ZrO$_2$, CeO$_2$, HAP, HT, MgO, and La$_2$O$_3$) on Ru(OH)$_x$ were tested along with various ILs. As shown in Table 4, Ru(OH)$_x$/La$_2$O$_3$ gives reasonable catalytic activity, with a FDCA yield of 48% and high HMF conversion (98%) at 100 °C and elevated pressure (30 bar O_2). $Ru(OH)_x$/HT also appeared to be active in ILs with 99% HMF conversion but FDCA yield obtained is very low (19%) at high temperature (140 °C) and ambient pressure after 24 h of reaction. High temperature (140 °C) is selected to decrease the influence of viscosity of the mixture formed in ILs and catalyst [80]. As the idea for using ionic liquids is flopped because of their high cost, low FDCA yield and unfeasible large scale production, researchers continue to search for base-free phase oxidation over modified support Ru catalysts. In another work, $MnCo_2O_4$ spinel supported Ru-catalyst (Ru/$MnCo_2O_4$) with 4% metal loading shows an exceptionally high FDCA yield of 99.1% in comparison with other modified support catalysts (Ru/$CoMn_2O_4$ and Ru/$MnCo_2CO_3$) under moderate reaction conditions (T = 120 °C, P_{air} = 24 bar and t = 10 h) with minor FFCA impurities [81]. In this study, it is concluded that catalyst supports structure plays vital role in catalytic activity. The high FDCA yield is due to Lewis and BrØnsted active acid sites on catalytic surface which is confirmed by NH_3-TPD results (Figure 4). Variation in supports to $CoMn_2O_4$ and $MnCo_2CO_3$ adversely affects the catalytic activity and FDCA yield is decreased to 82.2% and 69.9%, respectively (Table 4).

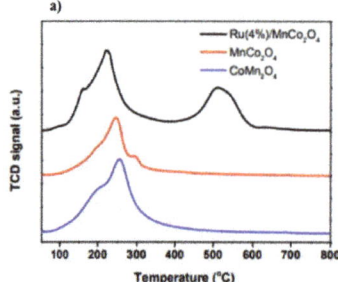

Samples	Lewis Acid Sites (mmol.g^{-1})	BrØnsted Acid Sites (mmol.g^{-1})	Total Acid Sites (mmol.g^{-1})
Ru (4%)/$MnCo_2O_4$	7.3	10.7	18.0
$MnCo_2O_4$	-	8.2	8.2
$CoMn_2O_4$	-	11.7	11.7

Figure 4. (a) NH_3-TPD and (b) summarized data obtained from NH_3-TPD for $CoMn_2O_4$, $MnCo_2O_4$ and Ru/$MnCo_2O_4$ catalyst (reproduced from [81] with permission from the Royal Society of Chemistry, copyright 2017).

Ru/HAP also shows good catalytic activity with 100% of HMF conversion and more than 99% of FDCA yield in base-free conditions [82]. This reaction is carried out in severe reaction conditions (T = 140 °C, P = 10 bar and t = 24 h) which is very challenging for practical applications. High catalytic performance of Ru/HAP catalyst is credited to formation of acidic-basic sites due to well dispersed Ru nanoparticles on HAP support. Christian and coworkers studied FDCA production via HMF oxidation over Ru catalyst supported on high surface area zirconia (ZrO_2) in a base-free environment [83]. High FDCA yield (97%) is achieved with 100% HMF conversion after 16 h reaction at similar severe reaction conditions (T = 120 °C and P = 10 bar). The catalytic tests in this investigation revealed that the small size of Ru particles due to utilization of high surface area ZrO_2 is crucial reason for better catalytic performance.

Rhodium (Rh) metal, in comparison with other noble metals (Ru, Pd, Au, and Pt), also has similar potential to act as catalytic active site in heterogeneous catalysis. However, this metal is not broadly explored by the researchers for catalytic oxidation of HMF to FDCA. This could be because of its high cost and less availability. Vuyyuru et al. have taken first step to use Rh to prepare a Rh/C catalyst for catalytic oxidation of HMF to FDCA [86]. The catalytic activity is compared for HMF oxidation reaction using different noble metals on carbon support at mild temperature (50 °C) and oxygen pressure (10 bar). The FDCA yield obtained in this work is 12.62%, which is comparatively low, with HMF conversion of 82%, after 4 h of reaction time over Rh/C catalysts. Ag-based catalysts were also studied for the aerobic oxidation of HMF to FDCA [87]. Inferior activities and selectivity to FDCA were shown with Ag-based catalysts, with HMFCA as the primary product. In addition, leaching of Ag was also demonstrated to be another issue during catalysis. Therefore, rooms are available for the improvement of catalytic performance of those noble-metal based catalysts in the aerobic oxidation of HMF into FDCA.

To sum up the overall results for the aerobic oxidation of HMF to FDCA over noble metal catalysts, a lot of progress has been achieved recently. Au catalysts are more effective catalysts in terms of stability and selectivity in comparison with Pt, Pd, Ru, and Rh based catalysts owing to a better ability in resistance to water and oxygen. However, deactivation of Au catalysts, and intermediates depositing on catalytic active sites are still observed for Au catalysts. Most of the studied noble metal catalysts use base additives to achieve high FDCA yield, to facilitate the oxidation of aldehyde part of HMF and keep the formed FDCA to dissolve into the solutions, in which the strong adsorption of products on the catalysts can be prevented. The aerial oxidation of HMF in base-free environment is a more green process, and more appropriate for sustainable chemistry. This entails a more research focus on base-free catalytic systems for the oxidation of HMF to FDCA. Applying an appropriate support and using an alloy strategy might make the catalysts show high activity and stability for the base-free oxidation of HMF into FDCA.

3. Non-Noble Metal Catalysts for FDCA Production

Noble metal catalysts are generally considered as active and stable in the field of catalysis, but due to their higher costs and less availability, it is of interest to design non-noble metal catalysts with high efficiency and excellent stability. Therefore, research shifted towards non-precious metal catalysts for aerobic oxidation of HMF to FDCA with prominent catalytic performance, and a reasonable progress has been achieved until now to get active and stable catalysts [88] (Table 5).

Table 5. Results of the oxidation of HMF to FDCA over non-noble metal catalysts.

Catalysts	Additive	Reaction Conditions			HMF Conv. (%)	FDCA Yield (%)	Ref.
		T (°C)	Oxidant P (bar)	Time (h)			
Fe-POP [a]	-	100	O_2, 10	10	100	85	[89]
MR-Co-Py [b]	CH_3CN	100	t-BuOOH	24	95.6	90.4	[90]
$Li_2CoMn_3O_8$	CH_3COOH	150	Air, 55	8	100	80	[91]
Fe_3O_4-CoO_x	-	80	t-BuOOH	12	97.2	68.6	[92]
$Ce_{0.5}Fe_{0.5}O_2$	[Bmim]Cl	140	O_2, 20	24	98.4	13.8	[93]
$Ce_{0.5}Zr_{0.5}O_2$	[Bmim]Cl	140	O_2, 20	24	96.1	23.2	[93]
$Ce_{0.5}Fe_{0.15}Zr_{0.35}O_2$	[Bmim]Cl	140	O_2, 20	24	99.9	44.2	[93]
$Fe_{0.6}Zr_{0.4}O_2$	[Bmim]Cl	160	O_2, 20	24	99.7	60.6	[94]
MnO_2	$NaHCO_3$	100	O_2, 10	24	>99	91.0	[95]
MOF-Mn_2O_3	$NaHCO_3$	100	O_2, 14	24	100	99.5	[96]
MnO_x-CeO_2	$KHCO_3$	110	O_2, 20	15	98	91	[97]
$MnCo_2O_4$	$KHCO_3$	100	O_2, 20	24	99.5	70.9	[98]
Co_3O_4/Mn_xCo	Base free	140	O_2, 1	24	100	>99	[99]

[a] POP = Porous organic polymer; [b] MR-Co-Py = Merrifield Resin supported Co(II)-*meso*-tetra(4-pyridylporhyrin).

Saha et al. prepared a thermally stable, robust structured iron catalyst (Fe/POP) by the integration of Fe^{3+} on the center of porphyrin ring supported on porous organic polymer (POP) to study the catalytic performance of aerobic oxidation of HMF to FDCA in aqueous medium [89]. This inexpensive metal catalyst can be reused without any significant loss of activity because the oxidation state of Fe remains intact after the reaction over this catalyst. As a result, complete HMF conversion was achieved with high FDCA yield (85%) at 100 °C and 10 bar O_2 pressure for 10 h reaction. The metal active site of Fe^{3+}-POP catalyst plays a significant role and a plausible radical chain mechanism for HMF oxidation would involve thermal autoxidation of organic substrate (R–H) to peroxides (ROOH) which further lead to FDCA product through Fenton-type cleavage of RO–OH bond over Fe. Later on, a stable cobalt (II)-*meso*-tetra(4-pyridyl) porphyrin supported on Merrifield resin catalyst was developed (abbreviated as Merrified Resin-Co-Py) by Gao et al. and studied the effect of various oxidants [90]. This catalyst showed excellent catalytic activity (FDCA yield = 90.4%, and HMF conversion = 95.6%) at 100 °C in the presence of *tert*-butylhydroperoxide (*t*-BuOOH) as oxidant after 24 h reaction (Table 5). On the other hand, in the presence of O_2 as oxidant, no FDCA was detected after 24 h of reaction. Methyl nitrile (CH_3CN) was found to be the best solvent in this reaction system. Furthermore, Jain et al. introduced a low-cost $Li_2CoMn_3O_8$ (spinel-mixed metal oxide) catalyst prepared by gel pyrolysis method to study the catalytic oxidation of HMF to FDCA in the presence of sodium bromide and acetic acid [91]. Although a reasonable FDCA yield (80%) is obtained but use of high temperature (150 °C) along with acetic acid and sodium bromide additives are the main drawbacks for this route.

Similar to noble metal magnetic catalysts, Wang et al. prepared non-noble metal (nano-Fe_3O_4-CoO_x) catalysts with magnetic properties [92]. As demonstrated earlier, this catalyst can also be easily recovered using external magnet because of its magnetic properties. This magnetic catalyst showed 97.2% of HMF conversion with FDCA yield (68.6%) and reasonable reusability of catalyst with minor mass loss at 80 °C for 12 h reaction (Table 5). Experimental results in this study demonstrated that the first step of HMF oxidation to produce FFCA is initiated by BrØnsted base, even without the presence of catalyst, whereas, the second step to produce FDCA from FFCA, requires the presence of catalysts. Similar conclusions were reached in another study of the catalytic oxidation of HMF to produce FFCA over a low cost $Mn_{0.75}/Fe_{0.25}$ heterogeneous metal catalyst [100].

Ionic liquids or ionic fluids have also been studied as solvents for this reaction system together with low cost transition metal oxide catalysts [93,94]. Several combinations of iron oxide (Fe_2O_3) and zirconia (ZrO_2) have been used with different Fe to Zr ratios to develop a highly efficient catalyst. For this system, even though the attained FDCA yield is low, excellent conversion of HMF attracts researchers to dig more about its mechanism. The results illustrated that change in the reaction parameters and using different Fe to Zr proportions can hardly improve the catalytic activity (Table 5).

Manganese (Mn) based catalysts were also reported for the aerobic oxidation of HMF with base additives. $MnO_2/NaHCO_3$ system was reported by Hayashi et al. [95]. A FDCA yield of 91% with a complete HMF conversion could be obtained at 100 °C and 10 bar O_2 after 24 h of reaction. However, catalyst deactivation was observed in the third cycle of reuse runs, mainly owing to adsorbed humin species covering the active sites. Catalyst reactivation could be achieved by calcination at 300 °C in air. Hayashi et al. further studied the impact of the structure of MnO_2 crystal on the performance in the areobic oxidation of HMF to FDCA through combined computational and experimental studies [101]. They demonstrated that reaction rates per surface area for the slowest step, FFCA oxidation to FDCA step, decrease in the order of β-MnO_2 > λ-MnO_2 > γ-MnO_2 ≈ α-MnO_2 > δ-MnO_2 > ϵ-MnO_2 on the basis of good agreements achieved between experimental results with the DFT calculations. β-MnO_2 exceeds that of the previously reported activated MnO_2. The successful synthesis of high-surface-area β-MnO_2 could significantly improve the catalytic activity for the aerobic oxidation of HMF to FDCA. Notably, a porous 2D Mn_2O_3 nanoflakes was prepared by a facile thermal treatment of a Mn-based metal-organic framework (MOF) precursor and applied for oxidation of HMF at 100 °C and 14 bar O_2 with $NaHCO_3$ [96]. A FDCA yield of 99.5% at complete conversion of HMF could be achieved after a reaction time of 24 h. However, a slight catalyst deactivation was observed during the recycle experiments.

Mixed/binary oxides have also been applied for the aerobic oxidation of HMF and showed enhanced catalytic performances as compared with their mono-oxide counterparts. Han et al prepared a mixed oxide MnO_x-CeO_2 (Mn/Ce = 6) catalyst by co-precipitation method, which afforded a high FDCA yield of 91% with a HMF conversion of 98% at 110 °C with $KHCO_3$ after 15 h reaction [97]. Structural analysis of mixed oxide catalyst revealed that the Mn^{4+} and Ce^{3+} on catalytic surface played the pivotal role as the active sites for HMF oxidation to FDCA. A mechanism involving the Mn^{4+} active center, the lattice oxygen transfer from CeO_2 to Mn oxide and the activation of O_2 on CeO_2 was proposed for the enhanced the performance. Mn-Co binary oxides catalysts with different Mn/Co molar ratios were also studied by Zhang et al. for catalytic oxidation of HMF to FDCA [98]. The $MnCo_2O_4$ catalyst with a Mn/Co molar ratios of 1/2, showed a HMF conversion of 99.5% and a FDCA yield of 70.9% at 100 °C and 10 bar O_2 with $KHCO_3$ for 24 h, which was significantly better than Mn_3O_4,

Co$_3$O$_4$ and Mn-Co binary oxides with other Mn/Co molar ratios. The enhanced catalytic activity was attributed to the presence of Mn^{3+} ions and the high oxygen mobility and reducibility.

The aerobic base-free oxidation of HMF to FDCA over non-precious metal catalysts has been limited reported. The use of organic solvent, organic peroxide and base additives may promote the product yield but undoubtedly hamper the green footprint of renewable FEDA production. It is still rather difficult to use non-noble metal based catalysts for the aerobic oxidation of HMF to FDCA in water and without base additives. Recently, Gao et al. prepared a Mn-Co mixed oxide catalyst (Co$_3$O$_4$/Mn$_x$Co) with Co$_3$O$_4$ nanoparticles well-dispersed on amorphous Mn-Co-O solid solutions by co-precipitation method [99]. They claimed that a FDCA yield of >99% could be obtained with Co$_3$O$_4$/Mn$_{0.2}$Co where the Mn/Co ratio was of 0.2, at 1 bar O$_2$ and 140 °C after 24 h without base additives. The high content of both Lewis (Mn^{4+}) and Brønsted (Co–O–H$^+$) acid sites on the surface, leading to an excellent ability of HMF adsorption and COOH group formation, as well as the enhanced oxygen mobility. This catalyst was shown stable after a minor deactivation (≤8%) during six recycling uses and its activity could be entirely regenerated by calcination in air.

According to the results of the aerobic oxidation of HMF to FDCA over non-noble metal based catalysts (Fe-, Co-, and Mn-based), an inferior catalytic selectivity to FDCA is normally shown as compared to noble metal catalysts. The activation of O$_2$ is not very efficient (especially Fe- and Co-based) and a strong oxidant (e.g., t-BuOOH) is more normally required to obtain a good selectivity of FDCA. Especially, base additives and organic solvent are often required to improve the FDCA yield. The stability of non-noble metal based catalyst suffers from more issues, i.e., metal leaching (especially Ni- and Cu-based), change of the active phase, and coverage of active species. Still, owing to the advance in the cost, it is promising to further explore non-precious transition-metal catalysts considering the lower catalyst cost for the upcoming practical production of FDCA. Therefore, further effort should be continuously devoted to develop new approaches for designing efficient non-precious transition-metal catalysts for the aerobic oxidation of HMF to FDCA.

4. Conclusions and Perspectives

The catalytic aerobic oxidation of biomass-derived HMF to FDCA is currently a hot topic, especially since FDCA exhibits the potential to replace petrochemical-derived terephthalic acid, one of the most widely used monomers in polymers, for the production of a series of biopolymers. According to the recent results of the aerobic oxidation of HMF to FDCA over supported metal catalysts, noble-metal catalysts are the most studied. Much progress and numerous breakthroughs have already been made in catalyst design and understanding the reaction mechanism. Although the application of inexpensive transition-metal catalysts might offer promising prospects in the practical synthesis of FDCA, the main issue of the non-noble metal-based catalyst is the inferior selectivity for FDCA compared to the noble-metal analogues, based on currently reported methods. Additional research efforts should be devoted to develop new methods based on non-noble transition-metal catalysts that can improve the selectivity of FDCA, especially with O$_2$ as the oxidant and without using additional base. The performance of the catalyst and the reaction pathway are highly dependent on the properties of the catalyst (i.e., the active phase, support, particle size) and reaction conditions (i.e., oxygen pressure, oxygen flow rate, pH, and temperature). Among the noble metal-based catalysts, Au-based catalysts appear to show a better performance in catalyst selectivity and stability for the aerobic oxidation of HMF into FDCA in water, compared to the Pt-, Pd-, Ru-, and Rh-based catalysts, owing to the better resistance to water and O$_2$. Nevertheless, deactivation of the Au-based catalysts by the deposition of the byproducts or intermediates on its active sites is also observed, as mentioned in some examples. For further improvement of the Au-based catalyst, a bimetallic alloy approach, achieved by alloying of a second metal with Au, has been applied and shown to be effective, with a higher catalytic activity and improved stability as compared to the monometallic counterparts. Although many bi-functional combinations have shown promising outcomes with rate enhancement or a synergistic effect. More details into how such effect of different function sites comes from metals or other sites with an optimal site-balance, have been limitedly studied. Better insight of the reaction mechanisms involved needs to be provided on atomic levels for the oxidation of HMF into FDCA.

Normally, excess base is used to promote the oxidation of HMF, which can not only facilitate the reaction, but also transform the formed FDCA into a salt form dissolved in aqueous solution. Otherwise, the strong adsorption of the carboxylic acids on the catalyst can hinder the further process of this reaction. The use of excess base can however also lead to a more expensive process, which is also less green. Therefore, it is necessary to develop a base-free oxidation system, which is more cost-effective and environmentally benign approach, appreciated in sustainable chemistry. Selecting an appropriate support with basic sites for the catalysts has shown advantages in catalytic activity and stability for the base-free oxidation of HMF into FDCA in recent examples. However, catalyst deactivation owing to a loss in basic sites has been observed with the support during reactions, and the stability of the catalyst is found to be a challenge for the catalyst. More effort needs to be put into finding out how to stabilize the required functional sites on the catalyst, and advanced strategies of catalyst design for preparing specific structures, i.e., single-atom metal, core-shell, sub-cluster segregated, multi-shell and random homogeneous alloys, etc., deserve more thoroughly exploration in research.

Most reported studies with high FDCA yields in the aerobic oxidation of HMF, have been conducted in dilute HMF solutions (0.5–2.1 wt%), which is unreal for the practical production of FDCA on an industrial scale. The limitation is attributed to the highly reactive functional groups in HMF, which can lead to the formation of undesired solid byproducts, namely humins, via complex side reactions (i.e., condensation and polymerization). Many fewer examples of concentrated HMF substrates are studied in the oxidation of HMF to FDCA. Recently, an approach for stabilizing the active formyl group of HMF by the acetalization with 1,3-propanediol was reported, which enables production of a high yield of FDCA and low humin formation, even in solutions of up to 20 wt% HMF acetal, by aerobic oxidation in the presence of a base additive [38]. In addition, extremely low concentration of HMF, together with their short-term reaction time often applied in reported cases, may underestimate other issues, particularly catalyst stability. Catalyst deactivation might not be well recognized in the liquid phase with a higher concentration of HMF (polar, aqueous and even corrosive). Thus, research efforts are still needed to conduct the reactions under a more practical concentration of HMF, and further understand the deeper fundamentals of catalyst stability challenges, in order to develop innovative and creative approaches.

The reaction mechanisms involved in the aerobic oxidation of HMF were revealed with some metal-based catalysts, mainly by applying isotope labeling and mass spectrometric techniques. Still, more efforts need to be devoted to the development of modern in-situ characterization technologies, to provide deep insights into the intrinsic kinetics and mechanisms. In addition, the state-of-the-art *operando* characterization methods combined with various spectroscopy techniques are also necessary for understanding the deep fundamentals of the nature of the intrinsic active sites for each elemental step in HMF oxidation, which might dynamically evaluate during catalysis [102]. The adequate understanding of the reaction mechanism will elucidate a more detailed understanding of catalytic chemistry. The deep insights on the active site can greatly benefit the rational design of catalysts, even with the use of the non-noble metals to prepare more efficient and stable catalysts for the oxidation of HMF to FDCA.

Although many significant achievements have been made for the aerobic oxidation of HMF to FDCA over metal-based heterogeneous catalysts, further improvements are still required for scaling up to an industrially large-scale production of FDCA. The mass balances for the aerobic oxidation of HMF to FDCA based on laboratory data need to be accurate and correct. The mass balance for the oxidation of HMF was not always mentioned or even not fully closed in many cases. Those unknown parameters might result in a significant amount of economic loss during scaling up processes if not properly done. Current approaches for the pioneering processes for the production of FDCA from HMF are technically feasible but not economically viable, mainly owing to the high price of HMF and its limited availability. Development of energetically and economically viable processes is a long-term task which requires extensive time, efforts and normally involves interdisciplinary knowledge of process engineering, chemistry, material science, etc. Only with a full grasp of the knowledge and reorganization of the fundamental details and catalytic challenges, we may develop an economically feasible approach for realizing industrially large-scale production of FDCA in the near future, which will alleviate our society's dependence on the traditional fossil resources.

Author Contributions: Conceptualization, S.H. and W.L.; validation, W.L.; formal analysis, L.L. and W.L.; investigation, S.H., L.L. and W.L.; resources A.W. and W.L.; writing—original draft preparation, S.H.; writing—review and editing, L.L.; supervision, A.W. and W.L. All authors have read and agreed to the published version of the manuscript.

Funding: This research was funded by the National Key Projects for Fundamental Research and Development of China (2018YFB1501602) and the National Natural Science Foundation of China (21703238, 21690084).

Acknowledgments: The authors acknowledge the National Key Projects for Fundamental Research and Development of China (2018YFB1501602), the National Natural Science Foundation of China (21703238, 21690084) and the CAS-TWAS President's Fellowship Program between Chinese Academy of Sciences (CAS) and The World Academy of Sciences (TWAS) for financial support.

Conflicts of Interest: The authors declare no conflict of interest.

References

1. Bozell, J.J. Connecting Biomass and Petroleum Processing with a Chemical Bridge. *Science* **2010**, *329*, 522–523. [CrossRef] [PubMed]
2. Corma, A.; Iborra, S.; Velty, A. Chemical routes for the transformation of biomass into chemicals. *Chem. Rev.* **2007**, *107*, 2411–2502. [CrossRef] [PubMed]
3. Gallezot, P. Conversion of biomass to selected chemical products. *Chem. Soc. Rev.* **2012**, *41*, 1538–1558. [CrossRef] [PubMed]
4. H. Clark, J.; EI Deswarte, F.; J. Farmer, T. The integration of green chemistry into future biorefineries. *Biofuels Bioprod. Biorefin.* **2009**, *3*, 72–90. [CrossRef]
5. Roy Goswami, S.; Dumont, M.-J.; Raghavan, V. Starch to value added biochemicals. *Starch Stärke* **2016**, *68*, 274–286. [CrossRef]
6. Morais, A.R.; da Costa Lopes, A.M.; Bogel-Łukasik, R. Carbon dioxide in biomass processing: Contributions to the green biorefinery concept. *Chem. Rev.* **2014**, *115*, 3–27. [CrossRef]
7. Van Putten, R.-J.; van der Waal, J.C.; de Jong, E.; Rasrendra, C.B.; Heeres, H.J.; de Vries, J.G. Hydroxymethylfurfural, A Versatile Platform Chemical Made from Renewable Resources. *Chem. Rev.* **2013**, *113*, 1499–1597. [CrossRef]
8. Fukuoka, A.; Dhepe, P.L. Catalytic Conversion of Cellulose into Sugar Alcohols. *Angew. Chem. Int. Ed.* **2006**, *45*, 5161–5163. [CrossRef]
9. Deng, W.; Zhang, Q.; Wang, Y. Polyoxometalates as efficient catalysts for transformations of cellulose into platform chemicals. *Dalton Trans.* **2012**, *41*, 9817–9831. [CrossRef]
10. Rinaldi, R.; Schüth, F. Acid Hydrolysis of Cellulose as the Entry Point into Biorefinery Schemes. *ChemSusChem* **2009**, *2*, 1096–1107. [CrossRef]
11. Pal, P.; Saravanamurugan, S. Recent Advances in the Development of 5-Hydroxymethylfurfural Oxidation with Base (Nonprecious)-Metal-Containing Catalysts. *ChemSusChem* **2019**, *12*, 145–163. [CrossRef]
12. Werpy, T.; Petersen, G. *Top Value Added Chemicals from Biomass: Volume I—Results of Screening for Potential Candidates from Sugars and Synthesis Gas*; National Renewable Energy Lab.: Golden, CO, USA, 2004.
13. Haworth, W.N.; Jones, W.G.M.; Wiggins, L.F. 1. The conversion of sucrose into furan compounds. Part II. Some 2: 5-disubstituted tetrahydrofurans and their products of ring scission. *J. Chem. Soc.* **1945**, *10*, 1–4. [CrossRef]
14. Chen, M.Y.; Ike, M.; Fujita, M. Acute toxicity, mutagenicity, and estrogenicity of bisphenol-A and other bisphenols. *Environ. Toxicol. Int. J.* **2002**, *17*, 80–86. [CrossRef] [PubMed]
15. Swan, S.H. Environmental phthalate exposure in relation to reproductive outcomes and other health endpoints in humans. *Environ. Res.* **2008**, *108*, 177–184. [CrossRef] [PubMed]
16. Ravindranath, K.; Mashelkar, R.A. Polyethylene terephthalate-I. Chemistry, thermodynamics and transport properties. *Chem. Eng. Sci.* **1986**, *41*, 2197–2214. [CrossRef]

17. Chen, G.; van Straalen, N.M.; Roelofs, D. The ecotoxicogenomic assessment of soil toxicity associated with the production chain of 2,5-furandicarboxylic acid (FDCA), a candidate bio-based green chemical building block. *Green Chem.* **2016**, *18*, 4420–4431. [CrossRef]
18. Lancefield, C.S.; Teunissen, L.W.; Weckhuysen, B.M.; Bruijnincx, P.C. Iridium-catalysed primary alcohol oxidation and hydrogen shuttling for the depolymerisation of lignin. *Green Chem.* **2018**, *20*, 3214–3221. [CrossRef]
19. De Jong, E.; Dam, M.; Sipos, L.; Gruter, G.-J. Furandicarboxylic acid (FDCA), a versatile building block for a very interesting class of polyesters. In *Biobased Monomers, Polymers, and Materials*; ACS Symposium Series; American Chemical Society: Washington, DC, USA, 2012; pp. 1–13.
20. Pan, T.; Deng, J.; Xu, Q.; Zuo, Y.; Guo, Q.X.; Fu, Y. Catalytic Conversion of Furfural into a 2,5-Furandicarboxylic Acid-Based Polyester with Total Carbon Utilization. *ChemSusChem* **2013**, *6*, 47–50. [CrossRef]
21. Ball, G.L.; McLellan, C.J.; Bhat, V.S. Toxicological review and oral risk assessment of terephthalic acid (TPA) and its esters: A category approach. *Crit. Rev. Toxicol.* **2012**, *42*, 28–67. [CrossRef]
22. Eerhart, A.J.J.E.; Faaij, A.P.C.; Patel, M.K. Replacing fossil based PET with biobased PEF; process analysis, energy and GHG balance. *Energy Environ. Sci.* **2012**, *5*, 6407–6422. [CrossRef]
23. Fittig, R.; Heinzelmann, H. Production of 2,5-furandicarboxylic acid by the reaction of fuming hydrobromic acid with mucic acid under pressure. *Chem. Ber.* **1876**, *9*, 1198.
24. Rose, M.; Weber, D.; Lotsch, B.V.; Kremer, R.K.; Goddard, R.; Palkovits, R. Biogenic metal–organic frameworks: 2,5-Furandicarboxylic acid as versatile building block. *Microporous Mesoporous Mater.* **2013**, *181*, 217–221. [CrossRef]
25. Chadderdon, D.J.; Xin, L.; Qi, J.; Qiu, Y.; Krishna, P.; More, K.L.; Li, W. Electrocatalytic oxidation of 5-hydroxymethylfurfural to 2,5-furandicarboxylic acid on supported Au and Pd bimetallic nanoparticles. *Green Chem.* **2014**, *16*, 3778–3786. [CrossRef]
26. Wang, K.F.; Liu, C.L.; Sui, K.Y.; Guo, C.; Liu, C.Z. Efficient Catalytic Oxidation of 5-Hydroxymethylfurfural to 2,5-Furandicarboxylic Acid by Magnetic Laccase Catalyst. *ChemBioChem* **2018**, *19*, 654–659. [CrossRef] [PubMed]
27. Ban, H.; Pan, T.; Cheng, Y.; Wang, L.; Li, X. Solubilities of 2,5-Furandicarboxylic Acid in Binary Acetic Acid + Water, Methanol + Water, and Ethanol + Water Solvent Mixtures. *J. Chem. Eng. Data* **2018**, *63*, 1987–1993. [CrossRef]
28. Sajid, M.; Zhao, X.; Liu, D. Production of 2, 5-furandicarboxylic acid (FDCA) from 5-hydroxymethylfurfural (HMF): Recent progress focusing on the chemical-catalytic routes. *Green Chem.* **2018**, *20*, 5427–5453. [CrossRef]
29. Xuan, Y.; He, R.; Han, B.; Wu, T.; Wu, Y. Catalytic Conversion of Cellulose into 5-Hydroxymethylfurfural Using [PSMIM] HSO_4 and $ZnSO_4·7H_2O$ Co-catalyst in Biphasic System. *Waste Biomass Valoriz.* **2018**, *9*, 401–408. [CrossRef]
30. Dijkman, W.P.; Groothuis, D.E.; Fraaije, M.W. Enzyme-Catalyzed Oxidation of 5-Hydroxymethylfurfural to Furan-2,5-dicarboxylic Acid. *Angew. Chem. Int. Ed.* **2014**, *53*, 6515–6518. [CrossRef]
31. Chen, C.T.; Nguyen, C.V.; Wang, Z.Y.; Bando, Y.; Yamauchi, Y.; Bazziz, M.T.S.; Fatehmulla, A.; Farooq, W.A.; Yoshikawa, T.; Masuda, T. Hydrogen Peroxide Assisted Selective Oxidation of 5-Hydroxymethylfurfural in Water under Mild Conditions. *ChemCatChem* **2018**, *10*, 361–365. [CrossRef]
32. Nam, D.-H.; Taitt, B.J.; Choi, K.-S. Copper-Based Catalytic Anodes To Produce 2,5-Furandicarboxylic Acid, a Biomass-Derived Alternative to Terephthalic Acid. *ACS Catal.* **2018**, *8*, 1197–1206. [CrossRef]
33. Xu, S.; Zhou, P.; Zhang, Z.; Yang, C.; Zhang, B.; Deng, K.; Bottle, S.; Zhu, H. Selective Oxidation of 5-Hydroxymethylfurfural to 2,5-Furandicarboxylic Acid Using O_2 and a Photocatalyst of Co-thioporphyrazine Bonded to g-C_3N_4. *J. Am. Chem. Soc.* **2017**, *139*, 14775–14782. [CrossRef] [PubMed]
34. Hutchings, G.J. Vapor phase hydrochlorination of acetylene: Correlation of catalytic activity of supported metal chloride catalysts. *J. Catal.* **1985**, *96*, 292–295. [CrossRef]
35. Masatake, H.; Tetsuhiko, K.; Hiroshi, S.; Nobumasa, Y. Novel Gold Catalysts for the Oxidation of Carbon Monoxide at a Temperature far Below 0 °C. *Chem. Lett.* **1987**, *16*, 405–408.
36. Casanova, O.; Iborra, S.; Corma, A. Biomass into Chemicals: Aerobic Oxidation of 5-Hydroxymethyl-2-furfural into 2,5-Furandicarboxylic Acid with Gold Nanoparticle Catalysts. *ChemSusChem* **2009**, *2*, 1138–1144. [CrossRef] [PubMed]

37. Lolli, A.; Amadori, R.; Lucarelli, C.; Cutrufello, M.G.; Rombi, E.; Cavani, F.; Albonetti, S. Hard-template preparation of Au/CeO$_2$ mesostructured catalysts and their activity for the selective oxidation of 5-hydroxymethylfurfural to 2,5-furandicarboxylic acid. *Microporous Mesoporous Mater.* **2016**, *226*, 466–475. [CrossRef]
38. Kim, M.; Su, Y.; Fukuoka, A.; Hensen, E.J.M.; Nakajima, K. Aerobic Oxidation of 5-(Hydroxymethyl)furfural Cyclic Acetal Enables Selective Furan-2,5-dicarboxylic Acid Formation with CeO$_2$-Supported Gold Catalyst. *Angew. Chem. Int. Ed.* **2018**, *57*, 8235–8239. [CrossRef]
39. Gorbanev, Y.Y.; Klitgaard, S.K.; Woodley, J.M.; Christensen, C.H.; Riisager, A. Gold-Catalyzed Aerobic Oxidation of 5-Hydroxymethylfurfural in Water at Ambient Temperature. *ChemSusChem* **2009**, *2*, 672–675. [CrossRef]
40. Cai, J.; Ma, H.; Zhang, J.; Song, Q.; Du, Z.; Huang, Y.; Xu, J. Gold Nanoclusters Confined in a Supercage of Y Zeolite for Aerobic Oxidation of HMF under Mild Conditions. *Chem. Eur. J.* **2013**, *19*, 14215–14223. [CrossRef]
41. Pasini, T.; Piccinini, M.; Blosi, M.; Bonelli, R.; Albonetti, S.; Dimitratos, N.; Lopez-Sanchez, J.A.; Sankar, M.; He, Q.; Kiely, C.J.; et al. Selective oxidation of 5-hydroxymethyl-2-furfural using supported gold–copper nanoparticles. *Green Chem.* **2011**, *13*, 2091–2099. [CrossRef]
42. Villa, A.; Schiavoni, M.; Campisi, S.; Veith, G.M.; Prati, L. Pd-modified Au on Carbon as an Effective and Durable Catalyst for the Direct Oxidation of HMF to 2,5-Furandicarboxylic Acid. *ChemSusChem* **2013**, *6*, 609–612. [CrossRef]
43. Gupta, N.K.; Nishimura, S.; Takagaki, A.; Ebitani, K. Hydrotalcite-supported gold-nanoparticle-catalyzed highly efficient base-free aqueous oxidation of 5-hydroxymethylfurfural into 2,5-furandicarboxylic acid under atmospheric oxygen pressure. *Green Chem.* **2011**, *13*, 824–827. [CrossRef]
44. Gao, T.; Gao, T.; Fang, W.; Cao, Q. Base-free aerobic oxidation of 5-hydroxymethylfurfural to 2,5-furandicarboxylic acid in water by hydrotalcite-activated carbon composite supported gold catalyst. *Mol. Catal.* **2017**, *439*, 171–179. [CrossRef]
45. Wan, X.; Zhou, C.; Chen, J.; Deng, W.; Zhang, Q.; Yang, Y.; Wang, Y. Base-Free Aerobic Oxidation of 5-Hydroxymethyl-furfural to 2,5-Furandicarboxylic Acid in Water Catalyzed by Functionalized Carbon Nanotube-Supported Au–Pd Alloy Nanoparticles. *ACS Catal.* **2014**, *4*, 2175–2185. [CrossRef]
46. Bonincontro, D.; Lolli, A.; Villa, A.; Prati, L.; Dimitratos, N.; Veith, G.M.; Chinchilla, L.E.; Botton, G.A.; Cavani, F.; Albonetti, S. AuPd-nNiO as an effective catalyst for the base-free oxidation of HMF under mild reaction conditions. *Green Chem.* **2019**, *21*, 4090–4099. [CrossRef]
47. Gao, Z.; Xie, R.; Fan, G.; Yang, L.; Li, F. Highly Efficient and Stable Bimetallic AuPd over La-Doped Ca–Mg–Al Layered Double Hydroxide for Base-Free Aerobic Oxidation of 5-Hydroxymethylfurfural in Water. *ACS Sustain. Chem. Eng.* **2017**, *5*, 5852–5861. [CrossRef]
48. Davis, S.E.; Zope, B.N.; Davis, R.J. On the mechanism of selective oxidation of 5-hydroxymethylfurfural to 2,5-furandicarboxylic acid over supported Pt and Au catalysts. *Green Chem.* **2012**, *14*, 143–147. [CrossRef]
49. Davis, S.E.; Benavidez, A.D.; Gosselink, R.W.; Bitter, J.H.; de Jong, K.P.; Datye, A.K.; Davis, R.J. Kinetics and mechanism of 5-hydroxymethylfurfural oxidation and their implications for catalyst development. *J. Mol. Catal. A* **2014**, *388*, 123–132. [CrossRef]
50. Zope, B.N.; Hibbitts, D.D.; Neurock, M.; Davis, R.J. Reactivity of the gold/water interface during selective oxidation catalysis. *Science* **2010**, *330*, 74–78. [CrossRef]
51. Zope, B.N.; Davis, S.E.; Davis, R.J. Influence of reaction conditions on diacid formation during Au-catalyzed oxidation of glycerol and hydroxymethylfurfural. *Top. Catal.* **2012**, *55*, 24–32. [CrossRef]
52. Verdeguer, P.; Merat, N.; Gaset, A. Oxydation catalytique du HMF en acide 2,5-furane dicarboxylique. *J. Mol. Catal.* **1993**, *85*, 327–344. [CrossRef]
53. Rass, H.A.; Essayem, N.; Besson, M. Selective aqueous phase oxidation of 5-hydroxymethylfurfural to 2,5-furandicarboxylic acid over Pt/C catalysts: Influence of the base and effect of bismuth promotion. *Green Chem.* **2013**, *15*, 2240–2251. [CrossRef]
54. Ait Rass, H.; Essayem, N.; Besson, M. Selective Aerobic Oxidation of 5-HMF into 2,5-Furandicarboxylic Acid with Pt Catalysts Supported on TiO$_2$- and ZrO$_2$-Based Supports. *ChemSusChem* **2015**, *8*, 1206–1217. [CrossRef] [PubMed]
55. Sahu, R.; Dhepe, P.L. Synthesis of 2,5-furandicarboxylic acid by the aerobic oxidation of 5-hydroxymethyl furfural over supported metal catalysts. *React. Kinet. Mech. Catal.* **2014**, *112*, 173–187. [CrossRef]

56. Miao, Z.; Wu, T.; Li, J.; Yi, T.; Zhang, Y.; Yang, X. Aerobic oxidation of 5-hydroxymethylfurfural (HMF) effectively catalyzed by a $Ce_{0.8}Bi_{0.2}O_{2-\delta}$ supported Pt catalyst at room temperature. *RSC Adv.* **2015**, *5*, 19823–19829. [CrossRef]
57. Niu, W.; Wang, D.; Yang, G.; Sun, J.; Wu, M.; Yoneyama, Y.; Tsubaki, N. Pt Nanoparticles Loaded on Reduced Graphene Oxide as an Effective Catalyst for the Direct Oxidation of 5-Hydroxymethylfurfural (HMF) to Produce 2,5-Furandicarboxylic Acid (FDCA) under Mild Conditions. *Bull. Chem. Soc. Jpn.* **2014**, *87*, 1124–1129. [CrossRef]
58. Zhang, Y.; Xue, Z.; Wang, J.; Zhao, X.; Deng, Y.; Zhao, W.; Mu, T. Controlled deposition of Pt nanoparticles on Fe_3O_4@carbon microspheres for efficient oxidation of 5-hydroxymethylfurfural. *RSC Adv.* **2016**, *6*, 51229–51237. [CrossRef]
59. Vinke, P.; van Dam, H.E.; van Bekkum, H. Platinum Catalyzed Oxidation of 5-Hydroxymethylfurfural. In *Studies in Surface Science and Catalysis*; Centi, G., Trifiro, F., Eds.; Elsevier: Amsterdam, The Netherlands, 1990; Volume 55, pp. 147–158.
60. Chen, H.; Shen, J.; Chen, K.; Qin, Y.; Lu, X.; Ouyang, P.; Fu, J. Atomic layer deposition of Pt nanoparticles on low surface area zirconium oxide for the efficient base-free oxidation of 5-hydroxymethylfurfural to 2,5-furandicarboxylic acid. *Appl. Catal. A Gen.* **2018**, *555*, 98–107. [CrossRef]
61. Han, X.; Geng, L.; Guo, Y.; Jia, R.; Liu, X.; Zhang, Y.; Wang, Y. Base-free aerobic oxidation of 5-hydroxymethylfurfural to 2,5-furandicarboxylic acid over a Pt/C–O–Mg catalyst. *Green Chem.* **2016**, *18*, 1597–1604. [CrossRef]
62. Han, X.; Li, C.; Guo, Y.; Liu, X.; Zhang, Y.; Wang, Y. N-doped carbon supported Pt catalyst for base-free oxidation of 5-hydroxymethylfurfural to 2,5-furandicarboxylic acid. *Appl. Catal. A Gen.* **2016**, *526*, 1–8. [CrossRef]
63. Shen, J.; Chen, H.; Chen, K.; Qin, Y.; Lu, X.; Ouyang, P.; Fu, J. Atomic Layer Deposition of a Pt-Skin Catalyst for Base-Free Aerobic Oxidation of 5-Hydroxymethylfurfural to 2,5-Furandicarboxylic Acid. *Ind. Eng. Chem. Res.* **2018**, *57*, 2811–2818. [CrossRef]
64. Siankevich, S.; Savoglidis, G.; Fei, Z.; Laurenczy, G.; Alexander, D.T.; Yan, N.; Dyson, P.J. A novel platinum nanocatalyst for the oxidation of 5-Hydroxymethylfurfural into 2,5-Furandicarboxylic acid under mild conditions. *J. Catal.* **2014**, *315*, 67–74. [CrossRef]
65. Siankevich, S.; Mozzettini, S.; Bobbink, F.; Ding, S.; Fei, Z.; Yan, N.; Dyson, P.J. Influence of the Anion on the Oxidation of 5-Hydroxymethylfurfural by Using Ionic-Polymer-Supported Platinum Nanoparticle Catalysts. *ChemPlusChem* **2018**, *83*, 19–23. [CrossRef]
66. Davis, S.E.; Houk, L.R.; Tamargo, E.C.; Datye, A.K.; Davis, R.J. Oxidation of 5-hydroxymethylfurfural over supported Pt, Pd and Au catalysts. *Catal. Today* **2011**, *160*, 55–60. [CrossRef]
67. Siyo, B.; Schneider, M.; Radnik, J.; Pohl, M.-M.; Langer, P.; Steinfeldt, N. Influence of support on the aerobic oxidation of HMF into FDCA over preformed Pd nanoparticle based materials. *Appl. Catal. A Gen.* **2014**, *478*, 107–116. [CrossRef]
68. Siyo, B.; Schneider, M.; Pohl, M.-M.; Langer, P.; Steinfeldt, N. Synthesis, characterization, and application of PVP-Pd NP in the aerobic oxidation of 5-hydroxymethylfurfural (hmf). *Catal. Lett.* **2014**, *144*, 498–506. [CrossRef]
69. Rathod, P.V.; Jadhav, V.H. Efficient Method for Synthesis of 2,5-Furandicarboxylic Acid from 5-Hydroxymethylfurfural and Fructose Using Pd/CC Catalyst under Aqueous Conditions. *ACS Sustain. Chem. Eng.* **2018**, *6*, 5766–5771. [CrossRef]
70. Zhang, Z.; Zhen, J.; Liu, B.; Lv, K.; Deng, K. Selective aerobic oxidation of the biomass-derived precursor 5-hydroxymethylfurfural to 2,5-furandicarboxylic acid under mild conditions over a magnetic palladium nanocatalyst. *Green Chem.* **2015**, *17*, 1308–1317. [CrossRef]
71. Mei, N.; Liu, B.; Zheng, J.; Lv, K.; Tang, D.; Zhang, Z. A novel magnetic palladium catalyst for the mild aerobic oxidation of 5-hydroxymethylfurfural into 2,5-furandicarboxylic acid in water. *Catal. Sci. Technol.* **2015**, *5*, 3194–3202. [CrossRef]
72. Liu, B.; Ren, Y.; Zhang, Z. Aerobic oxidation of 5-hydroxymethylfurfural into 2,5-furandicarboxylic acid in water under mild conditions. *Green Chem.* **2015**, *17*, 1610–1617. [CrossRef]
73. Lolli, A.; Albonetti, S.; Utili, L.; Amadori, R.; Ospitali, F.; Lucarelli, C.; Cavani, F. Insights into the reaction mechanism for 5-hydroxymethylfurfural oxidation to FDCA on bimetallic Pd–Au nanoparticles. *Appl. Catal. A Gen.* **2015**, *504*, 408–419. [CrossRef]

74. Xia, H.; An, J.; Hong, M.; Xu, S.; Zhang, L.; Zuo, S. Aerobic oxidation of 5-hydroxymethylfurfural to 2,5-difurancarboxylic acid over Pd-Au nanoparticles supported on Mg-Al hydrotalcite. *Catal. Today* **2019**, *319*, 113–120. [CrossRef]
75. Gupta, K.; Rai, R.K.; Singh, S.K. Catalytic aerial oxidation of 5-hydroxymethyl-2-furfural to furan-2,5-dicarboxylic acid over Ni–Pd nanoparticles supported on Mg(OH)$_2$ nanoflakes for the synthesis of furan diesters. *Inorg. Chem. Front.* **2017**, *4*, 871–880. [CrossRef]
76. Wang, Y.; Yu, K.; Lei, D.; Si, W.; Feng, Y.; Lou, L.-L.; Liu, S. Basicity-Tuned Hydrotalcite-Supported Pd Catalysts for Aerobic Oxidation of 5-Hydroxymethyl-2-furfural under Mild Conditions. *ACS Sustain. Chem. Eng.* **2016**, *4*, 4752–4761. [CrossRef]
77. Zakrzewska, M.E.; Bogel-Łukasik, E.; Bogel-Łukasik, R. Ionic Liquid-Mediated Formation of 5-Hydroxymethylfurfural—A Promising Biomass-Derived Building Block. *Chem. Rev.* **2011**, *111*, 397–417. [CrossRef]
78. Yi, G.; Teong, S.P.; Zhang, Y. Base-free conversion of 5-hydroxymethylfurfural to 2,5-furandicarboxylic acid over a Ru/C catalyst. *Green Chem.* **2016**, *18*, 979–983. [CrossRef]
79. Kerdi, F.; Ait Rass, H.; Pinel, C.; Besson, M.; Peru, G.; Leger, B.; Rio, S.; Monflier, E.; Ponchel, A. Evaluation of surface properties and pore structure of carbon on the activity of supported Ru catalysts in the aqueous-phase aerobic oxidation of HMF to FDCA. *Appl. Catal. A Gen.* **2015**, *506*, 206–219. [CrossRef]
80. Ståhlberg, T.; Eyjólfsdóttir, E.; Gorbanev, Y.Y.; Sádaba, I.; Riisager, A. Aerobic oxidation of 5-(hydroxymethyl) furfural in ionic liquids with solid ruthenium hydroxide catalysts. *Catal. Lett.* **2012**, *142*, 1089–1097. [CrossRef]
81. Mishra, D.K.; Lee, H.J.; Kim, J.; Lee, H.-S.; Cho, J.K.; Suh, Y.-W.; Yi, Y.; Kim, Y.J. MnCo$_2$O$_4$ spinel supported ruthenium catalyst for air-oxidation of HMF to FDCA under aqueous phase and base-free conditions. *Green Chem.* **2017**, *19*, 1619–1623. [CrossRef]
82. Gao, T.; Yin, Y.; Fang, W.; Cao, Q. Highly dispersed ruthenium nanoparticles on hydroxyapatite as selective and reusable catalyst for aerobic oxidation of 5-hydroxymethylfurfural to 2,5-furandicarboxylic acid under base-free conditions. *Mol. Catal.* **2018**, *450*, 55–64. [CrossRef]
83. Pichler, C.M.; Al-Shaal, M.G.; Gu, D.; Joshi, H.; Ciptonugroho, W.; Schüth, F. Ruthenium Supported on High-Surface-Area Zirconia as an Efficient Catalyst for the Base-Free Oxidation of 5-Hydroxymethylfurfural to 2,5-Furandicarboxylic Acid. *ChemSusChem* **2018**, *11*, 2083–2090. [CrossRef]
84. Nie, J.; Xie, J.; Liu, H. Activated carbon-supported ruthenium as an efficient catalyst for selective aerobic oxidation of 5-hydroxymethylfurfural to 2,5-diformylfuran. *Chin. J. Catal.* **2013**, *34*, 871–875. [CrossRef]
85. Artz, J.; Mallmann, S.; Palkovits, R. Selective Aerobic Oxidation of HMF to 2,5-Diformylfuran on Covalent Triazine Frameworks-Supported Ru Catalysts. *ChemSusChem* **2015**, *8*, 672–679. [CrossRef] [PubMed]
86. Vuyyuru, K.R.; Strasser, P. Oxidation of biomass derived 5-hydroxymethylfurfural using heterogeneous and electrochemical catalysis. *Catal. Today* **2012**, *195*, 144–154. [CrossRef]
87. Schade, O.R.; Kalz, K.F.; Neukum, D.; Kleist, W.; Grunwaldt, J.-D. Supported gold- and silver-based catalysts for the selective aerobic oxidation of 5-(hydroxymethyl)furfural to 2,5-furandicarboxylic acid and 5-hydroxymethyl-2-furancarboxylic acid. *Green Chem.* **2018**, *20*, 3530–3541. [CrossRef]
88. Zhang, Z.; Deng, K. Recent advances in the catalytic synthesis of 2,5-furandicarboxylic acid and its derivatives. *ACS Catal.* **2015**, *5*, 6529–6544. [CrossRef]
89. Saha, B.; Gupta, D.; Abu-Omar, M.M.; Modak, A.; Bhaumik, A. Porphyrin-based porous organic polymer-supported iron (III) catalyst for efficient aerobic oxidation of 5-hydroxymethyl-furfural into 2,5-furandicarboxylic acid. *J. Catal.* **2013**, *299*, 316–320. [CrossRef]
90. Gao, L.; Deng, K.; Zheng, J.; Liu, B.; Zhang, Z. Efficient oxidation of biomass derived 5-hydroxymethylfurfural into 2,5-furandicarboxylic acid catalyzed by Merrifield resin supported cobalt porphyrin. *Chem. Eng. J.* **2015**, *270*, 444–449. [CrossRef]
91. Jain, A.; Jonnalagadda, S.C.; Ramanujachary, K.V.; Mugweru, A. Selective oxidation of 5-hydroxymethyl-2-furfural to furan-2,5-dicarboxylic acid over spinel mixed metal oxide catalyst. *Catal. Commun.* **2015**, *58*, 179–182. [CrossRef]
92. Wang, S.; Zhang, Z.; Liu, B. Catalytic Conversion of Fructose and 5-Hydroxymethylfurfural into 2,5-Furandicarboxylic Acid over a Recyclable Fe$_3$O$_4$–CoO$_x$ Magnetite Nanocatalyst. *ACS Sustain. Chem. Eng.* **2015**, *3*, 406–412. [CrossRef]

93. Yan, D.; Xin, J.; Shi, C.; Lu, X.; Ni, L.; Wang, G.; Zhang, S. Base-free conversion of 5-hydroxymethylfurfural to 2,5-furandicarboxylic acid in ionic liquids. *Chem. Eng. J.* **2017**, *323*, 473–482. [CrossRef]
94. Yan, D.; Xin, J.; Zhao, Q.; Gao, K.; Lu, X.; Wang, G.; Zhang, S. Fe–Zr–O catalyzed base-free aerobic oxidation of 5-HMF to 2,5-FDCA as a bio-based polyester monomer. *Catal. Sci. Technol.* **2018**, *8*, 164–175. [CrossRef]
95. Hayashi, E.; Komanoya, T.; Kamata, K.; Hara, M. Heterogeneously-Catalyzed Aerobic Oxidation of 5-Hydroxymethylfurfural to 2,5-Furandicarboxylic Acid with MnO_2. *ChemSusChem* **2017**, *10*, 654–658. [CrossRef] [PubMed]
96. Bao, L.; Sun, F.-Z.; Zhang, G.-Y.; Hu, T.-L. Aerobic Oxidation of 5-Hydroxymethylfurfural to 2,5-Furandicarboxylic Acid over Holey 2D Mn_2O_3 Nanoflakes from a Mn-based MOF. *ChemSusChem* **2019**. [CrossRef] [PubMed]
97. Han, X.; Li, C.; Liu, X.; Xia, Q.; Wang, Y. Selective oxidation of 5-hydroxymethylfurfural to 2,5-furandicarboxylic acid over MnO_x–CeO_2 composite catalysts. *Green Chem.* **2017**, *19*, 996–1004. [CrossRef]
98. Zhang, S.; Sun, X.; Zheng, Z.; Zhang, L. Nanoscale center-hollowed hexagon $MnCo_2O_4$ spinel catalyzed aerobic oxidation of 5-hydroxymethylfurfural to 2,5-furandicarboxylic acid. *Catal. Commun.* **2018**, *113*, 19–22. [CrossRef]
99. Gao, T.; Yin, Y.; Zhu, G.; Cao, Q.; Fang, W. Co_3O_4 NPs decorated Mn-Co-O solid solution as highly selective catalyst for aerobic base-free oxidation of 5-HMF to 2,5-FDCA in water. *Catal. Today* **2019**. [CrossRef]
100. Neaţu, F.; Marin, R.S.; Florea, M.; Petrea, N.; Pavel, O.D.; Pârvulescu, V.I. Selective oxidation of 5-hydroxymethyl furfural over non-precious metal heterogeneous catalysts. *Appl. Catal. B Environ.* **2016**, *180*, 751–757. [CrossRef]
101. Hayashi, E.; Yamaguchi, Y.; Kamata, K.; Tsunoda, N.; Kumagai, Y.; Oba, F.; Hara, M. Effect of MnO_2 Crystal Structure on Aerobic Oxidation of 5-Hydroxymethylfurfural to 2,5-Furandicarboxylic Acid. *J. Am. Chem. Soc.* **2019**, *141*, 890–900. [CrossRef]
102. Weckhuysen, B.M. Preface: Recent advances in the in-situ characterization of heterogeneous catalysts. *Chem. Soc. Rev.* **2010**, *39*, 4557–4559. [CrossRef]

© 2020 by the authors. Licensee MDPI, Basel, Switzerland. This article is an open access article distributed under the terms and conditions of the Creative Commons Attribution (CC BY) license (http://creativecommons.org/licenses/by/4.0/).

Article

Performance of a Ni-Cu-Co/Al₂O₃ Catalyst on In-Situ Hydrodeoxygenation of Bio-derived Phenol

Huiyuan Xue [1,2], Jingjing Xu [1,2], Xingxing Gong [1,2] and Rongrong Hu [1,2,*]

1. Key Laboratory of Applied Surface and Colloid Chemistry, Ministry of Education, Xi'an 710062, China; xuehy@snnu.edu.cn (H.X.); XXGong0712@163.com (X.G.); xjj315@snnu.edu.cn (J.X.)
2. School of Chemistry and Chemical Engineering, Shaanxi Normal University, Xi'an 710062, China
* Correspondence: rrhu@snnu.edu.cn; Tel.: +86-29-81530726

Received: 16 October 2019; Accepted: 9 November 2019; Published: 14 November 2019

Abstract: The in-situ hydrodeoxygenation of bio-derived phenol is an attractive routine for upgrading bio-oils. Herein, an active trimetallic Ni-Cu-Co/Al₂O₃ catalyst was prepared and applied in the in-situ hydrodeoxygenation of bio-derived phenol. Comparison with the monometallic Ni/Al₂O₃ catalyst and the bimetallic Ni-Co/Al₂O₃ and Ni-Cu/Al₂O₃ catalysts, the Ni-Cu-Co/Al₂O₃ catalyst exhibited the highest catalytic activity because of the formation of Ni-Cu-Co alloy on the catalyst characterized by using X-ray powder diffraction (XRD), temperature programmed reduction (TPR), N₂ physisorption, scanning electron microscope (SEM), and transmission electron microscope (TEM). The phenol conversion of 100% and the cyclohexane yield of 98.3% could be achieved in the in-situ hydrodeoxygenation of phenol at 240 °C and 4 MPa N₂ for 6 h. The synergistic effects of Ni with Cu and Co of the trimetallic Ni-Cu-Co/Al₂O₃ catalyst played a significant role in the in-situ hydrodeoxygenation process of phenol, which not only had a positive effect on the production of hydrogen but also owned an excellent hydrogenolysis activity to accelerate the conversion of cyclohexanol to cyclohexane. Furthermore, the catalyst also exhibited excellent recyclability and good potential for the upgrading of bio-oils.

Keywords: bio-derived phenol; Ni-Cu-Co/Al₂O₃; in-situ hydrodeoxygenation; cyclohexane; hydrogenolysis

1. Introduction

Biomass is one of the most abundant renewable resources on the earth. Bio-oil, produced from biomass, is recognized as a green feedstock for the production of chemicals and fuels [1,2]. Bio-oil has the advantages of a higher energy density than the original biomass, secure storage and transportation, and flexible use. However, the much higher oxygen content in bio-oils results in lower heating value and poorer stability compared to crude oil, which makes it difficult to use directly as engine fuels or even oil additives without further upgrading [3,4]. Therefore, the study of upgrading bio-oils has attracted much attention in recent years due to environmental and sustainable concerns. One of the promising routines to upgrading bio-oils is catalytic hydrodeoxygenation (HDO) [4,5]. However, the conventional HDO process needs excessive hydrogen to maintain high hydrogen pressure of 7–20 MPa [6,7], which inevitably increases the storage and transportation costs and safety risks of hydrogen. Although the HDO process is useful, it is unfavorable for the production of fuels [8–10] and estimated to need approximately 0.11 kg H₂/L oil [9] by a techno-economic analyses. Therefore, reducing external hydrogen supply, such as using in-situ generated hydrogen, might be one of the augmented approaches for improving the economic feasibility [11]. More importantly, compared with the conventional HDO process, the in-situ hydrodeoxygenation (in-situ HDO) process is more flexible without the complicated equipment and suited for the distributed upgrading of bio-oils [12], which helps to expand its industrial application.

In recent years, the in-situ HDO has become a trend in the bio-oil upgrading research [12–15]. Fisk et al. [12] reported that the Pt/Al$_2$O$_3$ catalyst had high activity for the in-situ HDO of a model bio-oil, and after upgrading the oxygen content of the model oil decreased from 41.4 wt% to 2.8 wt%. Other noble metal catalysts such as Pd/AC [13], Ru/MCM-41 [16] and Pd/C [11] also showed high activity for the in-situ HDO of some model components in bio-oils. Feng et al. [17] reported that the in-situ HDO process could not only reduce the usage of external hydrogen but also significantly improve the yield of target products in the bio-oil upgrading. Xiang et al. [18] used Raney Ni and a Pd/Al$_2$O$_3$ catalyst in the in-situ hydrogenation of phenol, o-cresol, and p-tert-butylphenol. They found that the activity of the in-situ generated hydrogen from aqueous-phase reforming (APR) was different from that of H$_2$ gas used in the hydrogenation of bio-oil. For example, the hydrogen generated from APR favored the production of cyclohexanone, while the hydrogen from H$_2$ gas favored the generation of cyclohexanol in the hydrogenation reaction of phenol.

Ni-based catalysts also showed excellent catalytic activity for the upgrading of bio-oils [14,19–21]. Putra et al. [14] reported phenol conversion could increase to 86.75% when in-situ glycerol aqueous reforming and phenol hydrogenation over Raney Ni catalyst. Xu et al. [19] used Raney Ni for the in-situ HDO of phenol and found more than 64% phenol could be converted to cyclohexanol at 220 °C. However, they had low selectivity for some oxygen-free products. The combination with acidic sites could help to increase its deoxygenation ratio of bio-oil. Wang et al. [22] found the total deoxygenation ratio of bio-oil could reach 99.6% over the ZrNi/Ir-ZSM-5 and Pd/C catalysts through in-situ HDO by using methanol as a liquid hydrogen donor. When using Raney Ni and Nafion/SiO2 catalysts in the in-situ hydrodeoxygenation of bio-derived phenol, the phenol conversion and the total yield of cyclohexane can reach respectively 100% and 87%, respectively [23]. Feng et al. [15] reported that the catalysts of Raney Ni and HZSM-5 also yielded 70–90% cyclohexanes and hydrocarbons in the in-situ hydrodeoxygenation of biomass-derived phenolic compounds. Besides, the production of hydrogen obtained from aqueous-phase reforming of liquid hydrogen donors also had a significant effect on the in-situ HDO of bio-oils. Zeng et al. [16] observed that although methanol, ethanol, formic acid, and acetic acid could generate hydrogen, the product distributions of bio-oils after upgrading were very different due to their different productivity of hydrogen. Hence, another important strategy is to improve the productivity of hydrogen in the in-situ HDO of bio-oils. Many researchers have focused on the effect of liquid hydrogen donors [11,16,18] or the ratio of the hydrogen donors to phenols [13,14,22–24] on the hydrogen availability. Unfortunately, few works have reported the effect of the multifunctional catalysts on hydrogen production in this process. Raney Ni [14] and the monometallic Ni catalyst [24] have an excellent hydrogen yield for the oxygenates aqueous-phase reforming. However, a considerable amount of work [25–30] has demonstrated that the bimetallic or trimetallic catalysts have positive effects on hydrogen production in the oxygenates reforming and inhibit the sintering of the active phase due to their excellent dispersion and electronic properties. Also, bimetallic catalysts such as Ni-Cu/Al2O3 [31], Ni-Fe/Al2O3 [32,33], Ni-Fe/MCSs [34], and Ni-Co/HZSM-5 [35] had excellent catalytic activity for converting bio-derived phenols into hydrocarbons by the conventional HDO with external hydrogen gas. Recently, Zhang et al. [36] prepared the Cu-Ni/ZrO2 catalysts for in-situ HDO of oleic acid to heptadecane and found the bimetallic catalysts had higher activities compared with the monometallic Ni catalysts because of the synergistic effect between Ni and Cu alloy. These results indicated that the use of the bimetallic or trimetallic catalysts could improve both the hydrogen availability and the deoxygenation ratio of bio-oil in the in-situ HDO process.

In this study, a trimetallic Ni-Cu-Co/Al$_2$O$_3$ catalyst was prepared and applied in the in-situ hydrodeoxygenation of phenol. Phenol derivatives comprise up 30 wt% of bio-oil [37], and therefore, phenol is suitable as a probe molecular to understand the primary pathways of in-situ HDO reaction. The various reaction conditions such as temperature, N$_2$ pressure, reaction time, the molar ratio of water/methanol, and liquid hydrogen donors on the in-situ HDO reaction were studied. The recyclability of the Ni-Cu-Co/Al$_2$O$_3$ catalyst was also tested. Furthermore, the relationships of the structure and catalytic activity of the Ni-Cu-Co/Al$_2$O$_3$ catalyst were studied by X-ray powder diffraction

(XRD), BET analysis, temperature programmed reduction (TPR), scanning electron microscope (SEM), and transmission electron microscope (TEM), and the synergistic effects of Ni with Co and Cu on the in-situ hydrodeoxygenation of phenol was discovered.

2. Results and Discussion

2.1. Structural Characterization

The XRD pattern of the fresh catalysts is presented in Figure 1. A nickel phase at $2\theta = 44.5$, 51.8 and 76.4° (JCPDS #04-0850), and a copper phase at $2\theta = 43.4$, 50.6 and 74.3° (JCPDS #65-9743) were observed in 20Ni/Al and 5Cu/Al, respectively. The 5Co/Al catalyst showed a very tiny diffraction peak attributed to Co, indicating that the Co species were highly dispersed on the surface of the Al_2O_3 support. In Figure 1, it is noted that the prepared 20Ni-5Co/Al catalyst exhibits a diffraction peak similar to that of the fresh 20Ni/Al at around $2\theta = 44.4$. However, detailed studies of the diffraction peaks of the sample showed that it shifted towards a higher angle, indicating that a bimetallic alloy was formed due to the incorporation of cobalt, which was different from the diffraction peaks of the physically mixed metal samples [38]. Besides, in the 20Ni-5Cu-5Co/Al sample, the sight shift of the diffraction peak of Ni-Cu-Co at 44.26 (111) was also noticeable, indicating the formation of trimetallic alloy nanoparticles. A similar observation was also made by Singh et al. [39]. The diffraction peak at 44.26° corresponding to a nickel fcc phase was attributed to the Ni-Cu-Co solid solution caused by the incorporation of cobalt and copper [40,41].

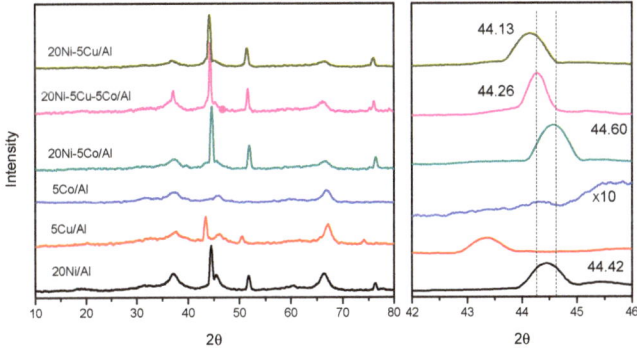

Figure 1. X-ray powder diffraction (XRD) patterns of the fresh catalysts.

The textural properties of the fresh catalysts are summarized in Table 1. The surface area of the Al_2O_3 support was 221 m²/g. After the incorporation of metal, the impregnated catalysts owned a lower surface area due to pore blockage during metal deposition. N_2 adsorption-desorption isotherms and pore size distributions are conducted to determine the structure of the samples, which is presented in Figure 2. All the isotherms obtained for these catalysts exhibit a type IV isotherm shape. The BJH pore size distribution (Figure 2b) indicates that the majority of pores were mesopores with pore sizes smaller than 10 nm.

Table 1. The textural properties of the fresh catalysts.

Catalyst	Surface Area	Pore Volume	Pore Size
	m²/g	cm³/g	nm
Al_2O_3	221	0.33	9.96
20Ni/Al	82	0.20	9.54
20Ni-5Cu/Al	80	0.18	8.71
20Ni-5Co/Al	82	0.18	8.63
20Ni-5Cu-5Co/Al	78	0.18	9.19

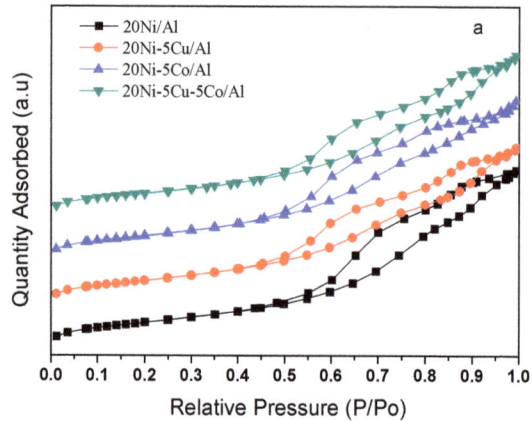
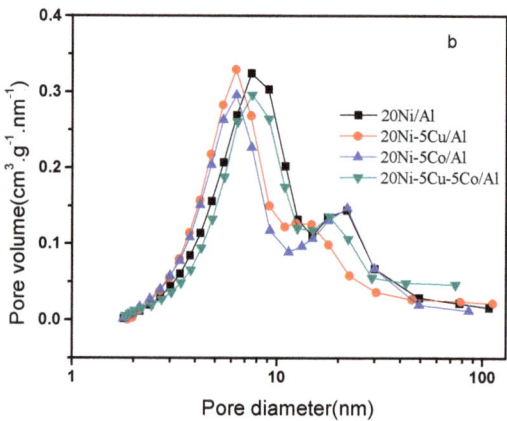

Figure 2. (**a**) N$_2$ adsorption-desorption isotherms and (**b**) pore size distribution calculated from the BJH desorption branch.

Figure 3 shows the TPR results of the fresh samples. The 20Ni/Al catalyst clearly showed multi broad peaks at 350 °C, 550 °C, and 700 °C. The first peak was associated with the reduction of Ni^{2+} to Ni^0, showing a weak interaction between NiO and Al$_2$O$_3$ support [29]. The other two peaks were broader reduction peaks from 480 °C to 800 °C. The wideness of the peaks suggested a stronger interaction between NiO and Al$_2$O$_3$. As reported by Shi et al. [42], there were octahedrally and tetrahedrally coordinated Ni^{2+} in the Ni supported catalysts, and the high temperature peak at 700 °C corresponded to the reduction of tetrahedrally coordinated Ni^{2+} since it was less reducible than the former. The 20Ni-5Co/Al catalyst clearly showed three distinct peaks. The reduction of Ni^{2+} to Ni^0 and Co^{3+} to Co^{2+} were responsible for the peaks at 300–410 °C, and the second weak peak at 500 °C corresponded to the complete reduction of cobalt [30]. For 20Ni-5Cu-5Co/Al, the area of the peak related to the reduction of tetrahedrally coordinated Ni^{2+} decreased, indicating that, for this sample, the amount of NiAl$_2$O$_4$ species was decreased [42]. The TPR results implied that the addition of cobalt and copper improved the reducibility of NiO, and hence more metallic Ni could be exposed in the trimetallic 20Ni-5Cu-5Co/Al catalyst.

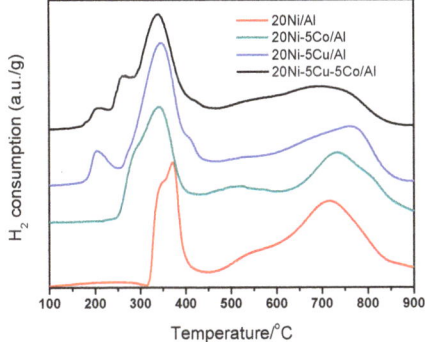

Figure 3. Spectra of the fresh catalysts.

The SEM images of the fresh samples are presented in Figure 4 and they displayed similar morphologies, i.e., rough surfaces and small grains. When compared with the 20Ni/Al catalysts, the bimetallic and trimetallic catalysts showed smaller particle sizes, which showed that the addition of Co or Cu could restrain agglomeration and promote the dispersion of the nickel species on the Al_2O_3 support surface.

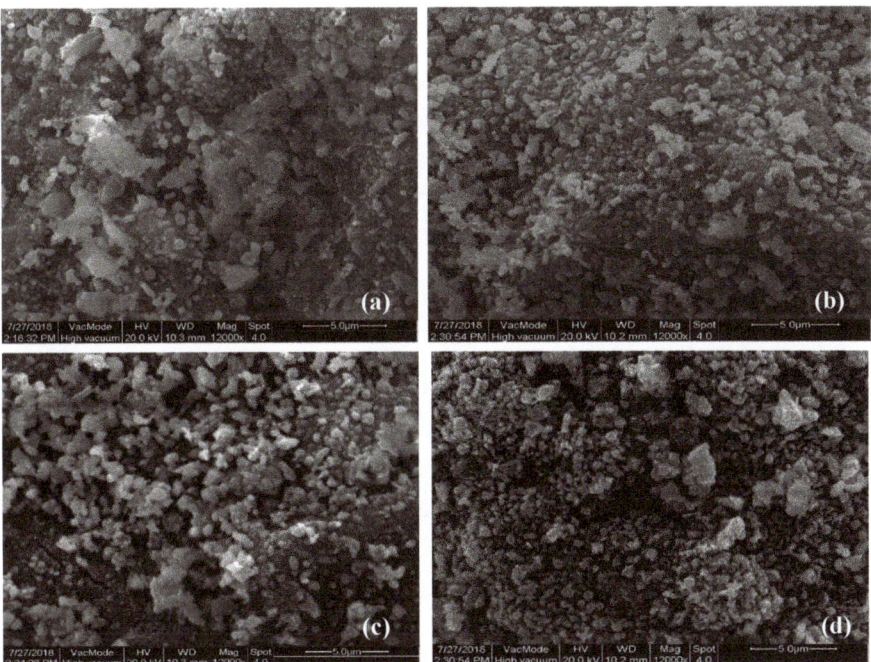

Figure 4. Scanning electron microscope (SEM) images of the fresh catalysts. 20Ni/Al (**a**), 20Ni-5Cu/Al (**b**), 20Ni-5Co/Al (**c**), 20Ni-5Cu-5Co/Al (**d**).

The TEM images and the particle size distribution for the fresh catalysts are presented in Figure 5. The particle distribution of the 20Ni/Al catalyst was 20–40 nm. When adding Co and Cu in 20Ni/Al, the particle distribution results indicated that it had a higher portion of smaller size in the 20Ni-5Cu-5Co/Al catalyst, which was attributed to the geometric and stabilizing effect of these metals [43]. The addition of cobalt or copper may show a "dilution effect" [44] or act as an inert spacer [45] on the surface of Ni species, forming much smaller metal alloy particles. The microstructure of 20Ni-5Cu-5Co/Al was also characterized by high-resolution transmission electron microscopy (HRTEM) and the crystalline fringe was measured to be 0.205 nm, which corresponds to the (111) interplanar spacing of the fcc structure of Ni-Cu-Co [46]. These images further confirmed the formation of the Ni-Cu-Co alloy, which was consistent with the XRD results.

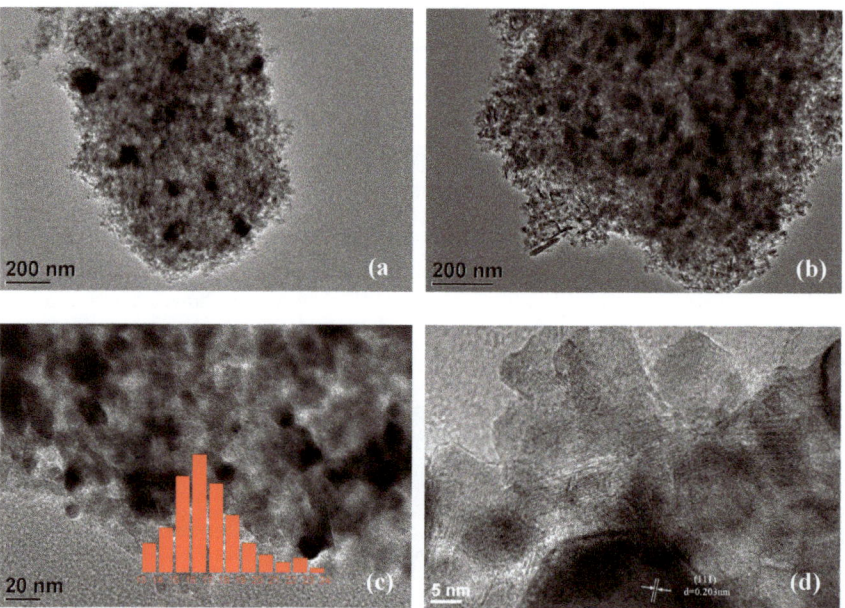

Figure 5. Images of the fresh catalysts. 20Ni/Al (**a**), 20Ni-5Cu-5Co/Al (**b–d**).

2.2. Performance of Ni-Based Catalysts in the in-situ Hydrodeoxygenation (HDO) of Phenol

According to the previous studies, the in-situ HDO process of phenol includes two main reactions: Hydrogen generation from the aqueous-phase reforming (APR) of methanol, and hydrodeoxygenation of phenol. Ni-based catalysts had been reported of excellent activities in both methanol aqueous-phase reforming and hydrodeoxygenation of phenols.

2.2.1. Hydrogen Production from Methanol Aqueous-Phase Reforming over Ni-Based Catalysts

The hydrogen used in the in-situ HDO, obtained from methanol aqueous-phase reforming, had an important effect on the HDO of phenol. For the over-all methanol aqueous-phase reforming, the hydrogen production is always accompanied by several side reactions, such as methanol decomposition, water gas shift reaction, methanation, and F-T reactions [29]. Therefore, the hydrogen yield is highly related to the performance of the catalysts. In order to further improve the catalytic activity and product selectivity, the second metal such as Co or Cu was often used to add into the Ni-based catalysts to tune surface active Ni by changing metal particle sizes or strengthen metal-support interaction. They could help to enhance water gas shift reaction or inhibit methanation reaction for higher hydrogen yield in the reforming reactions [26,27,29,30]. Table 2 shows the conversion and yield of gas products during the APR of methanol over the different Ni-based catalysts. From it, for all these catalysts, only H_2, CO_2, CO, and CH_4 could be detected in significant concentrations in the gas products. For the 20Ni/Al catalyst, the hydrogen yield was about 1.05 mol/mol CH_3OH and the water/methanol consumption was only 0.37, indicating that the methanol APR was part of the reaction. Besides, the amount of CO formed over this catalyst was more than CO_2, which confirmed that it was favorable for the decomposition reaction of methanol rather than the APR reaction itself and the decomposition reaction on Ni species was in charge of generating more CO. After adding a small amount of Co or Cu into the 20Ni/Al catalyst, the hydrogen yield rapidly increased to 1.31 and 1.75 mol/mol CH_3OH, respectively. It was also observed that with the 20Ni-5Cu/Al and 20Ni-5Co/Al catalyst, the water/methanol consumption increased and the amounts of CO decreased, which implied both Cu and Co could promote the methanol reforming and water gas shift reaction and produce more H_2 and CO_2. Notably, CH_4 due to CO or/and CO_2 hydrogenation over the 20Ni/Al catalyst in the reforming was not observed in the 20Ni-5Cu/Al and 20Ni-5Co/Al catalyst, indicating that the Ni-Cu or Ni-Co alloy phase may be responsible for reducing the effects of the methanation reaction. On the 20Ni-5Cu-5Co/Al catalyst, the H_2 yield was 2.15 mol/mol CH_3OH. Such a high amount of hydrogen showed the APR reaction and water gas shift reaction may dominate over the 20Ni-5Cu-5Co/Al catalyst, evidenced by an increase in both CO_2 production and water/methanol consumption. Indeed, for the 20Ni-5Cu-5Co/Al catalyst, the adding of Co and Cu in the Ni/Al catalyst affected in a positive way for the production of hydrogen due to the synergistic effects of Ni-Cu-CO alloy.

Table 2. The conversion and yield of gas products during the aqueous-phase reforming of methanol over the different Ni-based catalysts.

Sample	Conversion (%)		Water/Methanol Consumption	Yield (mol/mol CH_3OH)			
	CH_3OH	H_2O		H_2	CO	CH_4	CO_2
20Ni/Al	52	5.55	0.37	1.05	0.34	0.09	0.09
20Ni-5Co/Al	53.4	6.9	0.45	1.31	0.26		0.27
20Ni-5Cu/Al	64.1	13.5	0.74	1.75	0.17		0.47
20Ni-5Cu-5Co/Al	75.4	18.3	0.85	2.15	0.13		0.62

Reaction conditions: 30 g deionized water, 15 g methanol, and 0.5 g catalyst, 240 °C, 4 MPa N_2, 6 h.

2.2.2. In-situ HDO of Phenol over Ni-based Catalysts

The performances of the prepared catalysts in the HDO of phenol with H_2 and the in-situ HDO of phenol with methanol are shown in Figure 6a,b, respectively. From Figure 6a, all of the Ni-based catalysts exhibited pretty high activity in the phenol conversion. At high initial pressure of 4 MPa H_2, more than 50% phenol could yield the deep hydrogenated component of cyclohexane over these catalysts. Especially for the 20Ni-5Cu-5Co/Al catalyst, it could produce nearly 99% cyclohexane, showing its excellent activity for the conventional HDO of phenol. Interestingly, when the H_2 initial pressure decreased to 0.5 MPa, less than 2% cyclohexane could be yielded and benzene, cyclohexanone and cyclohexanol were the main productions. Obviously, the changes might be related to the amount of hydrogen. The low hydrogen pressure of 0.5 MPa led to a decrease of H_2 solubility in the liquid phase and stabilized the formation of cyclohexanone and cyclohexanol.

Figure 6b is the result of the performances of the prepared catalysts in the in-situ HDO of phenol with methanol as the liquid hydrogen donor. Among the catalysts, 20Ni-5Cu-5Co/Al was the most active and the phenol conversion was nearly 100%. In contrast, monometallic 20Ni/Al and bimetallic 20Ni-5Cu/Al and 20Ni-5Co/Al exhibited approximately the same conversions of less than 80% at the conditions of this work. Regarding product distribution, cyclohexanol and cyclohexane were the main products. Cyclohexanone was also detected in low amounts on some catalysts. It is important to highlight that no benzene was formed, which showed that the direct deoxygenation of phenol to benzene hardly occurs in the in-situ HDO process of phenol in these conditions. For the monometallic 20Ni/Al catalysts, cyclohexanol mainly formed on it and its selectivity reached 78%. However, when adding a small amount of Co or Cu into 20Ni/Al, the selectivity of cyclohexanol decreased significantly and more than 60% phenol was converted to cyclohexane. After both Cu and Co was added into 20Ni/Al, the phenol conversion and cyclohexane selectivity further increased and reached 100% and

95% on 20Ni-5Cu-5Co/Al, respectively, which means the addition of Co and Cu had an essential effect on the product distribution in the in-situ HDO of phenol. In general, there were two main pathways of the hydrogenation of phenol [16]: One is the direct deoxygenation to benzene; the other one is the consecutive hydrogenation to cyclohexanone, cyclohexanol, cyclohexene, and cyclohexane, successively. Mortensen et al. [47] and Han et al. [34] quantified the rates of these steps over Ni/ZrO2 and Ni-Fe/MCSs, respectively, and found that the hydrogenation rate of cyclohexanone to cyclohexanol was much higher than the hydrogenolysis rate of cyclohexanol to cyclohexane over these Ni-based catalysts. Therefore, the high selectivity for cyclohexane might imply a high activity of the catalyst in the conversion of cyclohexanol. It is worth noting that the 20Ni-5Co/Al catalyst yielded more cyclohexane than the 20Ni-5Cu/Al catalyst, indicating Co had a higher hydrogenolysis activity.

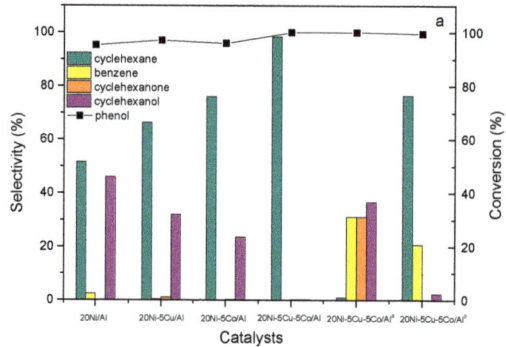

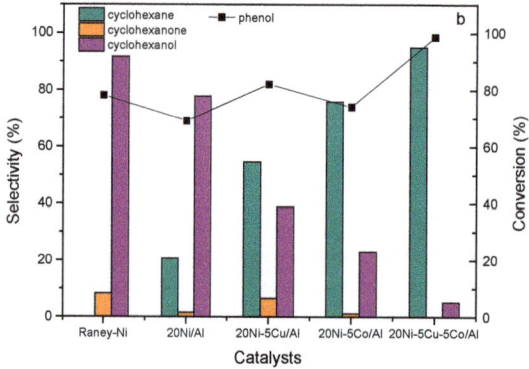

Figure 6. Catalytic performance of the Ni-based catalysts in the hydrodeoxygenation (HDO) of phenol with H_2 (**a**) and the in-situ HDO of phenol with methanol (**b**). Reaction conditions: (a) 3 g phenol and 0.5 g catalyst, 240 °C, 4 MPa H_2, 6 h; (b) 30 g deionized water, 15 g methanol, 3 g phenol and 0.5 g catalyst, 240 °C, 4 MPa N_2, 6 h.

To further clarify it, the hydrogenolysis of cyclohexanol to cyclohexane on the 20Ni-5Cu-5Co/Al catalyst was evaluated under 240 °C and 4 MPa H_2 and the results are summarized in Table 3.

Table 3. The hydrogenolysis of cyclohexanol over Ni-based catalysts.

Catalysts	Solvent	Conversion (%)	Selectivity (%)	
			Cyclohexene	Cyclohexane
20Ni/Al	None	72.4	–	>99.9
20Ni/Al	H$_2$O	48.7	–	>99.9
5Cu/Al	None	23.6	–	>99.9
5Cu/Al	H$_2$O	14.5	–	>99.9
5Co/Al	None	99.1	3.7	96.3
5Co/Al	H$_2$O	90.2	22.6	77.4
20Ni-5Cu/Al	H$_2$O	53.1	–	>99.9
20Ni-5Co/Al	H$_2$O	98.7	–	>99.9
20Ni-5Cu-5Co/Al	H$_2$O	99.3	–	>99.9

Reaction conditions: Solvent (30 g), 3 g cyclohexanol and 0.5 g catalyst, 240 °C, 4 MPa H$_2$, 6 h.

From Table 3, the conversion of cyclohexanol over 5Co/Al was more than 90% even in water. However, it seemed the hydrogenation rate of cyclohexene over monometallic Co metal sites was slower than the dehydration rate of cyclohexanol, which led to the accumulation of cyclohexene even after 6 h. For the 5Cu/Al and 20Ni/Al catalysts, they showed much lower cyclohexanol conversion, while both of them had higher selectivity of cyclohexane. These results implied that the hydrogenation rates of cyclohexene in water over Ni and Cu metal sites were much faster than the dehydration rate of cyclohexanol. Zhao [48] had investigated the detailed kinetics of cyclohexanol dehydration and cyclohexene hydrogenation over Ni/Al$_2$O$_3$-HZSM-5 and they found that the hydrogenation rate of the latter showed up nearly four times higher than the dehydration rate of the former. As a consequence, only cyclohexane formed during the cyclohexanol hydrogenolysis reaction over the Ni or Cu catalysts. It must be mentioned that though Cu had a weak activity for the hydrogenolysis of cyclohexanol, once it was added to the 20Ni/Al catalyst, the yield of cyclohexane could be increased. Furthermore, when both Cu and Co were added into the 20Ni/Al catalyst, nearly 100% cyclohexane was obtained from cyclohexanol.

Therefore, the main reason for the high selectivity to cyclohexane on the 20Ni-5Cu-5Co/Al catalyst in the phenol in-situ HDO could be summarized as follows: (1) The 20Ni-5Cu-5Co/Al catalyst had a substantial effect on the amount of producing hydrogen due to the different selectivity towards methanol decomposition reaction and the water gas shift reaction. The monometallic Ni catalyst was favorable for methanol decomposition reaction and it also promoted CH$_4$ formation. The 20Ni-5Cu-5Co/Al catalyst produced more hydrogen because of its excellent activity for the water gas shift reaction. Cu and Co alloying in Ni catalyst also had a negative effect on methanation reaction compared to 20Ni/Al. Since the 20Ni-5Cu-5Co/Al catalyst had the highest amount of hydrogen among all prepared Ni-based catalysts, it could make the converted phenol yield more cyclohexane than the other catalysts. (2) The 20Ni-5Cu-5Co/Al catalyst had an excellent hydrogenolysis activity to accelerate the conversion of cyclohexanol to cyclohexane because of the synergistic effect of Ni-Cu-Co. The monometallic Co catalyst had a good activity for the hydrogenolysis of cyclohexanol. After Co was added into 20Ni-5Cu/Al catalysts, both cyclohexanol conversion and cyclohexane yield were significantly increased even in water. Therefore, the formation of Ni-Co-Cu alloy on 20Ni-5Cu-5Co/Al was responsible for its high selectivity to cyclohexane in the in-situ HDO of phenol. (3) Nickel particle size also had an essential influence on the HDO of phenol. The relatively small Ni particles were active for a high yield of cyclohexane by increasing the deoxygenation rate [49]. For the 20Ni-5Cu-5Co/Al catalyst, the formation of Cu-Co-Ni alloy showed a dilution effect on Ni species, resulted in smaller particle sizes, evidenced by XRD and TEM. As a consequent, a higher yield of cyclohexane would be achieved on this trimetallic 20Ni-5Cu-5Co/Al catalyst.

2.2.3. Effect of Reaction Temperature and Initial N2 Pressure

The effect of temperature (120–260 °C) on the in-situ HDO of phenol at different initial N_2 pressure of 1–4 MPa over the 20Ni-5Cu-5Co/Al catalyst was also investigated. As displayed in Figure 7, when temperature increased from 120 °C to 260 °C, the phenol conversion increased and reached 52% and 100% at 1 and 4 MPa, respectively. At the lower temperature, it is unfavorable for the hydrogen production from methanol reforming due to an endothermic reaction, resulting in insufficient hydrogen, a lower conversion and a slower hydrogenation rate. The increase in temperature has not only an advantage to the conversion of phenol but also the selectivity of cyclohexane. When the temperature was lower than 180 °C, no cyclohexane could be detected in the products even in high initial pressure of 4 MPa. However, it showed a rapidly increasing trend with the temperature increasing from 180 °C to 260 °C, reaching 96.2% and 97.4% at 3 and 4 MPa, respectively. As mentioned previously, the formation of cyclohexane in the hydrogenation of phenol was originally from the hydrogenolysis reaction of cyclohexanol. Higher temperatures not only contributed to cyclohexanol transforming into cyclohexane [34,48] but also helped the production of more hydrogen, which could promote the hydrogenation of cyclohexene to cyclohexane. It must be mentioned that at low initial pressure, the highest selectivity of cyclohexane was less than 25% even at the high temperature, which means that the initial pressure is also one of the critical factors in the production distribution. High initial pressure increased the solubility of H_2 in the liquid phase, rendering more H_2 accessible the deep hydrogenation reaction, ultimately achieving high selectivity of cyclohexane.

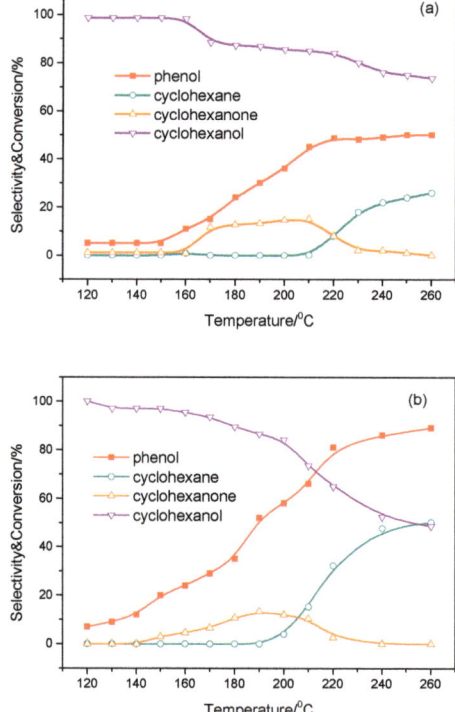

Figure 7. Cont.

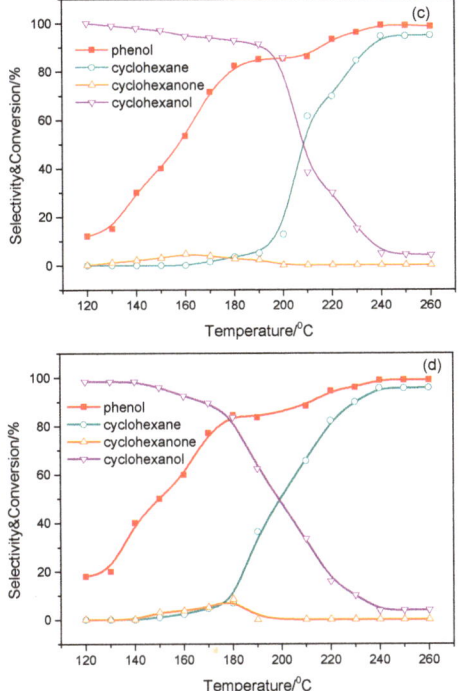

Figure 7. Effect of temperature on the in-situ HDO of phenol at different initial N2 pressure over the 20Ni-5Cu-5Co/Al catalyst. Reaction conditions: 30 g deionized water, 15 g methanol, 3 g phenol and 0.5 g catalyst, 4 h. (**a**) 1 MPa initial N_2 pressure, (**b**) 2 MPa initial N_2 pressure, (**c**) 3 MPa initial N_2 pressure, (**d**) 4 MPa initial N_2 pressure.

2.2.4. Effect of Reaction Time

The effect of the reaction time on the in-situ HDO of phenol over the 20Ni-5Cu-5Co/Al catalyst was investigated at 240 °C and 4 MPa N_2 over the 20Ni-5Cu-5Co/Al catalyst. From Figure 8, the conversion of phenol increased rapidly from 55.1% (1 h) to 99.2% (4 h) at the beginning and then slightly increased to 100% (5 h). Moreover, the selectivity of cyclohexane improved at the cost of the selectivity of cyclohexanol with increased reaction time. More than 95% of cyclohexane could be achieved by increasing the reaction time to 4 h. These results indicated that cyclohexanol, as an intermediate, would be converted to cyclohexane with prolonged reaction time. The hydrogenation rate of phenol to cyclohexanol might be faster than the hydrogenolysis rate of cyclohexanol to cyclohexane over the 20Ni-5Co-5Cu/Al catalyst. The hydrogenolysis of cyclohexanol was the rate-determining step of the overall reaction of phenol HDO over the catalyst with high activity for hydrogenation [34], which, as a consequence, would lead to the high selectivity of cyclohexanol in the initial phenol conversion process. After prolonging the reaction time, cyclohexane dominated the product distribution (95%) at 100% conversions, and cyclohexanol decreased to less than 5% selectivity eventually.

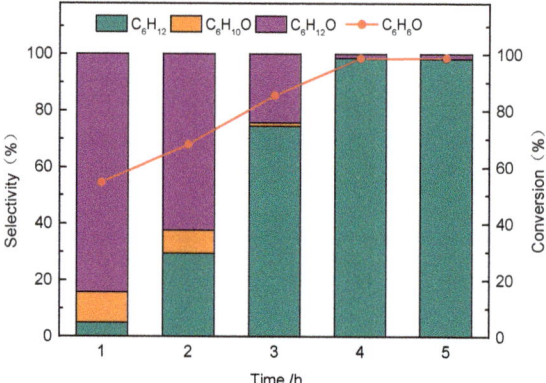

Figure 8. Effect of reaction time on the in-situ HDO of phenol over the 20Ni-5Cu-5Co/Al catalyst. Reaction conditions: 30 g deionized water, 15 g methanol, 3 g phenol and 0.5 g catalyst, 240 °C, 4 MPa initial N_2 pressure.

2.2.5. Effect of the Molar Ratio of Water/Methanol

Figure 9 presents the effect of the different molar ratios of water/methanol on the in-situ HDO of phenol over the 20Ni-5Co-5Cu/Al catalyst at 240 °C and 4 MPa initial N_2 pressure. Since the hydrogen of the in-situ hydrogenation reaction was obtained from methanol APR, the H_2O/methanol ratio had an essential effect on the hydrogenation reaction. From Figure 9, the conversion of phenol increased from 40.1% to 100%, with the increasing water/methanol ratio of 10:1 to 2.5:1. After further increasing the water/methanol ratio to 1.8:1, the conversion of phenol decreased slightly. These results showed that higher methanol concentration favored hydrogen production, thus leading to higher conversion of phenol. However, when the methanol concentration in the reactants was too high, it effectively competed for the active sites of the catalyst to block the access of reacted phenol, resulting in a decrease in phenol conversion [16,22].

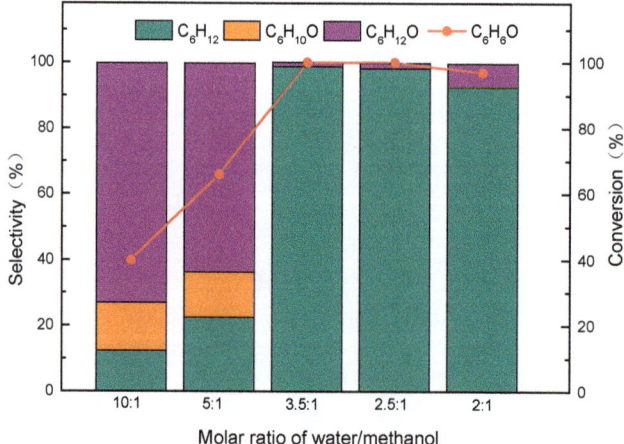

Figure 9. Effect of molar ratio of water/methanol on the in-situ HDO of phenol over the 20Ni-5Cu-5Co/Al catalyst. Reaction conditions: 30 g deionized water, 3 g phenol and 0.5 g catalyst, 240 °C, 4 MPa initial N_2 pressure, 4 h.

2.2.6. Effect of Liquid Hydrogen Donors

Methanol, ethanol, propanol and acetic acid were chosen as the liquid hydrogen donors and their effects on the in-situ HDO of phenol was evaluated over the 20Ni-5Co-5Cu/Al catalyst, as presented in Table 4. When ethanol was the liquid hydrogen donor, more than 90% conversion of phenol and selectivity of cyclohexane were achieved. However, when acetic acid was the liquid hydrogen donor, phenol conversion decreased to 39% under the same conditions. Meanwhile, the selectivity of cyclohexane was less than 5% and the main products of phenol in the in-situ HDO over the 20Ni-5Co-5Cu/Al catalyst were cyclohexanol and cyclohexanone. These results proved that methanol and ethanol were better than acetic acid for the in-situ HDO of phenol to cyclohexane over the 20Ni-5Co-5Cu/Al catalyst. It could be explained that the hydrogen produced by the APR of acetic acid was insufficient, and the low hydrogen pressure stabilized the formation of cyclohexanol and cyclohexanone intermediates, which was consistent with Tan et al.'s work [11]. Therefore, methanol is the best liquid hydrogen donor in the in-situ HDO process of phenol under these conditions.

Table 4. Effect of liquid hydrogen donors on the in-situ HDO of phenol.

Liquid Hydrogen Donors	Conversion (%)	Selectivity (%)		
		Cyclohexane	Cyclohexanone	Cyclohexanol
Methanol	100	98.6	–	1.4
Ethanol	92	93.8	–	6.2
Propanol	44	11.3	15.7	73
Acetic acid	39	2.6	21.8	75.6

Reaction conditions: 30 g deionized water, 15 g liquid hydrogen donor, 3 g phenol and 0.5 g catalyst, 240 °C, 4 MPa initial N_2 pressure, 4 h.

2.2.7. Recyclability

The recyclability of the sample is vital for practical industrial applications. The catalytic maintenance of the 20Ni-5Co-5Cu/Al catalyst was also evaluated at 240 °C, 4 MPa initial N_2 pressure for 4 h. After the reaction, the catalyst was then separated with the assistance of a magnet, washed with methanol several times and dried for the next cycle. Reaction results shown in Figure 10 revealed that the 20Ni-5Co-5Cu/Al catalyst could be used at least 10 times without significant decreases in catalytic conversion and selectivity, which indicated that the 20Ni-5Co-5Cu/Al catalyst exhibited excellent recyclability.

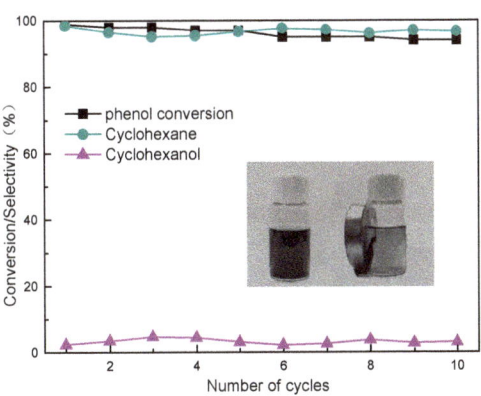

Figure 10. The recyclability of the 20Ni-5Co-5Cu/Al sample for in-situ HDO of phenol. Reaction conditions: 30 g deionized water, 15 g methanol, 3 g phenol and 0.5 g catalyst, 240 °C, 4 MPa initial N_2 pressure, 4 h.

3. Experimental

3.1. Catalyst Synthesis

A series of Ni-based catalysts were synthesized using an impregnation method. The monometallic catalysts (20 wt% Ni/Al_2O_3, 5 wt% Cu/Al_2O_3 and 5 wt% Co/Al_2O_3), bimetallic catalysts (20 wt% Ni-5 wt% Cu/Al_2O_3 and 20 wt% Ni-5 wt% Co/Al_2O_3) and trimetallic catalyst (20 wt% Ni-5 wt% Cu-5 wt% Co/Al_2O_3) were prepared. Briefly, the Al_2O_3 support was obtained through the calcination of pseudoboehmite precursor in the air at 550 °C for 4 h. The required amount of $Ni(NO_3)_2 \cdot 6H_2O$, $u(NO_3)_2 \cdot 3H_2O$ and $Co(NO_3)_3 \cdot 6H_2O$, obtained from Aladdin Industrial Corporation (Shanghai, China), were dissolved in distilled water under ultrasound for 30 min. Then 5 g Al_2O_3 support was added and mixed for 3 h using an ultrasonic mixer at 35 °C. The mixture was then dried at 100 °C for 12 h. Next, the sample was heated for calcination from 20 °C to 600 °C in a furnace at a rate of 1 °C/min, held at 600 °C for 5 h, then finally cooled to room temperature. Before the experiments, the catalyst was reduced in a 5% H_2/95% N_2 mixed gas with a flow rate of 50 mL/min at 500 °C for 5 h. The prepared Ni-based catalysts are denoted as 20Ni/Al, 5Cu/Al, 5Co/Al, 20Ni-5Co/Al, 20Ni-5Cu/Al, and 20Ni-5Cu-5Co/Al for brevity.

3.2. Catalyst Characterization

X-ray powder diffraction (XRD) patterns of these catalysts were measured by a Rigaku D/max2550VB3+/PC diffractometer (Rigaku International Corporation, Tokyo, Japan) using Cu K radiation at 40 KV and 40 mA. BET surface areas and average pore size of the catalysts were carried out by N_2 adsorption at 77 K with a Micrometric ASAP 2020 apparatus (Micromeritics GmbH, Aachen, Germany). Before measurements, the samples were degassed at 150 for 12 h. Temperature programmed reduction (TPR) of the samples was done in a Micromeritics AutoChem II 2920 apparatus equipped with TCD (Micromeritics GmbH, Aachen, Germany). A 50 mg sample was heated from room temperature to 800 °C in 10% H_2/90%N_2. Scanning electron microscope (SEM) images were recorded using an FEI Quanta 200 instrument (FEI, Eindhoven, Netherlands). High-resolution transmission electron microscopy (HRTEM) images were recorded using a JEM-2100 TEM Field Emission Electron Microscope (JEOL GmbH, Freising, Germany).

3.3. Activity Tests

The experiments were performed in a 100 mL high-pressure autoclave. The typical experimental conditions are as follows: 30 g deionized water, 15 g methanol, 3 g phenol, and 0.5 g catalyst were added to the autoclave. Before the reaction, the autoclave was washed five times with 5 MPa of N_2 to remove air. The autoclave was then heated to 130–250 °C and kept for 1–4 h with the stirring speed of 600 rpm. After finishing the experiment, the autoclave was cooled to room temperature. All the gas and liquid products were collected and analyzed by using a gas chromatography (Jingketianmei Instrument, Shanghai, China) equipped with thermal conductivity detector (TCD) and gas chromatography equipped with flame ionization detector (FID), respectively. The conversion of reactants, the selectivity to liquid products, the yield of gas products were defined as follows:

$$\text{Conversion} = \frac{(\text{moles of reactants})_{in} - (\text{moles of reactants})_{out}}{(\text{moles of reactants})_{in}} \times 100\%$$

$$\text{Selectivity of liquid product}_i = \frac{\text{moles of product}_i}{(\text{moles of phenol})_{in} - (\text{moles of phenol})_{out}} \times 100\%$$

$$\text{Yield of gas product}_i = \frac{\text{moles of gas product}_i}{\text{moles of methsnol}_{in}}$$

4. Conclusions

In this study, we prepared a series of Ni-based catalysts for the in-situ HDO process of biomass-derived phenol to cyclohexane. The trimetallic 20Ni-5Co-5Cu/Al catalyst presented the highest activity compared to the other monometallic catalysts or bimetallic catalysts. The phenol conversion of 100% and the cyclohexane yield of 98.3% could be obtained in the in-situ HDO of phenol at 240 °C and 4 MPa initial N_2 pressure. The high catalytic activity of the 20Ni-5Co-5Cu/Al catalyst could be attributed to the formation of Ni-Cu-Co alloy, which had the strong positive synergistic effects on the generation of hydrogen from the methanol aqueous-phase reforming and hydrodeoxygenation of phenol. The catalyst also showed excellent recyclability and exhibited good potential for upgrading the bio-oil to reduce the oxygen content by the in-situ hydrodeoxygenation.

Author Contributions: Conceptualization, H.X.; validation, H.X., X.G. and J.X.; writing—original draft preparation, H.X.; writing—review and editing, R.H.; funding acquisition, R.H.

Funding: Gratitude for the support from the National Natural Science Foundation of China (21276254, 21636006) and the Fundamental Research Funds for the Central Universities (GK201603051, GK201601005).

Conflicts of Interest: The authors declare no conflict of interest.

References

1. Kunkes, E.L.; Simonetti, D.A.; West, R.M.; Serrano-Ruiz, J.C.; Gartner, C.A.; Dumesic, J.A. Catalytic conversion of biomass to monofunctional hydrocarbons and targeted liquid-fuel classes. *Science* **2008**, *322*, 417–421. [CrossRef]
2. Li, C.Z.; Zhao, X.C.; Wang, A.Q.; Huber, G.W.; Zhang, T. Catalytic Transformation of Lignin for the Production of Chemicals and Fuels. *Chem. Rev.* **2015**, *115*, 11559–11624. [CrossRef]
3. Mortensen, P.M.; Grunwaldt, J.D.; Jensen, P.A.; Knudsen, K.G.; Jensen, A.D. A review of catalytic upgrading of bio-oil to engine fuels. *Appl. Catal. A Gen.* **2011**, *407*, 1–19. [CrossRef]
4. Wang, H.M.; Male, J.; Wang, Y. Recent Advances in Hydrotreating of Pyrolysis Bio-Oil and Its Oxygen-Containing Model Compounds. *Acs Catal.* **2013**, *3*, 1047–1070. [CrossRef]
5. Yoosuk, B.; Tumnantong, D.; Prasassarakich, P. Amorphous unsupported Ni-Mo sulfide prepared by one step hydrothermal method for phenol hydrodeoxygenation. *Fuel* **2012**, *91*, 246–252. [CrossRef]
6. Lu, Q.; Chen, C.J.; Luc, W.; Chen, J.G.G.; Bhan, A.; Jiao, F. Ordered Mesoporous Metal Carbides with Enhanced Anisole Hydrodeoxygenation Selectivity. *Acs Catal.* **2016**, *6*, 3506–3514. [CrossRef]
7. Anand, M.; Farooqui, S.A.; Kumar, R.; Joshi, R.; Kumar, R.; Sibi, M.G.; Singh, H.; Sinha, A.K. Kinetics, thermodynamics and mechanisms for hydroprocessing of renewable oils. *Appl. Catal. A Gen.* **2016**, *516*, 144–152. [CrossRef]
8. Huber, G.W.; Corma, A. Synergies between bio- and oil refineries for the production of fuels from biomass. *Angew. Chem. Int. Ed.* **2007**, *46*, 7184–7201. [CrossRef]
9. Singh, N.R.; Delgass, W.N.; Ribeiro, F.H.; Agrawal, R. Estimation of Liquid Fuel Yields from Biomass. *Environ. Sci. Technol.* **2010**, *44*, 5298–5305. [CrossRef]
10. Venderbosch, R.H.; Ardiyanti, A.R.; Wildschut, J.; Oasmaa, A.; Heeresb, H.J. Stabilization of biomass-derived pyrolysis oils. *J. Chem. Technol. Biot.* **2010**, *85*, 674–686. [CrossRef]
11. Tan, Z.; Xu, X.; Liu, Y.; Zhang, C.; Zhai, Y.; Li, Y.; Zhang, R. Upgrading Bio-Oil Model Compounds Phenol and Furfural with In Situ Generated Hydrogen. *Environ. Prog. Sustain.* **2014**, *33*, 751–755. [CrossRef]
12. Fisk, C.A.; Morgan, T.; Ji, Y.Y.; Crocker, M.; Crofcheck, C.; Lewis, S.A. Bio-oil upgrading over platinum catalysts using in situ generated hydrogen. *Appl. Catal. A Gen.* **2009**, *358*, 150–156. [CrossRef]
13. Zhang, D.M.; Ye, F.Y.; Xue, T.; Guan, Y.J.; Wang, Y.M. Transfer hydrogenation of phenol on supported Pd catalysts using formic acid as an alternative hydrogen source. *Catal. Today* **2014**, *234*, 133–138. [CrossRef]
14. Putra, R.D.D.; Trajano, H.L.; Liu, S.D.; Lee, H.; Smith, K.; Kim, C.S. In-situ glycerol aqueous phase reforming and phenol hydrogenation over Raney Ni (R). *Chem. Eng. J.* **2018**, *350*, 181–191. [CrossRef]
15. Feng, J.F.; Hse, C.Y.; Yang, Z.Z.; Wang, K.; Jiang, J.C.; Xu, J.M. Liquid phase in situ hydrodeoxygenation of biomass-derived phenolic compounds to hydrocarbons over bifunctional catalysts. *Appl. Catal. A Gen.* **2017**, *542*, 163–173. [CrossRef]

16. Zeng, Y.; Wang, Z.; Lin, W.G.; Song, W.L. In situ hydrodeoxygenation of phenol with liquid hydrogen donor over three supported noble-metal catalysts. *Chem. Eng. J.* **2017**, *320*, 55–62. [CrossRef]
17. Feng, J.; Yang, Z.; Hse, C.Y.; Su, Q.; Wang, K.; Jiang, J.; Xu, J. In situ catalytic hydrogenation of model compounds and biomass-derived phenolic compounds for bio-oil upgrading. *Renew. Energy* **2017**, *105*, 140–148. [CrossRef]
18. Xiang, Y.Z.; Li, X.N.; Lu, C.S.; Ma, L.; Yuan, J.F.; Feng, F. Reaction Performance of Hydrogen from Aqueous-Phase Reforming of Methanol or Ethanol in Hydrogenation of Phenol. *Ind. Eng. Chem. Res.* **2011**, *50*, 3139–3144. [CrossRef]
19. Xu, Y.; Long, J.; Liu, Q.; Li, Y.; Wang, C.; Zhang, Q.; Wei, L.; Xinghua, Z.; Songbai, Q.; Tiejun, W. In situ hydrogenation of model compounds and raw bio-oil over Raney Ni catalyst. *Energy Convers. Manag.* **2015**, *89*, 188–196. [CrossRef]
20. Xu, Y.; Qiu, S.; Long, J.; Wang, C.; Chang, J.; Tan, J.; Liu, Q.; Ma, L.; Wang, T.; Zhang, Q. In situ hydrogenation of furfural with additives over a RANEY (R) Ni catalyst. *Rsc Adv.* **2015**, *5*, 91190–91195. [CrossRef]
21. Huang, Y.H.; Xia, S.Q.; Ma, P.S. Effect of zeolite solid acids on the in situ hydrogenation of bio-derived phenol. *Catal. Commun.* **2017**, *89*, 111–116. [CrossRef]
22. Wang, L.; Liu, Q.; Jing, C.Y.; Mominou, N.; Li, S.Z.; Wang, H.W. In-situ hydrodeoxygenation of a mixture of oxygenated compounds with hydrogen donor over ZrNi/Ir-ZSM-5+Pd/C. *J. Alloy Compd.* **2018**, *753*, 664–672. [CrossRef]
23. Wang, L.; Ye, P.J.; Yuan, F.; Li, S.Z.; Ye, Z.X. Liquid phase in-situ hydrodeoxygenation of bio-derived phenol over Raney Ni and Nafion/SiO_2. *Int. J. Hydrog. Energy* **2015**, *40*, 14790–14797. [CrossRef]
24. Xu, Y.; Li, Y.; Wang, C.; Wang, C.; Ma, L.; Wang, T.; Zhang, X.; Zhang, Q. In-situ hydrogenation of model compounds and raw bio-oil over Ni/CMK-3 catalyst. *Fuel Process. Technol.* **2017**, *161*, 226–231. [CrossRef]
25. Carrero, A.; Calles, J.A.; Vizcaino, A.J. Hydrogen production by ethanol steam reforming over Cu-Ni/SBA-15 supported catalysts prepared by direct synthesis and impregnation. *Appl. Catal. A Gen.* **2007**, *327*, 82–94. [CrossRef]
26. Liao, P.H.; Yang, H.M. Preparation of catalyst Ni-Cu/CNTs by chemical reduction with formaldehyde for steam reforming of methanol. *Catal. Lett.* **2008**, *121*, 274–282. [CrossRef]
27. Khzouz, M.; Wood, J.; Pollet, B.; Bujalski, W. Characterization and activity test of commercial Ni/Al_2O_3, Cu/ZnO/Al_2O_3 and prepared Ni-Cu/Al_2O_3 catalysts for hydrogen production from methane and methanol fuels. *Int. J. Hydrog. Energy* **2013**, *38*, 1664–1675. [CrossRef]
28. Mrad, M.; Hammoud, D.; Gennequin, C.; Aboukais, A.; Abi-Aad, E. A comparative study on the effect of Zn addition to Cu/Ce and Cu/Ce-Al catalysts in the steam reforming of methanol. *Appl. Catal. A Gen.* **2014**, *471*, 84–90. [CrossRef]
29. Khzouz, M.; Gkanas, E.I.; Du, S.F.; Wood, J. Catalytic performance of Ni-Cu/Al2O3 for effective syngas production by methanol steam reforming. *Fuel* **2018**, *232*, 672–683. [CrossRef]
30. Zhao, X.X.; Lu, G.X. Modulating and controlling active species dispersion over Ni-Co bimetallic catalysts for enhancement of hydrogen production of ethanol steam reforming. *Int. J. Hydrog. Energy* **2016**, *41*, 3349–3362. [CrossRef]
31. Ardiyanti, A.R.; Khromova, S.A.; Venderbosch, R.H.; Yakovlev, V.A.; Heeres, H.J. Catalytic hydrotreatment of fast-pyrolysis oil using non-sulfided bimetallic Ni-Cu catalysts on a delta-Al_2O_3 support. *Appl. Catal. B Environ.* **2012**, *117*, 105–117. [CrossRef]
32. Leng, S.; Wang, X.; He, X.; Liu, L.; Liu, Y.E.; Zhong, X.; Zhuang, G.; Wang, J.G. NiFe/gamma-Al_2O_3: A universal catalyst for the hydrodeoxygenation of bio-oil and its model compounds. *Catal. Commun.* **2013**, *41*, 34–37. [CrossRef]
33. Nie, L.; de Souza, P.M.; Noronha, F.B.; An, W.; Sooknoi, T.; Resasco, D.E. Selective conversion of m-cresol to toluene over bimetallic Ni-Fe catalysts. *J. Mol. Catal. A Chem.* **2014**, *388*, 47–55. [CrossRef]
34. Han, Q.; Rehman, M.U.; Wang, J.; Rykov, A.; Gutiérrez, O.Y.; Zhao, Y.; Wang, X.; Ma, X.; Lercher, J.A. The synergistic effect between Ni sites and Ni-Fe alloy sites on hydrodeoxygenation of lignin-derived phenols. *Appl. Catal. B Environ.* **2019**, *253*, 348–358. [CrossRef]
35. Huynh, T.; Armbruster, U.; Kreyenschulte, C.; Nguyen, L.; Phan, B.; Nguyen, D.; Martin, A. Understanding the Performance and Stability of Supported Ni-Co-Based Catalysts in Phenol HDO. *Catalysts* **2016**, *6*, 176. [CrossRef]

36. Zhang, Z.; Pei, Z.; Chen, H.; Chen, K.; Hou, Z.; Lu, X.; Ouyang, P.; Fu, J. Catalytic in-Situ Hydrogenation of Furfural over Bimetallic Cu-Ni Alloy Catalysts in Isopropanol. *Ind. Eng. Chem. Res.* **2018**, *57*, 4225–4230. [CrossRef]
37. Moraes, M.S.A.; Migliorini, M.V.; Damasceno, F.C.; Georges, F.; Almeida, S.; Zini, C.A.; Jacquesa, R.A.; Caramão, E.B. Qualitative analysis of bio oils of agricultural residues obtained through pyrolysis using comprehensive two dimensional gas chromatography with time-of-flight mass spectrometric detector. *J. Anal. Appl. Pyrol.* **2012**, *98*, 51–64. [CrossRef]
38. Takanabe, K.; Nagaoka, K.; Nariai, K.; Aika, K.I. Titania-supported cobalt and nickel bimetallic catalysts for carbon dioxide reforming of methane. *J. Catal.* **2005**, *232*, 268–275. [CrossRef]
39. Singh, S.; Srivastava, P.; Singh, G. Synthesis, characterization of Co-Ni-Cu trimetallic alloy nanocrystals and their catalytic properties, Part-91. *J. Alloy Compd.* **2013**, *562*, 150–155. [CrossRef]
40. Ao, M.; Pham, G.H.; Sage, V.; Pareek, V.; Liu, S.M. Perovskite-derived trimetallic Co-Ni-Cu catalyst for higher alcohol synthesis from syngas. *Fuel Process. Technol.* **2019**, *193*, 141–148. [CrossRef]
41. Lua, A.C.; Wang, H.Y. Hydrogen production by catalytic decomposition of methane over Ni-Cu-Co alloy particles. *Appl. Catal. B Environ.* **2014**, *156*, 84–93. [CrossRef]
42. Shi, T.B.; Li, H.; Yao, L.H.; Ji, W.J.; Au, C.T. Ni-Co-Cu supported on pseudoboehmite-derived Al$_2$O$_3$: Highly efficient catalysts for the hydrogenation of organic functional groups. *Appl. Catal. A Gen.* **2012**, *425*, 68–73. [CrossRef]
43. Miranda, B.C.; Chimentao, R.J.; Szanyi, J.; Braga, A.H.; Santos, J.B.; Gispert-Guirado, F.; Llorca, J.; Medina, F. Influence of copper on nickel-based catalysts in the conversion of glycerol. *Appl. Catal. B Environ.* **2015**, *166*, 166–180. [CrossRef]
44. Chen, L.; Zhu, Q.S.; Wu, R.F. Effect of Co-Ni ratio on the activity and stability of Co-Ni bimetallic aerogel catalyst for methane Oxy-CO$_2$ reforming. *Int. J. Hydrog. Energy* **2011**, *36*, 2128–2136. [CrossRef]
45. Choi, H.Y.; Lee, W.Y. Effect of second metals and Cu content on catalyst performance of Ni-Cu/SiO$_2$ in the hydrodechlorination of 1,1,2-trichloroethane into vinyl chloride monomer. *J. Mol. Catal. A Chem.* **2001**, *174*, 193–204. [CrossRef]
46. Chen, T.; Sun, Y.; Guo, M.; Zhang, M. Hydrothermal synthesis of Ni-Co-Cu alloy nanoparticles from low nickel matte. *J. Alloy Compd.* **2018**, *766*, 229–240. [CrossRef]
47. 47. Mortensen, P.M.; Grunwaldt, J.D.; Jensen, P.A.; Jensen, A.D. Screening of Catalysts for Hydrodeoxygenation of Phenol as a Model Compound for Bio-oil. *Acs Catal.* **2013**, *3*, 1774–1785. [CrossRef]
48. Zhao, C.; Kasakov, S.; He, J.Y.; Lercher, J.A. Comparison of kinetics, activity and stability of Ni/HZSM-5 and Ni/Al$_2$O$_3$-HZSM-5 for phenol hydrodeoxygenation. *J. Catal.* **2012**, *296*, 12–23. [CrossRef]
49. Mortensen, P.M.; Grunwaldt, J.D.; Jensen, P.A.; Jensen, A.D. Influence on nickel particle size on the hydrodeoxygenation of phenol over Ni/SiO$_2$. *Catal. Today* **2016**, *259*, 277–284. [CrossRef]

© 2019 by the authors. Licensee MDPI, Basel, Switzerland. This article is an open access article distributed under the terms and conditions of the Creative Commons Attribution (CC BY) license (http://creativecommons.org/licenses/by/4.0/).

Article

Response Surface Methodology for the Optimization of Keratinase Production in Culture Medium Containing Feathers by *Bacillus* sp. UPM-AAG1

Aa'ishah Abdul Gafar, Mohd Ezuan Khayat, Siti Aqlima Ahmad, Nur Adeela Yasid and Mohd Yunus Shukor *

Department of Biochemistry, Faculty of Biotechnology and Biomolecular Sciences, Universiti Putra Malaysia, Serdang 43400, Malaysia; aishababa93@gmail.com (A.A.G.); m_ezuan@upm.edu.my (M.E.K.); aqlima@upm.edu.my (S.A.A.); adeela@upm.edu.my (N.A.Y.)
* Correspondence: mohdyunus@upm.edu.my; Tel.: +603-97696722

Received: 17 June 2020; Accepted: 7 July 2020; Published: 29 July 2020

Abstract: Keratinase is a type of proteolytic enzyme with broad application in industry. The main objective of this work is the optimization of keratinase production from *Bacillus* sp. strain UPM-AAG1 using Plackett-Burman (PB) and central composite design (CCD) for parameters, such as pH, temperature, feather concentration, and inoculum size. The optimum points for temperature, pH, and inoculum and feather concentrations were 31.66 °C, 6.87, 5.01 (w/v), and 4.53 (w/v), respectively, with an optimum keratinase activity of 60.55 U/mL. The keratinase activity was further numerically optimized for commercial application. The best numerical solution recommended a pH of 5.84, temperature of 25 °C, inoculums' size of 5.0 (v/v), feather concentration of 4.97 (w/v). Optimization resulted an activity of 56.218 U/mL with the desirability value of 0.968. Amino acid analysis profile revealed the presence of essential and non-essential amino acids. These properties make *Bacillus* sp. UPM-AAG1 a potential bacterium to be used locally for the production of keratinase from feather waste.

Keywords: RSM; numerical optimization; keratinase; feather; *Bacillus* sp.; amino acids

1. Introduction

Keratinase (EC 3.4.99.11) is a type of protease enzyme that started to gain interest due to its broad application in industry. They are commonly extracellular inducible enzymes secreted by various microorganism in the medium containing keratin showing high substrate specificity toward keratin [1]. They are widely used in most of the biotechnological processing industry, mainly in feed formulation, nitrogen fertilizer, leather processing, and pharmaceutical industry [2–4].

In food and feed supplements, keratinase-treated feather is increasingly seen as a viable source of dietary protein, as the enzyme-treated final product preserved good nutritional value. Keratinases are expected to develop a significant total global demand comparable to other commercial proteases. Diverse class of keratinase have been isolated from various microbial populations, such as bacteria [5,6] actinomycetes [7,8], and fungi [9,10]. However, among bacteria, keratinase from *Bacillus* genera has been widely reported as keratinase from this genera and appears to be the most promising keratinase producer for commercial application [11,12]. In general, the reasons why *Bacillus* spp. are preferred in bioremediation and industrial biotechnology are due to their generally regarded as safe (GRAS) property and the capacity of selected *Bacillus* strains to produce and secrete large quantities (20–25 g/L) of extracellular enzymes [13]. This is also the reason as to why an increase in the number of reports on the isolation of keratin-degrading *Bacillus* spp. is on the rise.

Numerous commercial keratinases are from *Bacillus* spp., such as Versazyme from *B. licheniformis* PWD-1 (Odetallah et al. 2005); Prionzyme (Genencor) and Cibenza DP100™, both also from *B. licheniformis* PWD-1; Esperase and Savinase (Novozymes A/S), both from *Bacillus* spp.; and Alcalse (Novozymes A/S) from *B. licheniformis* [14]. We have been approached by a small feather-processing company interested in feather-degrading technology with the main target in producing keratinase at ambient temperature (25 to 32 °C) without utilizing additional nitrogen sources and heating process to lower the cost. There are many *Bacillus* spp. keratin-degrading bacteria reported in the literature, but most require additional nitrogen sources, such as yeast extract, peptone, soybean meal, ammonium ions, and soy flour [4,13,15–25], that may elevate the cost. Scouring through the literature shows that only two *Bacillus* spp. keratin-degrading bacteria fits the bill with chicken feather as the sources of carbon and nitrogen, but both required elevated temperatures 37 °C [26] and 50 °C [27] for optimum activity. In light of the current Covid-19 pandemic, where imported products have great difficulties in being available, sometimes months at a time, this mean that local sources need to be developed.

The objectives of this work are to optimize keratinase production using feather as the only source of carbon and nitrogen and to numerically select the best conditions to maximize keratinase activity under ambient temperature and maximum feather concentration as a substrate. In this work, we report the optimization via response surface method (RSM) followed by a numerical optimization of a *Bacillus* sp. keratin-degrading bacterium having optimum ambient temperature for growth with chicken feather as the sole carbon and nitrogen sources.

2. Results

2.1. Isolation and Screening of Keratinase Producing Bacterium

In the present study, five prevalent colonies are that competent for sustainable growth on feather meal agar (FMA) were successfully isolated based on hydrolysis zone on FMA indicate the use of feather keratin as both carbon and nitrogen sources. The four isolates were able to utilize keratin in FMA for its growth. The morphology of each isolate is shown in Table 1. For further analysis, all pure strains were subjected to endospore screening for the best keratinase producing *Bacillus*. For this purpose, endospore-forming species was confirmed by the formation of green-colored spore after staining with malachite green and safranin. Among the tested isolate, three isolates were spore positive within 2 days of incubation in sporulation media signifying a potential member of *Bacillus* sp. The isolate was isolated UPM-AAG1, UPM-AAG6, and UPM-AAG14. A further screening process to select the highest keratinase producer was conducted based on bacterial growth and keratinase activity in 1% feather as sole carbon and nitrogen sources. The result suggests that the highest keratinolytic activity was isolate UPM-AAG1 (35.23 U/mL), followed by isolate UPM-AAG14 (33.97 U/mL), while isolate UPM-AAG6 resulted in the lowest keratinase production at only 25.56 U/mL for the same incubation time. The bacterial growth showed the same pattern where isolate UPM-AAG1 gave the highest bacterial count at 7.771 Log Colony Forming Unit or CFU/mL followed by isolate UPM-AAG6 at 7.628 Log CFU/mL and isolate UPM-AAG14 at 7.573 Log CFU/mL (Figure 1). Based on the results, isolate UPM-AAG1 was subjected further for identification study.

Table 1. Morphology of isolated microorganism.

Isolate	Morphology
UPM-AAG1	Circular, White, Dry, Flat
UPM-AAG6	Irregular, Dry, White, Flat
UPM-AAG14	Irregular, Dry, White, Flat
UPM-AAG15	Irregular, Dry, White, Flat
UPM-AAG16	Irregular, Dry, White, Flat

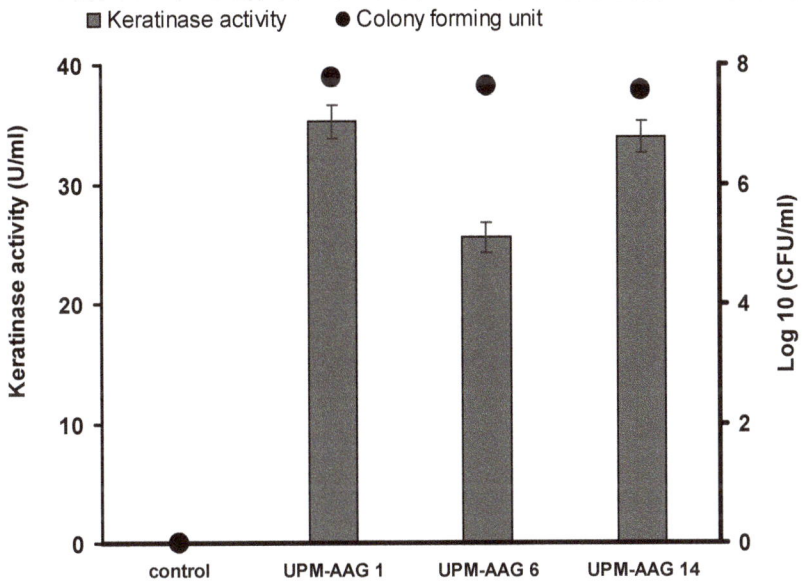

Figure 1. Screening result of three isolates with 1% feather. Error bars represent mean ± standard deviation (n = 3).

2.2. Identification of Keratinolytic Microorganism

Micromorphology of isolate AAG1 was examined microscopically and demonstrated rod-shaped blue color bacterial cells, signifying their Gram-positive characteristic. Biochemical analysis showed positive results towards oxidase, catalase, Voges-Proskauer, and citrate test but negative result towards nitrate production. Further identification was supported by the 16S rRNA sequencing. BLASTn result showed that isolate AAG1 belonged to the *Bacillus* genus with high similarity percentage of (>99%). The phylogenetic tree constructed using partial 16S rRNA sequence and *Escherichia coli* strain U5/41 as the outgroup demonstrated that isolate AAG1 was not attached to any know species in the clade. However, bootstrap result AAG1 shows sequence similarity to *Bacillus safensis* strain FO-36b, *Bacillus pumilis* strain ATCC 7061, *Bacillus pumilis* strain SBMP2, and *Bacillus stratosphericus* strain 41KF2a with a bootstrap value of 78% (Figure 2). Therefore, UPM-AAG1 isolate was identified as *Bacillus* sp. strain UPM-AAG1 and deposited in the GenBank with the Accession No. MK285608.1.

2.3. Optimization of Keratinase Activity Using Plackett Burman and Response Surface Methodology

2.3.1. Pre-Screening of Significant Parameters Using Plackett-Burman

Four independent factors (i.e., inoculum size (v/v), feather concentration (w/v), pH, and temperature) were screened to evaluate their effects on keratinase production using Plackett-Burman design. A total of 12 experimental variables generated using software screening for keratinase production, and their corresponding responds, as shown in Table 2a. The adequacy of the model was calculated using ANOVA analysis and presented in Table 2b. The model F value 70.33 indicates the model is significant with only 0.25% chance that a "Model F-value" this large could occur due to noise. The factors with $p < 0.05$ were considered to have a significant effect on the response. As presented in the table, all factors—temperature, inoculum size (v/v), pH, and feather concentration (w/v)—exert a positive effect on the model. Therefore, all four significant factors screened were further brought into the central composite design.

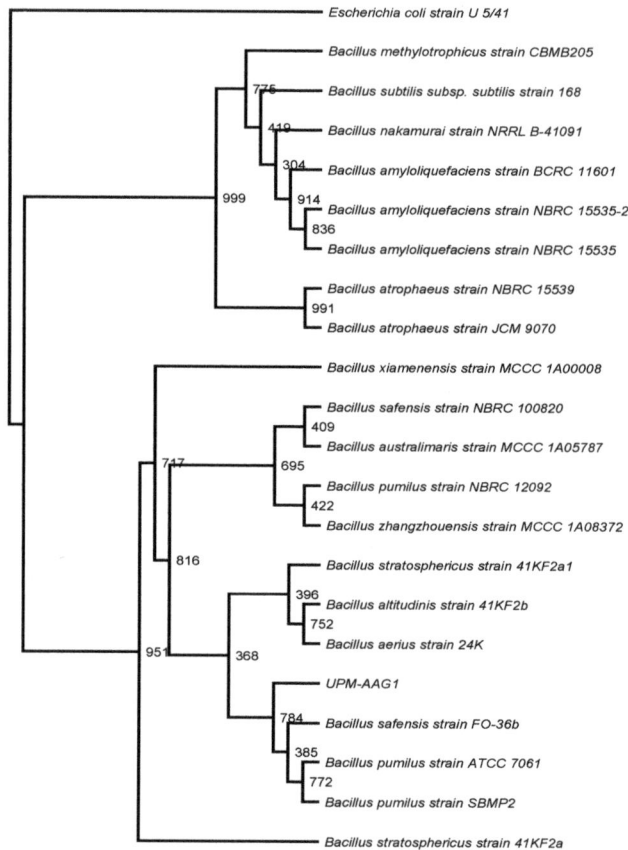

Figure 2. Phylogram (neighbor-joining method) showing the genetic relationship between strain UPM-AAG1 and other related reference micro-organisms based on the 16S rRNA gene sequence analysis. Species names are followed by the strain of their 16S rRNA sequences. The numbers at branching points or nodes refer to bootstrap values, based on 1000 resamplings (GenBank MK285608.1).

Table 2. Prescreening of significant parameters using Plackett-Burman design matrix with keratinase activity as the response (±standard deviation, n = 3).

Run	Factors				Keratinase Activity (U/mL)
	Temperature (°C)	Inoculum (v/v)	pH	Feather Concentration (w/v)	
	A	B	C	D	
1	25.00	10.00	5.00	1.00	6.4 ± 0.25
2	25.00	10.00	8.00	5.00	8.7 ± 0.07
3	35.00	5.00	8.00	1.00	7.6 ± 0.13
4	35.00	10.00	5.00	5.00	9.7 ± 0.26
5	35.00	10.00	8.00	1.00	7.7 ± 0.01
6	35.00	10.00	5.00	5.00	9.3 ± 0.11
7	25.00	5.00	5.00	1.00	8.3 ± 0.14
8	25.00	5.00	5.00	5.00	12.5 ± 0.28
9	25.00	10.00	8.00	1.00	5.8 ± 0.1
10	35.00	5.00	5.00	1.00	12 ± 0.16
11	25.00	5.00	8.00	5.00	10.5 ± 0.01
12	35.00	5.00	8.00	5.00	7.8 ± 0.26

Table 2. Cont.

Source	Factors	F-Value	p-Value		Remarks
Model		70.33	0.0025		Significant
A	Temperature	16.98	0.0259		Significant
B	Inoculum	63.75	0.0041		Significant
C	pH	179.38	0.0009		Significant
D	Feather	72.24	0.0034		Significant
				Value	
R^2				0.9947	
Adjusted R^2				0.9806	
Predicted R^2				0.8955	
Adeq Precision				26.980	

2.3.2. Optimization of Significant Variables Using Central Composite Design (CCD)

CCD was used to determine the optimum condition of the four selected significant variables (temperature, inoculum, pH, and feather concentration) for keratinase production using keratinase activity as the output response. A total of 30 experiments with different combinations of the four selected variables were performed. The experimental designs used are shown in Table 3.

Table 3. Optimization of keratinase activity by strain AAG-1 using central composite design (CCD) with six center points showing observed and predicted values (±standard deviation, n = 3).

Run Order	X1: Temperature	X2: Inoculum	X3: pH	X4: Feather (w/v)	Keratinase Activity (U/mL)	
					Experimental Value	Predicted Value
1	35.00	5.00	5.50	5.00	42 ± 1.87	42.42
2	25.00	5.00	5.50	5.00	42.3 ± 1.75	46.18
3	30.00	7.50	6.75	3.00	41.1 ± 1.74	38.85
4	30.00	7.50	6.75	3.00	48 ± 0.28	48
5	35.00	5.00	5.50	1.00	22.5 ± 1.62	22.56
6	35.00	10.00	8.00	1.00	31.9 ± 1.57	29.97
7	35.00	10.00	5.50	5.00	38.4 ± 0.76	40.99
8	30.00	7.50	4.25	3.00	55.7 ± 0.49	53.8
9	25.00	10.00	8.00	5.00	50.5 ± 1.81	54.68
10	35.00	5.00	8.00	1.00	56.6 ± 0.51	54.34
11	25.00	5.00	8.00	1.00	14.9 ± 0.76	17.15
12	35.00	10.00	8.00	5.00	20 ± 1.02	22.21
13	25.00	5.00	8.00	5.00	33.1 ± 0.43	33.42
14	20.00	7.50	6.75	3.00	32.2 ± 0.86	36.73
15	40.00	7.50	6.75	3.00	19.5 ± 0.54	17.9
16	35.00	5.00	8.00	5.00	26.7 ± 0.86	26.6
17	30.00	12.50	6.75	3.00	20.2 ± 1.91	18.52
18	25.00	10.00	5.50	5.00	31.9 ± 0.11	30.98
19	30.00	7.50	6.75	-1.00	60.1 ± 1	56.8
20	25.00	10.00	8.00	1.00	42.4 ± 0.61	43.1
21	25.00	5.00	5.50	1.00	47.4 ± 0.84	44.48
22	30.00	2.50	6.75	3.00	28.7 ± 0.54	29.02
23	30.00	7.50	6.75	3.00	41.6 ± 1.12	42.47
24	35.00	10.00	5.50	1.00	31 ± 1.11	27.53
25	30.00	7.50	6.75	3.00	45.1 ± 0.75	49.2
26	30.00	7.50	6.75	3.00	44 ± 1.87	49.2
27	30.00	7.50	6.75	3.00	50.6 ± 0.8	49.2
28	30.00	7.50	9.25	3.00	52.4 ± 1.12	49.2
29	30.00	7.50	6.75	7.00	51.1 ± 1.69	49.2
30	25.00	10.00	5.50	1.00	52 ± 1.67	49.2

The responses were studied using four independent variables with six center point showing both observed and predicted values for keratinase activity. The multiple regression analysis of the observed responses resulted in the below quadratic equation:

Keratinase Activity = + 47.10 + 3.65*A + 0.65* B − 9.70*C − 8.72*D − 6.11*A2 − 0.78*B2 + 0.75*C2 − 3.55*D2 + 0.46*A*B + 2.70*A*C − 1.02*A*D + 5.50*B*C − 0.18*B*D − 16.97*C*D,

where A, B, C, and D, each represent concentrations (coded values) of temperature, pH, inoculum, and feather concentrations, respectively. From Table 4, it can be observed that all four linear terms (A, B, C, D), three squared terms (A^2, C^2, D^2), and two quadratic terms (BC and CD) of the model were significant to the response, suggesting that keratinase production highly depends on the interactions between these factors.

Table 4. ANOVA analysis of CCD for optimization of keratinase activity by *Bacillus* sp. strain UPM-AAG1.

Source	Sum of Squares	DF	Mean Square	F Value	Prob > F	
Model	4308.841	14	307.7743	23.79177	<0.0001	significant
A	191.1947	1	191.1947	14.77986	0.0016	
B	21.89327	1	21.89327	1.692408	0.2129	
C	235.7161	1	235.7161	18.22148	0.0007	
D	1089.714	1	1089.714	84.23776	<0.0001	
A2	1024.804	1	1024.804	79.22009	<0.0001	
B2	265.7186	1	265.7186	20.54075	0.0004	
C2	0.964286	1	0.964286	0.074542	0.7886	
D2	345.6686	1	345.6686	26.7211	0.0001	
AB	13.3225	1	13.3225	1.029865	0.3263	
AC	29.16	1	29.16	2.254145	0.1540	
AD	16.81	1	16.81	1.299458	0.2722	
BC	484	1	484	37.41448	<0.0001	
BD	1.96	1	1.96	0.151513	0.7026	
CD	1152.603	1	1152.603	89.09923	<0.0001	
Residual	194.0425	15	12.93617			
Lack of Fit	126.5425	10	12.65425	0.937352	0.5669	not significant
Pure Error	67.5	5	13.5			
Cor Total	4502.883	29				
Std. Dev.	3.596688		R-Squared	0.956907		
Mean	39.13		Adj R-Squared	0.916687		
C.V.	9.191639		Pred R-Squared	0.816543		
PRESS	826.0848		Adeq Precision	15.58869		

Based on the coded value below, the effects of inoculum and feather concentrations outweigh the effect of other factors.

Final equation in terms of actual factors:

Keratinase Activity = −114.67125 + 13.59533*Temperature − 7.93667*pH − 13.85500*Inoculum + 29.97458 * Feather − 0.24450*Temperature2 − 0.49800*pH2 + 0.12000*Inoculum2 − 0.88750*Feather 2 + 0.073000*Temperature*pH + 0.21600*Temperature*Inoculum − 0.10250*Temperature*Feather + 1.7600*pH*Inoculum − 0.070000*pH*Feather − 3.39500*Inoculum*Feather.

The predicted model was assessed further by RSM analysis. The 3D response plot for keratinase activity represents the interaction between two parameters at a time, while fixing the other parameter at zero levels (constant) for maximum keratinase production (Figure 3a–f). The predicted optimum points for temperature, pH, and inoculum and feather concentrations were 31.66 °C, 6.87, 5.01 (w/v), and 4.53 (w/v), respectively, with an optimum keratinase activity of 60.5539 U/mL. Verification of the value obtained showed a close value of 60.02 U/mL indicating good agreement.

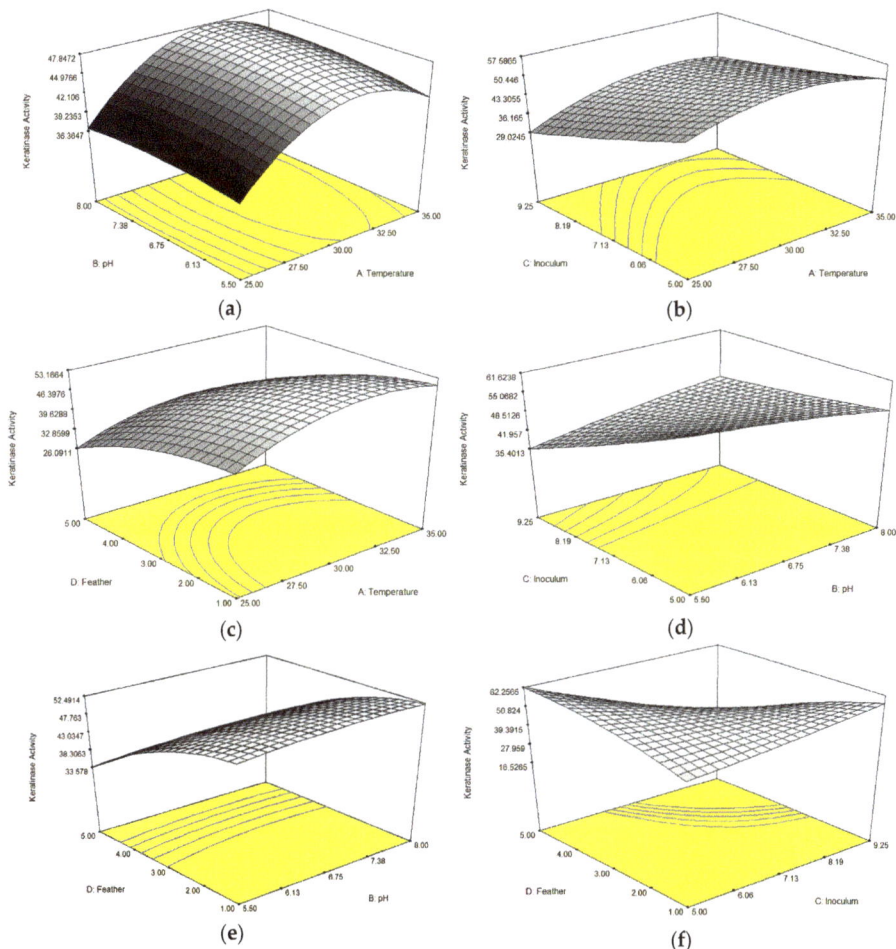

Figure 3. Response surface 3D plot showing the interaction of factors affecting keratinase production (a) pH and temperature, (b) inoculum size and temperature, (c) feather concentration and temperature, (d) pH and inoculum size, (e) pH and feather concentration, and (f) feather concentration and inoculum size.

Data fitness into the selected model was examined using diagnostic model plots (Supplementary Figure S2a–d). The plots are especially important in the evaluation of data error which varies from model predictions, which helps to assess and improve model adequacy. The actual versus predicted response plot obtained from the experiment (Figure S2a) showed a similar relationship between the predicted and actual values as the data points were clustered near the line dividing the plot into identical halves (45°). Plotting the predicted values and studentized residuals (Figure S2b) further verified the suitability of the model. Studentized residues are utilized to indicate differences between the predicted value and the actual model responses. The experimental data exhibit slight or no abnormality based on visual observation of the normal probability plot (Figure S2c). To visualize the distantly standout standard deviation, an outlier plot (Figure S2d) can show the presence of outlier(s). The result shows that the data falls between 3.5 and −3.5, suggesting the absence of outlier.

2.3.3. Numerical Optimization

As the company requested minimum costs to for the developed system, a numerical optimization was calculated so that the best conditions under the following criteria (Table 5)—minimum temperature, pH within range, minimum inoculum, maximum substrate (feather), and maximum keratinase activity—were obtained. Under the constraint criteria selected, ten solutions were obtained, and the best solution recommended was as follows.

Table 5. Numerical optimization for selected criteria for keratinase activity by *Bacillus* sp. strain UPMAGG-1.

Name	Goal	Lower Limit	Upper Limit	Lower Weight	Upper Weight	Importance
Temperature	minimize	25	35	1	1	3
pH	is in range	5.5	8	1	1	3
Inoculum	minimize	5	10	1	1	3
Feather	maximize	1	5	1	1	3
Keratinase Activity	maximize	14.9	60.1	1	1	5

For verification purposes, a series of validation experiment was conducted based on the conditions provided by CCD for optimum keratinase production (Table 6). Based on the provided solution, the highest keratinase activity obtained through solution 1 with a pH of 7.00, temperature 25.00, inoculums' size of 5.0 (v/v), feather concentration 4.97 (w/v) resulted in an activity of 56.218 U/mL with the desirability value of 0.968.

Table 6. Suggested value for each variable for optimum keratinase activity by *Bacillus* sp. strain UPMAGG-1.

Number	Temperature	pH	Inoculum	Feather	Keratinase Activity	Desirability
1	25.05	7.00	5.00	5.00	56.218	0.968

2.4. Amino Acid Profile of Hydrolysate of Bacillus sp. UPM-AAG1 Using High-Performance Liquid Chromatography (HPLC)

Amino acid analysis profile of the keratinase from *Bacillus* sp. strain UPM-AAG1 (Figure 4) revealed the presence of 17 different amino acids, including essential amino acids histidine, isoleucine, leucine, lysine, methionine, phenylalanine, threonine and valine, and non-essentials amino acids, like aspartic acid, glutamic, glycine, alanine, cysteine, tyrosine, arginine, serine, and proline, as evident from the HPLC chromatogram (Figure 5). The sequence is largely composed of phenylalanine (65.73 µmol/mL), isoleucine (24.04 µmol/mL), and lysine (20.14 µmol/mL) as essential amino acids and glutamine (32.48 µmol/mL), glycine (60.47 µmol/mL), and serine (158.42 µmol/mL) as a non-essential amino acid.

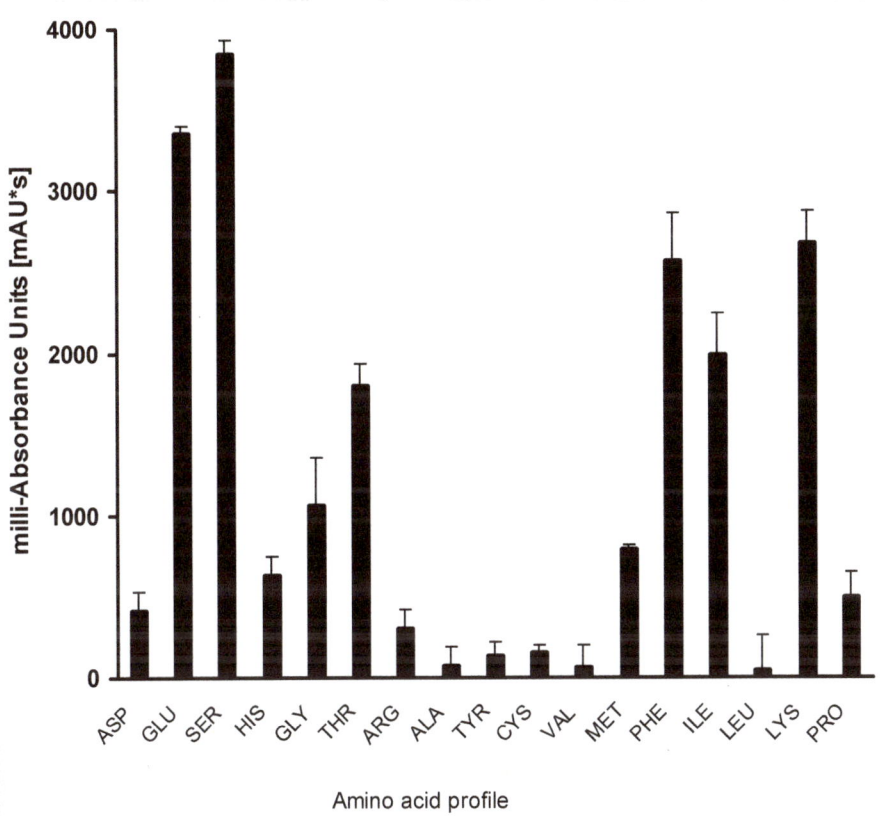

Figure 4. Amino acid profile of *Bacillus* sp. UPM-AAG1 hydrolysate. Error bars represent mean ± standard deviation (n = 2).

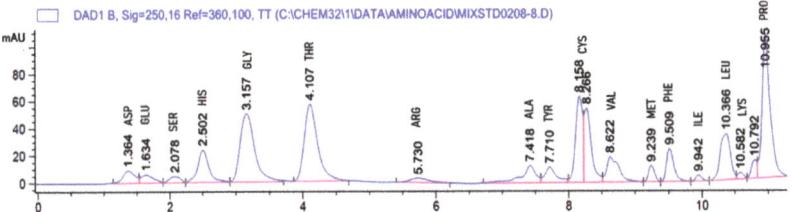

Figure 5. High-performance liquid chromatography (HPLC) chromatogram of chicken feather degradation lysate by *Bacillus* sp. strain UPMAGG-1.

3. Discussion

Keratinase is a protease that is very robust with broad application in industry. In this work, isolation of a new potential keratinase producer from the *Bacillus* genera due to its high keratinolytic activity [28]. Five colonies were successfully isolated from poultry waste and soil in Selangor using heat treatment method for the selection of spore-forming *Bacillus* species. As one of the exclusivities of *Bacillus* species is characterized by endospore formation, heat treatment is the most common and simplest method use to infer the presence of spore-forming *Bacillus* [29,30]. Under extreme environment, spore-forming *Bacillus* spp. develop endospore; a metabolically inactive dormant cell to protect themselves against the harsh environment [31]. Once the environment returns to favorable conditions for growth, the cell will proceed with its vegetative life cycle and continue to germinate. The purpose of sporulation in this experiment is to kill other vegetative cells, leaving only dormant endospore cells to survive. Of all five cultures tested, only isolate UPM-AAG1 shows positive growth response in FMA, spore positive, and highest keratinase activity, as well as the highest bacterial growth count; hence, this isolate was selected for further studies. This method of isolation of keratinase producing organism has been reported before [32], where feather as a substrate was used as sole carbon and nitrogen source. A few keratinolytic *Bacillus* spp. that utilize feather solely as carbon and nitrogen sources include *Bacillus pumilus* GRK [26] and *Bacillus cereus* LAU 08 [27].

Physiological and biochemical identification of isolate AAG1 revealed a rod-shaped structure indicating their Gram-positive characteristic. Gram-positive keratinase producer is not exclusive to *Bacillus* spp. as other Gram-positive bacteria have been reported, including *BreviBacillus parabrevis* [33], *Micrococcus luteus*, and *Actinobacter* sp. [34]. On the basic of 16srDNA sequencing and phylogenetic analysis, the keratinolytic isolate UPM-AAG1 was tentatively identified as *Bacillus* sp. strain UPM-AAG1. Many of the major commercial keratinolytic bacteria come from the genus *Bacillus* spp. [28,35], chiefly due to its generally regards as safe (GRAS) property [36]. The keratinolytic bacteria in this study were isolated from a poultry farm environment, and numerous keratinolytic bacteria have been isolated from poultry farm soils, such as *Bacillus subtilis* DP [25] and five keratinolytic strains of *Bacillus* spp. [37], to name a few, making them predictable as the feather-contaminated soils offer rich sources of keratin for selective enrichment [25,38].

Despite being the most studied and widely documented, the major drawback in keratinase study is to optimize keratinase production using feather keratin as the sole carbon and nitrogen sources. Addition of supplements will incur a high cost when production is scaled up (Table 7).

Table 7. Summary of keratinase production by *Bacillus* spp.

Organism	Optimization Model	Optimum Temperature (°C)	Keratinase Activity (U/mL)	Substrate during Optimization	Carbon Sources during Optimization	Nitrogen Sources during Optimization	Time (h)	References
Bacillus pumilus A1	OFAT Plackett-Burman CCD	30	87.73 U/mL	heat-treated chicken feather meal	heat-treated chicken feather meal	peptone	24	[39]
Bacillus subtilis P13	OFAT Plackett-Burman Box–Behnken	room	2.07 U/mL	soybean meal	soybean meal	soybean meal	24	[17]
Bacillus sp. RKY3	Plackett-Burman CCD	-ns-	939 U/mL	corn starch, corn steep liquor	corn starch	corn steep liquor	24	[16]
Bacillus thuringiensis TS2	OFAT	50	90.78 U/mL	feather meal	starch	yeast extract	96	[22]
B. Subtilis KD-N2	OFAT	23	60.9 U/mL	feather	sucrose	feather	30	[40]
Bacillus subtilis DP1	OFAT	37	379.65 U/mL	feather coffee	feather coffee	feather coffee	96	[25]
Bacillus subtilis (MTCC9102)	OFAT	37	15.972 U/mL	horn meal	dextrose	peptone	48	[41]
Bacillus licheniformis ALW1	OFAT	65	72.2 U/mL	native feather	galactose	corn steep liquor	96	[4]
Bacillus pumilus GRK	OFAT	37	373 U/mL	feather	feather	feather	24	[26]
Bacillus subtilis AMR	OFAT	50	163 U/mL	human hair	yeast extract	yeast extract	192	[42]
Bacillus cereus LAU 08	OFAT	50	51.7 U/mL	feather powder	feather powder	feather powder	72	[27]
Bacillus licheniformis	OFAT	50	11 U/mL	feather meal	feather meal	feather meal	32	[20]
B. subtilis BLBC17	CCD	33	170 U/mL	soybean meal	soybean meal	yeast extract	48	[18]

Table 7. Cont.

Organism	Optimization Model	Optimum Temperature (°C)	Keratinase Activity (U/mL)	Substrate during Optimization	Carbon Sources during Optimization	Nitrogen Sources during Optimization	Time (h)	References
Bacillus licheniformis ER-15:	OFAT PB CCD	70	1962 U/mL	feather	glucose	soy flour	48	[43]
Bacillus subtilis S14	OFAT CCD	50	5.5 U/mL	feather meal	feather meal	feather meal	24	[23]
Bacillus cereus Wu2	OFAT	30	1750 U/mL	chicken feather powder	chicken feather powder	nh4cl	96	[19]
Bacillus weihenstephanensis	OFAT	40	15.3 U/mL	chicken feather	cellulose	(NH4)2 SO4	168	[21]
Bacillus pumilus FH9	OFAT	37	647 U/mL	chicken feather	chicken feather	NH4CL yeast extract	48	[15]
Bacillus sp. 5 MG-MASC-BT	BBD	55	1075 U/mL	alkali-treated soluble keratin	alkali-treated soluble keratin	alkali-treated soluble keratin	60	[24]
Bacillus licheniformis RPk	OFAT	60	37.35 U/mL	chicken feather	chicken feather	yeast extract	-	[44]
B. subtilis 1273	-	-	412 U/mL	feather meal	feather meal	feather meal	168	[45]
Bacillus sp. UPM-AAG1	PB CCD	30	60.1 U/mL	chicken feather	chicken feather	chicken feather	24	Current study

In keratinase research, the main objective is to maximize keratinase production through manipulating external and internal parameters [46]. Most of the optimization studied involving keratinase involved conventional optimization through one factor-at-a-time (OFAT) [22,41,47] or both conventional and statistical approach [17,48] but with non-keratin carbon and nitrogen sources (Table 7). By carrying out optimization relying on no additional supplements, as well as optimizing process at ambient temperature suiting Malaysia (from 24 to 32 °C), will increase the chances of a successful keratinase production by local small and medium enterprise (SME) companies. To date, only very few keratinase-producing *Bacillus* spp. bacteria have been optimized using feather as the sole carbon and nitrogen sources (Table 7). Additional C and N sources supplementation may not work during actual feather degradation as the augmented bacterium may choose to utilize the easily assimilable C and N source rather than the feather itself. In addition, competition with the easily assimilable C and N sources by indigenous bacteria may outcompete the augmented bacterium resulting in a lower production of keratinase and poor degradation of feather waste [49]. Compared to many keratin-degrading *Bacillus* spp. (Table 7), *Bacillus* sp. UPM-AAG1 produce a relatively good keratinase activity (60.1 U/mL) in a shorter time (24 h) at 30 °C, features that make this bacterium suitable for the requirement of the local SME company where keratinase production should be optimum or acceptable activity at ambient temperature. However, the applicability of this strain in real world conditions need to be tested, and this remains the limit of this work.

RSM CCD's result showed that all factors—temperature, inoculum size (v/v), pH, and feather concentration (w/v)—exert positive effects to the model with feather concentration forming a major contributor. This result is similar to Yusuf et al. [48], where feather concentration was found to be the most significant substrates for keratinase production when Plackett-Burman (PB) was used in the screening process. Apart from that, Govarthanan et al. [24] also reported the same result where a significant increase in keratinase production was observed when feather was used as substrate. The inoculum size was reported to give significant effect towards keratinase production in *Bacillus licheniformis* ER-15 [43]. This is because inoculum size significantly affects the growth profile of aerobic microorganism. Further, a neutral to alkaline pH were reported to promote keratinase production in various microorganism [50,51] with the exception of a few including *Bacillus subtilis* [52] where the highest activity occurs at acid to neutral pH range (pH 5–7). Apart from that, temperature also plays an important role in the production of keratinase enzyme. Generally, most of the reported keratinases work optimally in between 28 to 50 °C [24,39]. The ANOVA analysis result of keratinase activity obtained through RSM indicated that the model is adequate with a correlation coefficient, R^2 of the model was 0.9569 and an adjusted R^2 value of 0.9167 showing a high correlation between the experimental data design (Table 5). The nearer the R^2 value to 1, the better the accuracy of the model. The "Pred R-Squared" of 0.8165 was in consistent agreement to the "Adj R-Squared" of 0.9167 indicating an acceptable degree of correlation between the observed value and predicted values [53], although "Pred R-Squared" value of >0.9 is more desirable in many cases [54]. A ratio > 4 for the Adeq Precision value is sought and the result from this study with a value of 15.586 indicates a good signal to noise ratio [55]. The large lack of fit F value is normally sough and, with a value of 0.94, suggests an insignificant lack of fit relative to the pure error [56,57]. The *p*-value for the lack of fit value was 0.5666, and this demonstrated the model appropriateness for the optimal region. The Model F-value of 23.79 implies the model is significant. There is only a 0.01% chance that "Model F-value" this large could occur due to noise. Values of "Prob > F" less than 0.0500 indicate model terms are significant [58]. Significant model terms in this case were A, B, C, D, A^2, C^2, D^2, BC, BD. Further 3D analysis of the model showed an escalated pattern in keratinase production when the temperature was increased, and pH maintained in the targeted range (Figure 3a). The same escalating pattern was also observed in Figure 3b and c where keratinase production increased only when the temperature was increased and could not increase further with increasing in inoculum size and feather concentrations, respectively. Verification of the model with a predicted value close to the actual value showed the reliability of the experiment to predicted precise condition, thus supporting the accuracy of the model over 95%.

The amino acid profile of hydrolysate of *Bacillus* sp. UPM-AAG1 revealed the presence of both essential amino acid and non-essential amino acids. The result is in accordance with Reference [19], that demonstrated the presence 17 different amino acid acquired from fermentation of *Bacillus cereus* utilizing chicken feather as sole their carbon and nitrogen sources. The fermented hydrolysate was rich with nutritionally essential amino acid particularly lysine, threonine, and methionine. Similarly, Ghosh et al. [59] reported on purified keratinolytic protease from feather waste hydrolysate by *Bacillus cereus* DCUW that comprises of 17 different amino acids.

4. Materials and Methods

4.1. Azokeratin and Keratinase Assay

Azo keratin substrate was prepared [60] with the modification where, instead of ball milling, the feather was cut into small pieces with a scissor. One gram of a finely cut white chicken feather, 20 mL deionized water, and 10% of $NaHCO_3$ were mixed in a 100 mL round bottom flask. Separately, 0.174 g of sulfanilic acid was dissolved in 5 mL of 0.2 M NaOH. Next 0.069 g of $NaNO_2$ was added to the suspension. The solution then was acidified with 0.4 mL of 5 M HCl for 2 min and neutralized by 0.4 mL of 5 M NaOH. The prepared solution was then added to finely cut feather keratin and mixed properly for 10 min. The reaction mixture then was filtered. Insoluble azo keratin was retrieved and rinsed with deionized water. The azo keratin was then suspended in water and shaken for 2 h at 50 °C. The pH of the filtrate and absorbance readings were taken periodically until the pH of the filtrate reached 6.0–7.0 and the absorbance value was less than 0.01 [60]. The resulting azokeratin (Supplementary Figure S1) is utilized for keratinase assay. All experiments were carried out three times unless otherwise stated.

The keratinase activity was determined using azo keratin as a substrate. 5 mg azo keratin substrate was added to 1.5 mL mini centrifuges tube together with 800 µL of 0.1 M phosphate buffer pH 8.0. Then, 200 µL of enzyme supernatant was added to the mixture. The mixture was vortexed thoroughly and incubated at 30 °C for 30 min in a water bath. The enzymatic reaction was stopped by 200 µL of 10% (w/v) trichloroacetic acid added to the mixture, and the absorbance was read at 450 nm (DTX 800-Multimode detector, Beckman Coulter, Brea, CA, USA). Control was prepared by adding trichloroacetic acid (TCA) to the mixture before the enzyme. One unit of keratinase activity was defined as 0.010 unit increase in the absorbance at 450 nm compared to control [48]. All experiments were carried out in triplicate, unless stated otherwise.

4.2. Isolation and Screening of Bacillus sp. with Keratinolytic Activity

Soil samples and poultry waste specimens were collected from a waste collection area of a poultry research farm in Universiti Putra Malaysia. One percent (w/v) of each soil samples and poultry waste were dissolved in 10 mL of sterilized phosphate buffer saline (PBS) and incubated in 80 °C water bath for 10 min to further biased the selection towards spore-forming *Bacillus* species. The PBS medium used was adopted from Dulbecco and Vogt [61]. The suspension (100 µL) was spread on nutrient agar (NA) supplemented with keratin substrate. The plates were incubated at room temperature (25 °C) for 24–48 h. Surviving bacteria that showed different morphology and high hydrolysis zone on NA were further re-streaked on NA until pure cultures were obtained [62].

All potentially isolated keratinase-producing bacteria were screened according to the ability to develop endospore under stress environment in a sporing medium (pH 7.0) composed of g/L: 1.6 NH_4Cl, 0.9 K_2HPO_4, 0.6 KH_2PO_4, 0.2 $MgSO_4·7H_2O$, and 0.07 $CaCl·2H_2O$, 0.01 $FeSO_4·7H_2O$ and 0.01 EDTA for two days at 25 °C under shaking condition at 150 rpm. All strains were spore stained with malachite green and safranin according to Reference [63]'s method and observed under a light microscope (Olympus BX.40F4, Japan) with 100× magnification [64]. Positive endospore-forming isolate were further screened based on bacterial growth (CFU/mL) on feather meal agar (FMA) composed of (g/L); 1.0 feather, 0.5 NaCl, 0.7 K_2HPO_4, 0.001 $MgSO_4·6H_2O$ and 15.0 bacteriological agar and keratinase assay with 1% feather as sole carbon and nitrogen sources [48].

4.3. Morphological, Biochemical and Molecular Identification of Keratinolytic Microorganism

The identity of the selected bacterium was further identified morphologically using Gram staining method and a series of biochemical test (oxidase test, catalase, Voges-Proskauer, nitrate, citrate, lipase, and gelatinase) [64]. Meanwhile, molecular identification confirmation was performed by 16S rDNA sequence analysis using a 24 h culture of the bacterial cell using InnuPREP Bacteria DNA Kit (Analytik Jena, Jena, Germany) according to the manufacturer's protocol. Amplification of the partial 16S rRNA gene was carried out using universal primers. The PCR mixtures comprise of mixtures of 1µL of 5 mM 27F (5′-AGA GTT TGATCC TGG CTC AG-3′) and 1429R (5′-TAC GGT TACCTT GTT ACG ACTT-3′) of forward and reverse primer, 1 µL of DNA sample, 12.5 µL of Master mix 2 × Taq (Vivantis Technologies Sdn. Bhd., Selangor, Malaysia) and 9.5 µL sterile deionized water for a final volume of 25 µL. The polymerase chain reaction was accomplished using a gradient thermocycler (Hercuvan, Milton, UK) under the following conditions: 3 min initial denaturation at 94 °C, 29 cycles denaturation for 1 min at 94 °C, 1 min of annealing at 58 °C, 2 min of extension for 10 min, and final extension at 72 °C for 10 min with incubation at 4 °C. Successfully amplified DNA fragments were analyzed on 1% (w/v) agarose gel [65].

4.4. Sequence Analysis and Phylogenetic Analysis

The selected sequence was analyzed using BLASTn [66]. Twenty sequence alignment with more than 95% similarity was selected for further analysis using neighbor-joining method, as in Ref. [67], fitting to the distances of Jukes-Cantor [68]. Phylogenetic analysis was done using PHYLIP software v3.696 (http://evolution.genetics.washington.edu/phylip.html). *E. coli* strain U5/41 was used as the outgroups in the cladogram for identification analysis. The confidence level of each branch was calculated by 1000 bootstraps replicates. The constructed tree was viewed using Tree View version 1.6.6.

4.5. Optimization of Keratinase Activity Using Response Surface Methodology

The effect of four factors namely temperature, inoculum size (v/v), pH, and feather concentration (w/v) on keratinase production was screened statistically using Plackett-Burman factorial design (PBFD) to verify the significance of the named factors in the production of keratinase. The experimental design and statistical analysis were performed using statistical software Design-Expert® 6.0.8 (Stat-Ease, Minneapolis, MN, USA). Each independent factor was evaluated at two different levels; minimum and maximum levels (+1, −1) as shown in Table 8. Keratinase activity was analyzed as the response. The independent factors that show significance by PBFD were optimized further for their interaction effects by composite design (CCD) of response surface methodology (RSM). Each independent factor was studied at five different level: −α, −1, 0, +1, +α. Keratinase activity was evaluated as a response based on 30 experimental design. All experiments were conducted in triplicate, and keratinase activity was examined as the response using a second-order polynomial equation as below:

$$y = \sum_{i=1}^{k} \beta_i X_i + \sum_{i}^{k} \beta_{ii} X_{i2} + \sum_{1 \leq i \leq j}^{k} \beta_{ij} X_i X_j$$

where Y is the predicted response, X is the independent factor that is affected by Y, k is the number of factors, β_0 is the constant term, β_i is the linear coefficient, βii is the i the quadratic coefficient, and β_{ij} is the ij the interaction coefficient, whereas i and j = 1,2,3 and $i \neq j$ are coefficient in the model. The significance of each coefficient in the equation was determined by Fisher's F test and analysis of variances ($p < 0.05$). The experimental design and statistical analysis were performed using statistical software Design-Expert® 6.0.8 (Stat-Ease, Minneapolis, MN, USA). All experiments were conducted in triplicate.

Table 8. Experimental factors and level of minimum and maximum range for statistical screening using Plackett-Burman factorial design (PBFD).

Factors	Independent Factor	Unit	Range Level	
			Minimum (−1)	Maximum (+1)
X_1	Temperature	(°C)	25	35
X_2	Inoculum	% (v/v)	1	5
X_3	pH	-	5	8
X_4	Feather (w/v)	% (w/v)	1	5

4.6. Amino Acid Profile of Hydrolysate

Amino acid profile of hydrolysate was performed according to a previous method [69], with slight modifications. The amino acid profile of sample hydrolysate was determined using an HPLC system (Agilent 1200, Agilent Technologies, Santa Clara, CA, USA). The sample was subjected to automated pre-column derivatization using orthopthalaldehyde (OPA) run through the injector program. The injector program protocols were as follows where 2.5 µL were drawn from a borate buffer vial (0.4 min, pH 10.2). Next, 0.5 µL of the sample was drawn from a sample vial, followed by mixing with 3 µL in a wash port five times and waiting for 0.2 min. Next, 0.5 µL of orthopthalaldehyde (OPA) was drawn, followed by mixing of 3.5 µL in wash port 6 times. Next, 32 µL of injection diluent (1 mL of mobile phase A + 15 µL of concentrated H_3PO_3) was mixed with 20 µL in seat 8 times. The sample was injected, then wait for 0.10 min and valve bypass. The mobile phase A consisted of 10 mM of Na_2HPO_4, 10 mM $Na_2B_4O_7$, pH 8.2, and mobile phase B (acetonitrile-methanol-water; 45:45:10, v/v). A programmed gradient elution was performed from 2% B to 57% B for 7 min, followed by 57% B to 100% B for 8.4 min, with a flow rate of 1.5 mL/min at 40 °C. Amino acid detection was detected with a 250 nm Diode Array Detector (DAD) detector using an amino acid standard solution (Sigma-Aldrich, St. Louis, MO, USA).

5. Conclusions

The reliability of statistical optimization of the external parameter in enhancing keratinase production *Bacillus* sp. UPM-AAG1 I was demonstrated in this work. The significant parameter required for the optimum keratinase production was screened using Plackett-Burman design. Optimization of keratinase by RSM allowed us to evaluate the effect of various parameter at different levels. The CCD design applied results 1.7-fold in keratinase yield. The acceptable degree of similarity between the predicted model and actual activity signifies the reliability of the statistical model in optimization of keratinase. The optimized parameters and characteristics of the bacterium include optimal growth at near neutrality, ambient temperature, and able to support growth and keratinase production without external supplementary requirements. Moreover, the hydrolysate of *Bacillus* sp. UPM-AAG1 obtained through statistical optimization is rich in amino acids. These properties make the bacterium an excellent choice for local commercial application where keratinase production should be optimum at ambient temperature and no additional C or N sources should be added to minimize cost. In the future, cheaper, or even waste, materials from the local agricultural industries, such as waste bagasse or Palm Mill Oil Effluent or POME, may be tested to improve keratinase production and feather degradation in general.

Supplementary Materials: The following are available online at http://www.mdpi.com/2073-4344/10/8/848/s1, Figure S1: Azokeratin formation from feather keratin treated with the azotization of sulfanilic acid. Figure S2: Model diagnostic plots; (a) predicted versus actual, (b) studentized residue versus predicted, (c) normal plots of residue and (d) outlier T versus run.

Author Contributions: Conceptualization, M.Y.S. and N.A.Y.; methodology, M.Y.S.; software, S.A.A.; validation, M.E.K., M.Y.S. and N.A.Y.; formal analysis, A.A.G. and S.A.A.; investigation, A.A.G.; resources, M.Y.S.; data curation, M.Y.S.; writing—original draft preparation, A.A.A.; writing—review and editing, M.Y.S.; visualization, S.A.A.; supervision, M.E.K., M.Y.S. and N.A.Y.; project administration, M.Y.S.; funding acquisition, M.Y.S. All authors have read and agreed to the published version of the manuscript.

Funding: The authors would like to thank the support of the Graduate Research Fellowship University Putra Malaysia (GRF-UPM) given to Aa'ishah Abd Gafar.

Conflicts of Interest: The authors declare the results obtained in this study will be used in discussion for a possible transfer of technology to a local small and medium enterprise (SME) company, of which the company did not sponsor the works carried out in this study.

References

1. Fang, Z.; Zhang, J.; Liu, B.; Du, G.; Chen, J. Biodegradation of wool waste and keratinase production in scale-up fermenter with different strategies by *Stenotrophomonas maltophilia* BBE11-1. *Bioresour. Technol.* **2013**, *140*, 286–291. [CrossRef]
2. Brandelli, A.; Daroit, D.J.; Riffel, A. Biochemical features of microbial keratinases and their production and applications. *Appl. Microbiol. Biotechnol.* **2010**, *85*, 1735–1750. [CrossRef] [PubMed]
3. Tiwary, E.; Gupta, R. Rapid Conversion of Chicken Feather to Feather Meal Using Dimeric Keratinase from *Bacillus licheniformis* ER-15. *J. Bioprocess. Biotech.* **2012**, *2*, 1000123. [CrossRef]
4. Abdel-Fattah, A.M.; El-Gamal, M.S.; Ismail, S.; Emran, M.; Hashem, A. Biodegradation of feather waste by keratinase produced from newly isolated *Bacillus licheniformis* ALW1. *J. Genet. Eng. Biotechnol.* **2018**, *16*, 311–318. [CrossRef] [PubMed]
5. Fang, Z.; Zhang, J.; Liu, B.; Jiang, L. Cloning, heterologous expression and characterization of two keratinases from *Stenotrophomonas maltophilia* BBE11-1. *Process Biochem.* **2014**, *49*, 647–654. [CrossRef]
6. Pereira, J.Q.; Lopes, F.C.; Petry, M.V.; da Costa Medina, L.F.; Brandelli, A. Isolation of three novel Antarctic psychrotolerant feather-degrading bacteria and partial purification of keratinolytic enzyme from *Lysobacter* sp. A03. *Int. Biodeterior. Biodegrad.* **2014**, *88*, 1–7. [CrossRef]
7. Mohamedin, A.H. Isolation, identification and some cultural conditions of a protease- producing thermophilic *Streptomyces* strain grown on chicken feather as a substrate. *Int. Biodeterior. Biodegrad.* **1999**, *43*, 13–21. [CrossRef]
8. Syeda, D.G.; Leeb, J.C.; Lic, W.-J.; Kimb, C.-J.; Agasard, D. Production, characterization and application of keratinase from *Streptomyces gulbargensis*. *Bioresour. Technol.* **2009**, *100*, 1868–1871. [CrossRef]
9. Malviya, H.K.; Rajak, R.C.; Hasija, S.K. Synthesis and regulation of extracellular keratinase in three fungi isolated from the grounds of a gelatin factory, Jabalpur, India. *Mycopathologia* **1992**, *120*, 1–4. [CrossRef]
10. Ramakrishnaiah, G.; Mustafa, S.M.; Srihari, G. Studies on Keratinase Producing Fungi Isolated from Poultry Waste and their Enzymatic Activity. *Microbiol. Res.* **2013**, *3*, 148–151. [CrossRef]

11. Lin, X.; Chung, G.L.; Casale, E.S.; Jason, C.H.S. Purification and characterization of a keratinase from a degrading *Bacillus licheniformis* strain. *Appl. Environ. Microbiol.* **1992**, *58*, 3271–3275. [CrossRef]
12. Suntornsuk, W.; Suntornsuk, L. Feather degradation by *Bacillus* sp. FK 46 in submerged cultivation. *Bioresour. Technol.* **2003**, *86*, 239–243. [CrossRef]
13. Schallmey, M.; Singh, A.; Ward, O.P. Developments in the use of Bacillus species for industrial production. *Can. J. Microbiol.* **2004**, *50*, 1–17. [CrossRef]
14. Huang, Y.; Busk, P.K.; Herbst, F.-A.; Lange, L. Genome and secretome analyses provide insights into keratin decomposition by novel proteases from the non-pathogenic fungus *Onygena corvina*. *Appl. Microbiol. Biotechnol.* **2015**, *99*, 9635–9649. [CrossRef] [PubMed]
15. El-Refai, H.A.; AbdelNaby, M.A.; Gaballa, A.; El-Araby, M.H.; Abdel Fattah, A.F. Improvement of the newly isolated *Bacillus pumilus* FH9 keratinolytic activity. *Process Biochem.* **2005**, *40*, 2325–2332. [CrossRef]
16. Reddy, L.V.A.; Wee, Y.J.; Yun, J.S.; Ryu, H.W. Optimization of alkaline protease production by batch culture of *Bacillus* sp. RKY3 through Plackett–Burman and response surface methodological approaches. *Bioresour. Technol.* **2008**, *99*, 2242–2249. [CrossRef]
17. Pillai, P.; Mandge, S.; Archana, G. Statistical optimization of production and tannery applications of a keratinolytic serine protease from *Bacillus subtilis* P13. *Process Biochem.* **2011**, *46*, 1110–1117. [CrossRef]
18. Dettmer, A.; Cavalli, É.; Ayub, M.A.Z.; Gutterres, M. Optimization of the unhairing leather processing with enzymes and the evaluation of inter-fibrillary proteins removal: An environment-friendly alternative. *Bioprocess Biosyst. Eng.* **2012**, *35*, 1317–1324. [CrossRef] [PubMed]
19. Lo, W.H.; Too, J.R.; Wu, J.Y. Production of keratinolytic enzyme by an indigenous feather-degrading strain *Bacillus cereus* WU2. *J. Biosci. Bioeng.* **2012**, *114*, 640–647. [CrossRef]
20. Okoroma, E.A.; Garelick, H.; Abiola, O.O.; Purchase, D. Identification and characterisation of a *Bacillus licheniformis* strain with profound keratinase activity for degradation of melanised feather. *Int. Biodeterior. Biodegrad.* **2012**, *74*, 54–60. [CrossRef]
21. Sahoo, D.K.; Das, A.; Thatoi, H.; Mondal, K.C.; Mohapatra, P.K.D. Keratinase production and biodegradation of whole chicken feather keratin by a newly isolated bacterium under submerged fermentation. *Appl. Biochem. Biotechnol.* **2012**, *167*, 1040–1051. [CrossRef] [PubMed]
22. Sivakumar, T.; Shankar, T.; Vijayabaskar, P.; Ramasubramanian, V. Optimization for Keratinase Enzyme Production Using *Bacillus thuringiensis* TS2. *Acad. J. Plant Sci.* **2012**, *5*, 102–109. [CrossRef]
23. E Silva, L.A.D.; Macedo, A.J.; Termignoni, C. Production of keratinase by *Bacillus subtilis* S14. *Ann. Microbiol.* **2014**, *64*, 1725–1733. [CrossRef]
24. Govarthanan, M.; Selvankumar, T.; Selvam, K.; Sudhakar, C.; Kamala-Kannan, S. Response surface methodology based optimization of keratinase production from alkali-treated feather waste and horn waste using *Bacillus* sp. MG-MASC-BT. *J. Ind. Eng. Chem.* **2015**, *27*, 25–30. [CrossRef]
25. Sanghvi, G.; Patel, H.; Vaishnav, D.; Oza, T.; Dave, G.; Kunjadia, P.; Sheth, N. A novel alkaline keratinase from *Bacillus subtilis* DP1 with potential utility in cosmetic formulation. *Int. J. Biol. Macromol.* **2016**, *87*, 256–262. [CrossRef]
26. Reddy, M.R.; Reddy, K.S.; Chouhan, Y.R.; Bee, H.; Reddy, G. Effective feather degradation and keratinase production by *Bacillus pumilus* GRK for its application as bio-detergent additive. *Bioresour. Technol.* **2017**, *243*, 254–263. [CrossRef]
27. Lateef, A.; Oloke, J.K.; Gueguim Kana, E.B.; Sobowale, B.O.; Ajao, S.O.; Bello, B.Y. Keratinolytic activities of a new feather-degrading isolate of *Bacillus cereus* LAU 08 isolated from Nigerian soil. *Int. Biodeterior. Biodegrad.* **2010**, *64*, 162–165. [CrossRef]
28. Kothari, D.; Rani, A.; Goyal, A. Keratinase. In *Current Developments in Biotechnology and Bioengineering: Production, Isolation and Purification of Industrial Products*; Elsevier: London, UK, 2017; pp. 447–469, ISBN 9780444636621.
29. Walker, R.; Powell, A.A.; Seddon, B. Bacillus isolates from the spermosphere of peas and dwarf French beans with antifungal activity against *Botrytis cinerea* and Pythium species. *J. Appl. Microbiol.* **1998**, *84*, 791–801. [CrossRef]
30. Monteiro, S.M.; Clemente, J.; Henriques, A.O.; Gomes, R.J.; Carrondo, M.J. A Procedure for High-Yield Spore Production by *Bacillus subtilis*. *Biotechnol. Prog.* **2005**, *21*, 1026–1031. [CrossRef]
31. Tan, I.S.; Ramamurthi, K.S. Spore formation in *Bacillus subtilis*. *Environ. Microbiol. Rep.* **2014**, *6*, 212–225. [CrossRef]

32. Riffel, A.; Brandelli, A. Keratinolytic bacteria isolated from feather waste. *Braz. J. Microbiol.* **2006**, *37*, 395–399. [CrossRef]
33. Zhang, R.X.; Gong, J.S.; Su, C.; Zhang, D.D.; Tian, H.; Dou, W.F.; Li, H.; Shi, J.S.; Xu, Z.H. Biochemical characterization of a novel surfactant-stable serine keratinase with no collagenase activity from *Brevibacillus parabrevis* CGMCC 10798. *Int. J. Biol. Macromol.* **2016**, *93*, 843–851. [CrossRef]
34. Laba, W.; Choinska, A.; Rodziewicz, A.; Piegza, M.; Laba, W.; Choinska, A.; Rodziewicz, A.; Piegza, M. Keratinolytic abilities of *Micrococcus luteus* from poultry waste. *Braz. J. Microbiol.* **2015**, *46*, 691–700. [CrossRef] [PubMed]
35. Arokiyaraj, S.; Varghese, R.; Ahmed, B.A.; Duraipandiyan, V.; Al-dhabi, N.A. Optimizing the fermentation conditions and enhanced production of keratinase from *Bacillus cereus* isolated from halophilic environment. *Saudi J. Biol. Sci.* **2019**, *26*, 378–381. [CrossRef] [PubMed]
36. Vidmar, B.; Vodovnik, M. Microbial Keratinases: Enzymes with Promising Biotechnological Applications. *Food Technol. Biotechnol.* **2018**, *56*, 312–328. [CrossRef]
37. Khodayari, S.; Kafilzadeh, F. Separating Keratinase Producer Bacteria from the Soil of Poultry Farms and Optimization of the Conditions for Maximum Enzyme Production. *Eur. J. Exp. Biol.* **2018**, *8*, 1–8. [CrossRef]
38. Fang, Z.; Zhang, J.; Liu, B.; Du, G.; Chen, J. Biochemical characterization of three keratinolytic enzymes from *Stenotrophomonas maltophilia* BBE11-1 for biodegrading keratin wastes. *Int. Biodeterior. Biodegrad.* **2013**, *82*, 166–172. [CrossRef]
39. Fakhfakh-Zouari, N.; Haddar, A.; Hmidet, N.; Frikha, F.; Nasri, M. Application of statistical experimental design for optimization of keratinases production by Bacillus pumilus A1 grown on chicken feather and some biochemical properties. *Process Biochem.* **2010**, *45*, 617–626. [CrossRef]
40. Cai, C.; Zheng, X. Medium optimization for keratinase production in hair substrate by a new *Bacillus subtilis* KD-N2 using response surface methodology. *J. Ind. Microbiol. Biotechnol.* **2009**, *36*, 875–883. [CrossRef]
41. Kumar, R.; Balaji, S.; Uma, T.S. Optimization of influential parameters for extracellular keratinase production by *Bacillus subtilis* (MTCC9102) in solid state fermentation using horn meal -A biowaste management. *Appl. Biochem. Biotechnol.* **2010**, *160*, 30–39. [CrossRef]
42. Mazotto, A.M.; Cedrola, S.M.L.; Lins, U.; Rosado, A.S.; Silva, K.T.; Chaves, J.Q.; Rabinovitch, L. Keratinolytic activity of *Bacillus subtilis* AMR using human hair. *Soc. Appl. Microbiol.* **2010**, *50*, 89–96. [CrossRef] [PubMed]
43. Tiwary, E.; Gupta, R. Medium optimization for a novel 58 kDa dimeric keratinase from *Bacillus licheniformis* ER-15: Biochemical characterization and application in feather degradation and dehairing of hides. *Bioresour. Technol.* **2010**, *101*, 6103–6110. [CrossRef] [PubMed]
44. Fakhfakh, N.; Kanoun, S.; Manni, L.; Nasri, M. Production and biochemical and molecular characterization of a keratinolytic serine protease from chicken feather-degrading *Bacillus licheniformis* RPk. *Can. J. Microbiol.* **2009**, *55*, 427–436. [CrossRef]
45. Mazotto, A.M.; Coelho, R.R.R.; Cedrola, S.M.L.; De Lima, M.F.; Couri, S.; Paraguai De Souza, E.; Vermelho, A.B. Keratinase production by three Bacillus spp. using feather meal and whole feather as substrate in a submerged fermentation. *Enzyme Res.* **2011**, *2011*. [CrossRef] [PubMed]
46. Gupta, R.; Beg, Q.; Lorenz, P. Bacterial alkaline proteases: Molecular approaches and industrial applications. *Appl. Microbiol. Biotechnol.* **2005**, *59*, 15–32. [CrossRef]
47. Mousavi, S.; Salouti, M.; Shapoury, R.; Heidari, Z. Optimization of keratinase production for feather degradation by *Bacillus subtilis*. *Jundishapur J. Microbiol.* **2013**, *6*. [CrossRef]
48. Yusuf, I.; Ahmad, S.A.; Phang, L.Y.; Syed, M.A.; Shamaan, N.A.; Abdul Khalil, K.; Dahalan, F.A.; Shukor, M.Y. Keratinase production and biodegradation of polluted secondary chicken feather wastes by a newly isolated multi heavy metal tolerant bacterium-Alcaligenes sp. AQ05-001. *J. Environ. Manag.* **2016**, *183*, 182–195. [CrossRef]
49. Covino, S.; D'Annibale, A.; Stazi, S.R.; Cajthaml, T.; Čvančarová, M.; Stella, T.; Petruccioli, M. Assessment of degradation potential of aliphatic hydrocarbons by autochthonous filamentous fungi from a historically polluted clay soil. *Sci. Total Environ.* **2015**, *505*, 545–554. [CrossRef]
50. Riffel, A.; Lucas, F.; Heeb, P.; Brandelli, A. Characterization of a new keratinolytic bacterium that completely degrades native feather keratin. *Arch. Microbiol.* **2003**, *179*, 258–265. [CrossRef]
51. Pillai, P.; Archana, G. Hide depilation and feather disintegration studies with keratinolytic serine protease from a novel *Bacillus subtilis* isolate. *Appl. Microbiol. Biotechnol.* **2008**, *78*, 643–650. [CrossRef]

52. Balaji, S.; Kumar, M.S.; Karthikeyan, R.; Kumar, R.; Kirubanandan, S.; Sridhar, R.; Sehgal, P.K. Purification and characterization of an extracellular keratinase from a hornmeal-degrading *Bacillus subtilis* MTCC (9102). *World J. Microbiol. Biotechnol.* **2008**, *24*, 2741–2745. [CrossRef]
53. Bansal, M.; Sudhakara Reddy, M.; Kumar, A. Optimization of cell growth and bacoside—A production in suspension cultures of *Bacopa monnieri* (L.) Wettst. using response surface methodology. *In Vitro Cell. Dev. Biol. Plant* **2017**, *53*, 527–537. [CrossRef]
54. Whitcomb, P.J.; Anderson, M.J. *RSM Simplified: Optimizing Processes Using Response Surface Methods for Design of Experiments*; Productivity Press: New York, NY, USA, 2004; ISBN 978-1-56327-297-4.
55. Manogaran, M.; Shukor, M.Y.; Yasid, N.A.; Johari, W.L.W.; Ahmad, S.A. Isolation and characterisation of glyphosate-degrading bacteria isolated from local soils in Malaysia. *Rend. Lincei* **2017**, *28*, 471–479. [CrossRef]
56. Roslan, M.A.H.; Abdullah, N.; Mustafa, S. Removal of shells in palm kernel cake via static cling and electrostatic separation. *J. Biochem. Microbiol. Biotechnol.* **2015**, *3*, 1–6.
57. Aziz, N.F.; Halmi, M.I.E.; Johari, W.L.W. Statistical optimization of hexavalent molybdenum reduction by *Serratia* sp. strain MIE2 using Central Composite Design (CCD). *J. Biochem. Microbiol. Biotechnol.* **2017**, *5*, 8–11.
58. Richa, K.; Bose, H.; K, S.; Loganathan, K.; Kumar, G.; Rao, B. Response surface optimization for the production of marine eubacterial protease and its application. *Res. J. Biotechnol.* **2013**, *8*, 78–85.
59. Ghosh, A.; Chakrabarti, K.; Chattopadhyay, D. Degradation of raw feather by a novel high molecular weight extracellular protease from newly isolated *Bacillus cereus* DCUW. *J. Ind. Microbiol. Biotechnol.* **2008**, *35*, 825–834. [CrossRef]
60. Joshi, S.G.; Tejashwini, M.M.; Revati, N.; Sridevi, R.; Roma, D. Isolation, identification and characterization of feather degrading bacteria. *Int. Journall Poult. Sci.* **2007**, *6*, 689–693. [CrossRef]
61. Dulbecco, R.; Vogt, M. Plaque formation and isolation of pure lines with poliomyelitis viruses. *J. Exp. Med.* **1954**, *99*, 167–182. [CrossRef]
62. Gajbhiye, A.; Rai, A.R.; Meshram, S.U.; Dongre, A.B. Isolation, evaluation and characterization of *Bacillus subtilis* from cotton rhizospheric soil with biocontrol activity against *Fusarium oxysporum*. *World J. Microbiol. Biotechnol.* **2010**, *26*, 1187–1194. [CrossRef]
63. Schaeffer, A.B.; Fulton, M.D. A simplified methode of staining endospores. *Science* **1933**, *77*, 1990. [CrossRef] [PubMed]
64. Cappuccino, J.; Sherman, N. *Microbiology: A Laboratory Manual*, 10th ed.; Pearson: London, UK, 2010.
65. Habib, S.; Ahmad, S.A.; Johari, W.L.W.; Shukor, M.Y.A.; Alias, S.A.; Khalil, K.A.; Yasid, N.A. Evaluation of conventional and response surface level optimisation of n-dodecane (n-C12) mineralisation by psychrotolerant strains isolated from pristine soil at Southern Victoria Island, Antarctica. *Microb. Cell Factories* **2018**, *17*, 1–21. [CrossRef] [PubMed]
66. Altschul, S.F.; Madden, T.L.; Schäffer, A.A.; Zhang, J.; Zhang, Z.; Miller, W.; Lipman, D.J. Gapped BLAST and PSI-BLAST: A new generation of protein database search programs. *Nucleic Acids Res.* **1997**, *25*, 3389–3402. [CrossRef] [PubMed]
67. Saitou, N.; Nei, M. The Neighbor-joining Method: A New Method for Reconstructing Phylogenetic Trees'. *Mol. Biol. Evol.* **1987**, *4*, 406–425. [PubMed]
68. Jukes, T.H.; Cantor, C.R. Evolution of protein molecules. In *Mammalian Protein Metabolism*; Munro, N., Ed.; Academic Press: New York, NY, USA, 1969; Volume 3, pp. 21–132.
69. Jones, B.N.; Gilligan, J.P. o-phthaldialdehyde precolumn derivatization and reversed-phase high-performance liquid chromatography of polypeptide hydrolysates and physiological fluids. *J. Chromatogr. A* **1983**, *266*, 471–482. [CrossRef]

© 2020 by the authors. Licensee MDPI, Basel, Switzerland. This article is an open access article distributed under the terms and conditions of the Creative Commons Attribution (CC BY) license (http://creativecommons.org/licenses/by/4.0/).

Article

Enzymatic Hydrolysis of Softwood Derived Paper Sludge by an In Vitro Recombinant Cellulase Cocktail for the Production of Fermentable Sugars

Samkelo Malgas [1], Shaunita H. Rose [2], Willem H. van Zyl [2] and Brett I. Pletschke [1,*]

[1] Enzyme Science Programme (ESP), Department of Biochemistry and Microbiology, Rhodes University, Grahamstown 6140, South Africa; samkelomalgas@yahoo.com
[2] Department of Microbiology, Stellenbosch University, Stellenbosch 7600, South Africa; shrose@sun.ac.za (S.H.R.); whvz@sun.ac.za (W.H.v.Z.)
* Correspondence: b.pletschke@ru.ac.za; Tel.: +27-46-603-8081

Received: 3 June 2020; Accepted: 19 June 2020; Published: 11 July 2020

Abstract: Paper sludge is an attractive biomass feedstock for bioconversion to ethanol due to its low cost and the lack of pretreatment required for its bioprocessing. This study assessed the use of a recombinant cellulase cocktail (mono-components: *S. cerevisiae*-derived *Pc*BGL1B (BGL), *Te*Cel7A (CBHI), *Cl*Cel6A (CBHII) and *Tr*Cel5A (EGII) mono-component cellulase enzymes) for the efficient saccharification of softwood-derived paper sludge to produce fermentable sugars. The paper sludge mainly contained 74.3% moisture and 89.7% (per dry mass (DM)) glucan with a crystallinity index of 91.5%. The optimal protein ratio for paper sludge hydrolysis was observed at 9.4: 30.2: 30.2: 30.2% for BGL: CBHI: CBHII: EGII. At a protein loading of 7.5 mg/g DW paper sludge, the yield from hydrolysis was approximately 80%, based on glucan, with scanning electron microscopy micrographs indicating a significant alteration in the microfibril size (length reduced from ≥ 2 mm to 93 µm) of the paper sludge. The paper sludge hydrolysis potential of the Opt CelMix (formulated cellulase cocktail) was similar to the commercial Cellic CTec2® and Celluclast® 1.5 L cellulase preparations and better than Viscozyme® L. Low enzyme loadings (15 mg/g paper sludge) of the Opt CelMix and solid loadings ranging between 1 to 10% (*w/v*) rendered over 80% glucan conversion. The high glucose yields attained on the paper sludge by the low enzyme loading of the Opt CelMix demonstrated the value of enzyme cocktail optimisation on specific substrates for efficient cellulose conversion to fermentable sugars.

Keywords: cellulase; cellulose; paper sludge; *Saccharomyces cerevisiae*; synergism

1. Introduction

The production of bioethanol from lignocellulosic biomass or lignocellulose-derived wastes is one of the most promising alternatives to conventional fossil fuels [1]. Lignocellulosic biomass is an ideal feedstock for bioethanol production because it is an abundant renewable source, obtainable at low cost and generally not considered to compete with food sources [1,2].

Paper sludge (waste fibre) is a residual stream produced by pulp mills [3]. Currently, it is utilised in either landfilling or incineration to generate energy for pulp mills [4]. Paper sludge has a high carbohydrate and low lignin content which eliminates the requirement of additional thermochemical pretreatment steps. Paper sludge has high potential and value as a suitable feedstock for the production of lactic acid and bioethanol [5]. The latter requires the enzymatic hydrolysis of paper sludge by cellulases to fermentable sugars which can subsequently be fermented by yeast into bioethanol [3]. Four key cellulase activities working synergistically are required for cellulose utilisation; exo-glucanases (CBHI and CBHII) target the crystalline regions of cellulose and produce cellobiose in a processive manner, while endo-glucanases randomly cleave internal sites at amorphous regions to produce long-chain cello-oligomers, β-glucosidases then process the products from the aforementioned enzymes into glucose [6,7].

In spite of the progress achieved by enzyme manufacturers in lowering enzyme production costs, cellulases still constitute a significant portion of the final bioethanol production costs. Many strategies have been developed to reduce enzyme cost, including enzyme recycling, improving enzyme synergism through synthetic cocktail design and enzyme engineering [8–10]. Synthetic cocktails with improved enzyme synergism is one of the strategies that can be implemented to lower the amount of enzyme required for hydrolysis [8].

Numerous studies have successfully produced cellulase mono-components; endo-glucanase, exo-glucanases (CBHI and CBHII) and β-glucosidases in *Saccharomyces cerevisiae* strains using heterologous gene expression [11–13]. More recent studies focused on using and improving *S. cerevisiae* wild type and industrial strains for second-generation (2G) bioethanol production by targeting ideal traits, such as the secretory capacity of heterologous cellulase enzymes through evolutionary and genetic engineering, and mutagenesis [14–17]. However, the optimum enzyme ratios for specific cellulosic substrates are crucial for developing more efficient consolidated bioprocessing (CBP) yeast strains. In this study, mono-component recombinant cellulases—*S. cerevisiae*-derived PcBGL1B (BGL), TeCel7A (CBHI), ClCel6A (CBHII) and TrCel5A (EGII)—were produced in *S. cerevisiae* Y294 strains and partially purified for use in formulating an enzyme cocktail for the efficient deconstruction of paper sludge. The performance of the formulated cellulase cocktail was compared to that of different commercially available cellulase cocktails in terms of its hydrolytic performance on paper sludge. Furthermore, the effect of higher amounts of solids and enzyme loadings on the hydrolysis efficiency of the cocktail was studied to optimise the conversion process.

2. Results and Discussion

2.1. Composition and Structural Analysis of Paper Sludge

The chemical composition and structural characterisation (crystallinity and cellulase accessibility) of paper sludge was assessed and is displayed in Table 1. The paper sludge contained a dry mass (DM) of 24.7% of the total mass, similar to that reported for paper sludge derived from a mixture of pine, cypress and eucalyptus (65% moisture content) [1]. The carbohydrate fraction was the major component of the dry mass (DM) of paper sludge, with glucan estimated at 89.7% followed by hemicellulose, lignin and ash on dry mass basis.

Table 1. Chemical composition as percentage of dry mass basis and structural analysis of paper sludge.

Component/Property	Content/Accessibility
Glucan *	89.7%
Mannan *	2.73%
Xylan *	1.65%
Galactan *	0.08%
Arabinan *	Nd
Lignin *	0.8%
Ash *	1.7%
Moisture	74.3%
Crystallinity index	91.5%
Substrate accessibility (mg/g)	87 mg/g

Not detected (Nd). * = percentage of DW.

X-ray diffraction (XRD) analysis was used to determine the type of cellulose and the crystallinity index of the paper sludge biomass. Based on the XRD pattern (four main peaks at 1–10, 110, 200 and 004) displayed by paper sludge, it could be deduced that the glucan content of the biomass is in the form of cellulose Iβ (Figure 1).

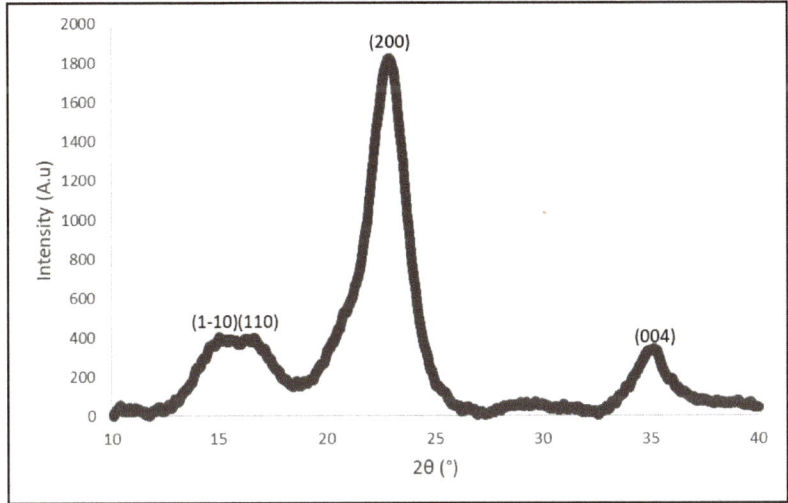

Figure 1. An X-ray diffraction pattern of paper sludge illustrating four main peaks (1–10, 110, 200 and 004) specific for cellulose Iβ.

The crystallinity index (CrI) of the paper sludge was estimated using the peak height method as previous described [18,19]. The CrI value of the paper sludge was estimated at approximately 91.5% (Table 1), similar to that of Avicel PH-101 (model cellulose Iβ biomass) with 89 to 92% reported by Park et al. [18] and Malgas et al. [20]. Biomass with high cellulosic contents have higher CrI values compared to biomasses with lower cellulosic contents [21].

Furthermore, the accessibility of the paper sludge biomass to cellulases was evaluated using a modified version of Simon's staining method. Paper sludge was shown to be highly accessible to the crystalline cellulose-specific direct orange (DO) dye (see Table 1). The accessibility of paper sludge to DO (87 mg/g) was slightly higher than that previously reported for Avicel PH-101 (67 mg/g) [20]. This coincided with the slightly higher crystallinity index observed in paper sludge compared to that of Avicel PH-101.

2.2. Enzyme Production and Substrate Specificities Using "Model" Substrates

In this study, the *S. cerevisiae*-derived *Pc*BGL1B (BGL), *Te*Cel7A (CBHI), *Cl*Cel6A (CBHII) and *Tr*Cel5A (EGII) mono-component cellulase enzymes were individually produced under the transcriptional control of the enolase 1 gene (*ENO1*) promoter (constitutive expression) and concentrated to approximately 1.5 mg/mL. Since the yeast strains were cultured in minimal media, very few other protein species were present in the concentrated culture supernatants containing the mono-component cellulase enzymes. To assess the relative purity of the cellulases upon ultrafiltration, they were separated by sodium dodecyl sulphate-polyacrylamide gel electrophoresis (SDS-PAGE) (Figure 2). The estimated molecular masses of the enzymes were approximately 80 and 140 kDa (60 kDa was expected as reported in the literature), 70 kDa, 75 kDa (heterogeneous in size due to hyper-glycosylation) and 54 kDa for BGL, CBHI, CBHII and EGII, respectively, and corresponded to molecular masses determined previously [22,23]. Migration on SDS-PAGE that does not correlate with formula molecular masses, termed "gel shifting", is common for some proteins [24], this may explain why the anticipated molecular mass of BGL was not observed in Figure 2.

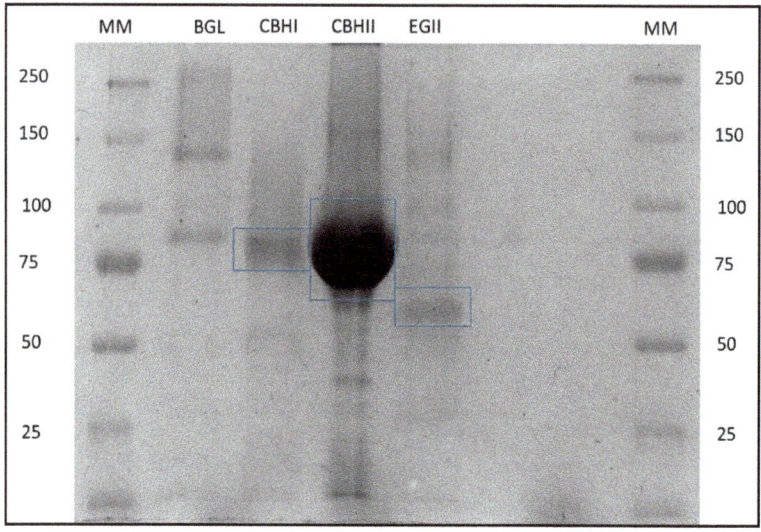

Figure 2. Denaturing 12% sodium dodecyl sulphate-polyacrylamide gel electrophoresis (SDS-PAGE) of ultrafiltration-concentrated mono-component cellulases (15 µL) visualised by Coomassie Brilliant Blue R-250 staining. Molecular mass marker (MM) (BioRad Precision Plus Protein Standards); the blue rectangles represent the identified protein molecular masses corresponding to the literature.

The cellulases were tested for their specific activities using different substrates at 50 °C in 50 mM sodium citrate buffer, pH 5.0, since these conditions assured sufficient and stable activity of all the enzymes assessed (data not shown). The specific activities of the individual recombinant cellulases on model cellulosic substrates is displayed in Table 2.

Table 2. Specific activities (U/mg protein) of the cellulase enzymes on model substrates. The underlined values are the specific activities expected (based on the literature) for each enzyme tested.

Enzyme	CMC-Na	Cellopentaitol	pNPC	pNPG	Reference
BGL	0.67	-	5.74	56	[10,14,25]
CBHI	0.24	Nd	0.20	0.35	[12,26]
CBHII	1.69	0.38	Nd	0.11	[26]
EGII	44	-	Nd	0.11	[10,12,14,26]

Not detected (Nd). Not determined (-).

The β-glucosidase, BGL, showed the highest activity (56 U/mg) on 4-nitrophenyl-β-D-glucopyranoside (pNPG), with some minor activity (5.74 U/mg) observed on 4-nitrophenyl-β-D-cellobioside (pNPC) (Table 2). CBHI displayed comparable activities (~0.3 U/mg) on carboxymethylcellulose sodium (CMC-Na), pNPC and pNPG. The β-glucosidase activity displayed by CBHI was unexpected as the enzymes are reported to require at least a minimum of three subsites to initiate catalysis [27,28]. CBHII, on the other hand, displayed activity on both CMC-Na and cellopentaitol. The endo-glucanase, EGII, displayed high activity (44 U/mg) on CMC-Na. Overall, the activities of the cellulases were generally in agreement with the literature [26].

2.3. Enzyme Cocktail Formulation

Paper sludge hydrolysis by CBHI or CBHII and/or EGII was performed in the presence of BGL and the concentration of glucose released was measured after 48 h. A BGL dosage of 10% of the overall cellulase loading was found to be sufficient for converting all the cello-oligosaccharides into glucose (data not shown), and as a result, this BGL loading was used in all subsequent paper sludge hydrolysis evaluations.

Numerous studies have evaluated synergism between cellulolytic enzymes and the hydrolysis of cellulose, particularly: (i) exo–exo synergism, whereby cellobiohydrolases (CBHs) hydrolyse cellulose at opposite cellulose fibril ends and (ii) endo–exo synergism between endo-glucanases (EGs) and CBHs, whereby endo-glucanases cleave amorphous regions of cellulose exposing more chain ends for CBHs [6,29].

In this study, enzyme synergism between the cellulolytic enzymes, BGL, CBHI, CBHII and EGII was assessed using paper sludge as a substrate. These experiments were conducted in order to formulate a "cellulolytic core set" for the efficient hydrolysis of paper sludge cellulose to glucose. The binary enzyme combinations between CBHI and CBHII, were synergistic at all enzyme combinations considered, with an increase from 2.5–2.9 mg/mL of glucose by the individual enzymes at 100% dosage to approximately 3.7 to 4.0 mg/mL of glucose release by the combinations of CBHI and CBHII (Figure 3). The yield obtained by the optimal binary combination (CBHI and CBHII at 50: 50%) corresponded to approximately 22% glucan conversion after 48 h.

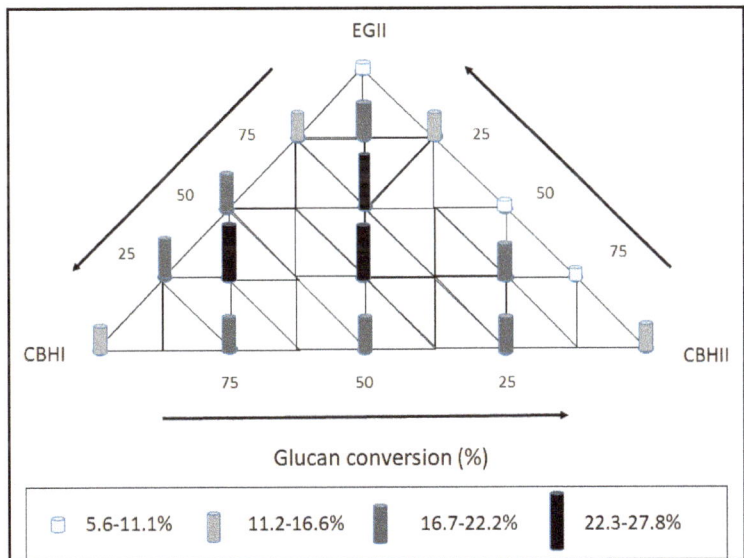

Figure 3. Glucose yields obtained from the optimisation of the formulated cellulase cocktail (TeCel7A (CBHI), ClCel6A (CBHII) and TrCel5A (EGII)) on the hydrolysis of 2% (w/v, dry mass basis) paper sludge for 48 h at 50 °C, at a protein loading of 1.875 mg/g biomass with PcBGL1B (BGL) at 10% protein loading. All experiments were performed in quadruplicate and the mean values were calculated. The different shades of the bars indicate data sets that exhibited differences which were statistically significant (one-way ANOVA, $p < 0.05$).

The classical exo–exo-cellulase model of synergistic enzymatic hydrolysis of cellulose proposes that CBHI starts catalysis at the reducing end of cellulose chains, whereas CBHII initiates activity on the non-reducing, opposite chain ends of cellulose, and has been confirmed [30,31].

Exo-cellulase (CBHI or CBHII) and endo-cellulase (EGII) synergism was also assessed during paper sludge hydrolysis. Figure 3 indicates that CBHI synergised with EGII and that the enzyme combinations where CBHI was present at a higher protein ratio to EGII, released the most glucose. At 75% CBHI to 25% EGII protein dosage, approximately 3.3 mg/mL of glucose was released from 2% (w/v) paper sludge, which corresponded to approximately 20% glucan conversion after 48 h.

According to Wood and McCrae (1979), the endo–exo-cellulase (s) model of synergism involves endo-cellulases (endo-glucanases) which attack the amorphous regions within cellulose, creating new and more chain ends for exo-cellulases (cellobiohydrolases I and or II) to attack [32,33]. In turn, exo-cellulases expose new amorphous regions within the bulk substrate and this then necessitates re-application of endo-cellulase activity. Data attesting to this interdependence between endo-glucanases and cellobiohydrolases during cellulose hydrolysis has been reported by others [7].

The CBHII to EGII combinations exhibited a lower level of glucose release than that exhibited by CBHII at 100% dosage (2.5 mg/mL), except for 25% CBHII to 75% EGII which released comparable glucose concentrations to 100% CBHII. It is possible that the CBHII could be competing with EGII for the same subsites (amorphous regions) on the paper sludge biomass since CBHII exhibited residual endo-glucanase activity when evaluated on CMC as substrate (see Table 2). As a result, this could have led to the anti-synergism (competition for the same binding sites) observed between the two cellulases during paper sludge hydrolysis (see Figure 3). Previous studies have reported that, in addition to the processive exo-activity from the non-reducing ends of cellulose chains, CBHIIs also occasionally have an endo-acting character [6,34,35].

The ternary combinations of CBHI, CBHII and EGII generally showed the highest synergistic effect compared to the binary combinations of the same enzymes, with equal amounts of the three enzymes at 33.3: 33.3: 33.3% or a higher proportion of cellobiohydrolases to EGII, particularly CBHI, releasing the highest glucose of approximately 4 and 5 mg/mL (22.2 and 27.8% glucan conversion), respectively, from paper sludge (Figure 3). Previous studies reported that the best enzyme combinations for cellulose hydrolysis were those with higher proportions of cellobiohydrolases (CBHI and CBHII) to the endo-glucanase, EGII [36,37]. Therefore, it seems that highly crystalline cellulosic biomass substrates such as paper sludge and Avicel cellulose require a higher proportion of the crystalline region-specific CBHs compared to the amorphous region-specific EGs (in this study, a ratio of 66.6% CBHI/II to 33% EGII was required). Two other substrate parameters; the degree of polymerisation (DP) and the fraction of β-glucosidic bonds accessible to cellulase, reaction time and enzyme loading seem to affect enzyme activity and the synergistic interactions between cellulase mono-components [38,39]. Väljamäe and co-workers [38] showed a correlation between the DP length and crystallinity of bacterial cellulose (BC) which was acid treated to various degrees, with a decrease in DP length being accompanied by an increase in CrI. In addition, the study showed that the relative activity of CBHI increased with an in increase in CrI and a decrease in the DP of BC, while the relative activity of EG decreased with an increase in CrI and decrease in the DP of BC [38]. Similarly, Zhang and Lynd [39] showed that DP length of cellulose negatively affects the relative activity of exo-cellulases. In addition, the study showed that the EG concentration required for maximal synergism was 1 EG molecule to 17 exo-cellulase molecules. Den Haan and co-workers [22] also showed that a 3.4 mg/g biomass mixture of cellulases in a ratio of CBHI: CBHII: EGII of 11:5:1 released more glucose from Avicel compared to the individual mono-component cellulases at >20 mg/g biomass. It has been postulated that the lower surface area limits EG adsorption on the surface of cellulose such that hydrolysis of many β-glucosidic bonds cannot occur until they are made accessible by the action of exo-cellulases such as CBHI or CBHII [40]. However, EGII was the largest contributing factor for successful synergism in the current study, as the enzyme was highly efficient even at very low dosages of 33.3%, leading to nearly a doubling effect in paper sludge glucan conversion (see Figure 3). Similarly, previous studies have indicated that the addition of minute quantities of EG to CBHI or a mixture of CBHI and CBHII induced an increase in the saccharification of cellulose [22,35,38].

CBHI appeared to be the only mono-component whereby an increase in its proportion in the ternary mixture was tolerated without affecting glucose yields (Figure 2). Similarly, a recent study observed that a minimum amount of surface area was needed for a single cellulase to be effective in hydrolysing cellulose and the hydrolytic efficiency will decrease when the total cellulase concentration exceeds saturation. Therefore, the use of excess enzymes should be avoided, even if their production costs are negligible [41]. The formulated cellulase cocktail consisting of 33.3:33.3:33.3% of CBHI, CBHII and EGII with supplementation of BGL at 10% protein dosage (henceforth referred to as Opt CelMix) was used in all subsequent studies.

2.4. Effect of Opt CelMix Loading on the Hydrolysis Yields of Paper Sludge

The Opt CelMix was used to hydrolyse the paper sludge (2% w/v, DM) at different enzyme loadings (Figure 4). Saccharification yields of approximately 90% was obtained at protein loadings exceeding 7.5 mg protein/g biomass. The yields are in agreement with the more than 80% yield obtained from softwood (pine pulp) derived paper sludge after 48 h of hydrolysis at 50 °C and pH 5.0 with the NS-22086 multi-enzyme preparation (Novozymes A/S) [42].

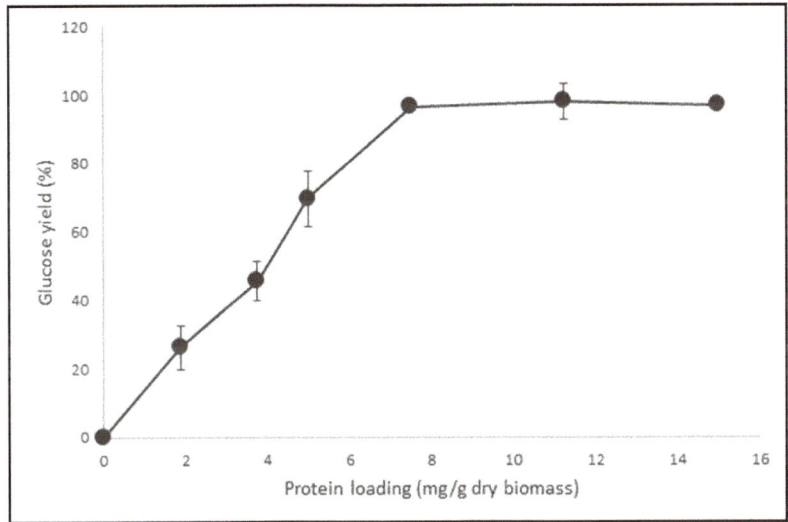

Figure 4. Glucose yield obtained from the hydrolysis of 2% (w/v, DM) paper sludge for 48 h at 50 °C using varying protein loadings (1.875 to 15 mg/g biomass) of the formulated cellulase cocktail with BGL at 10% protein loading. All experiments were performed in quadruplicate, and the mean values and error bars (±SD) were calculated. The statistical analysis showed that the differences were statistically significant (one-way ANOVA, $p < 0.05$).

The saccharification ability of the Opt CelMix (108 mg sugars/mg enzyme) on paper sludge was five-fold higher than that (approximately 27.3 mg sugars/mg enzyme) reported by Hu and colleagues with the hydrolysis of steam pretreated corn stover (SPCS) after 72 h [43] and by Malgas and co-workers with the hydrolysis of steam pre-treated hardwoods after 24 h [26]. This was not surprising, as some authors promote the amenability of paper sludge for enzymatic hydrolysis compared to raw wood or untreated plant materials since paper sludge generally has negligible quantities of hemicellulose and lignin that contribute to biomass recalcitrance.

2.5. Scanning Electron Microscopy (SEM)

SEM visualisation was used to analyse the changes in the size and surface morphology of the paper sludge microfibrils upon cellulase hydrolysis, wherein the 48 h hydrolysates by the Opt CelMix at 7.5 mg/g biomass were analysed (Figure 5). The untreated (with no enzyme addition) paper sludge microfibrils are present as thick and long ribbons with a smooth and regular surface (Figure 5A). In contrast, the surface and dimensions of the microfibrils changed significantly after enzyme treatment (Figure 5B). After cellulase treatment, the paper sludge fibrils appeared to have dislocations on the surface of the microfibrils. Enzyme treatment of paper sludge microfibrils led to a partial decrease in width, from 19.12 to 13.42 µm, and a significant reduction in length, from ≥2 mm to 93 µm.

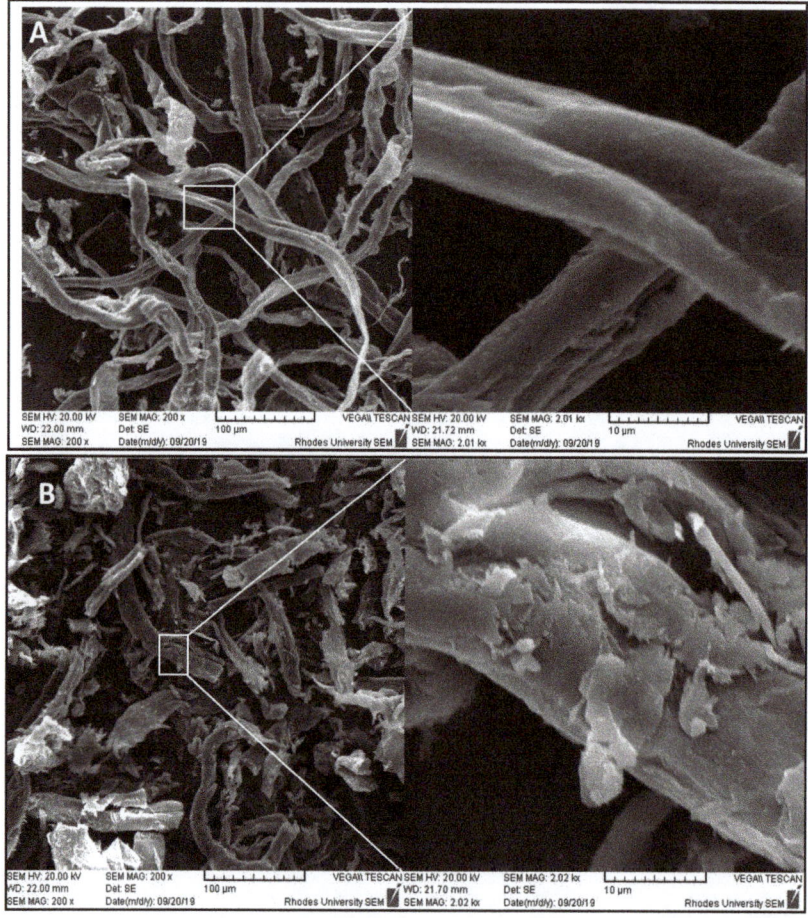

Figure 5. Scanning electron microscopic (SEM) analysis of paper sludge: surface morphology of (**A**) paper sludge before enzyme treatment and (**B**) solid residues after enzymatic hydrolysis by the Opt CelMix at a protein loading of 7.5 mg/g biomass. Two magnifications for each sample were used; 200× magnification allows the observer to see fibril length reduction while 2000× magnification allows the observer to see fibril surface erosion upon enzyme hydrolysis.

2.6. Evaluation of the Performance of the Opt CelMix at Varying Paper Sludge Loadings

Fermentation processes that produce ethanol at concentrations lower than 4% (w/v) require subsequent distillation of the product (with an extensive energy consumption) [44]. Therefore, increasing substrate loading during hydrolysis and fermentation steps is essential to make ethanol production economically feasible. Therefore, the performance of the Opt CelMix (at an enzyme dosage of 15 mg/g biomass) was investigated at various paper sludge loadings (1–20% w/v) (Figure 6).

The Opt CelMix enzyme cocktail released consistent levels of glucose (~80% yield) across the substrate loadings of 1 to 10% with a decrease in saccharification yield at higher substrate loadings (Figure 5). Similar to these findings, a study reported approximately 80% yield for two cocktails; (1) an optimised in vivo thermostable-cellulase mixture and (2) a Cellic CTec2® to Cellic HTec® mixture, during the hydrolysis of 8% (*w/v*) steam exploded bagasse [37]. The incomplete hydrolysis at higher paper sludge concentrations could be due to glucose inhibition of the enzymes. This is not a problem during fermentations since the glucose will be utilised by the yeast as its being produced. At 10 to 20% (*w/v*) paper sludge loadings, the saccharification yields obtained by Opt CelMix averaged at approximately 90 mg/mL glucose, this amount being enough to produce in excess of 40 mg/mL or 4% (*w/v*) ethanol.

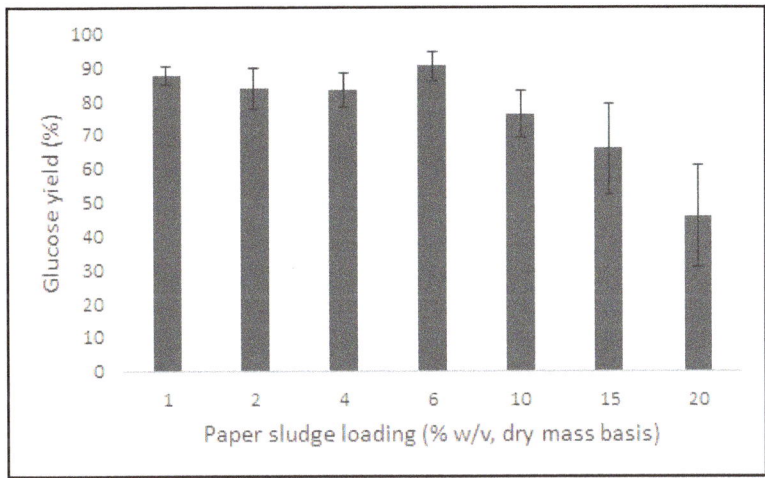

Figure 6. Glucose yield obtained by the action of the formulated cellulase cocktail at 15 mg/g biomass with BGL at 10% protein loading during the hydrolysis of varying substrate loadings (1 to 20% (*w/v*)) paper sludge at 50 °C for 48 h. All experiments were performed in quadruplicate, and the mean values and error bars (±SD) were calculated. The statistical analysis showed that the differences were statistically significant (one-way ANOVA, $p < 0.05$).

The initially high viscosity of lignocellulosic materials at high solid loadings prevents efficient mixing and is associated with poor mass transfer which leads to lowered saccharification of the biomass by the cellulases [45]. For high-water retention materials, such as paper sludge, whereby enzymes have a reduced mobility due to lower free liquid in suspension [46], the Opt CelMix maintained its performance at high solid loads. These results are encouraging with regards to the economic feasibility of using Opt CelMix industrially.

2.7. Comparison of the Hydrolytic Efficiency of the Opt CelMix to Commercial Cellulase Preparations

The hydrolytic performance of the Opt CelMix and the commercially available cellulase preparations (Celluclast® 1.5 L, Cellic® CTec2 and Viscozyme® L) were assessed on paper sludge under "optimal" enzymatic conditions (pH 5.0, 45 rpm, 50 °C) (Figure 7).

After the first 6 h of hydrolysis the Opt CelMix, Cellic® CTec2 and Celluclast® 1.5 L displayed a hydrolytic rate of approximately 1.2 mg/mL/h glucose, while Viscozyme® L displayed a rate of 0.35 mg/mL/h glucose (Figure 7A). The hydrolysis yields by Opt CelMix and the different commercial cellulase preparations were then evaluated after 48 h. Comparable yields of about 90% were obtained by the Opt CelMix, Cellic CTec2® and Celluclast® 1.5 L, while Viscozyme® L displayed a yield of 26.4% (Figure 7B). High yields from the conversion of paper sludge by the commercial enzymes were expected as a recent study showed that Celluclast® 1.5 L displayed a hydrolysis yield of 76.8% with the hydrolysis of deinking paper sludge [47].

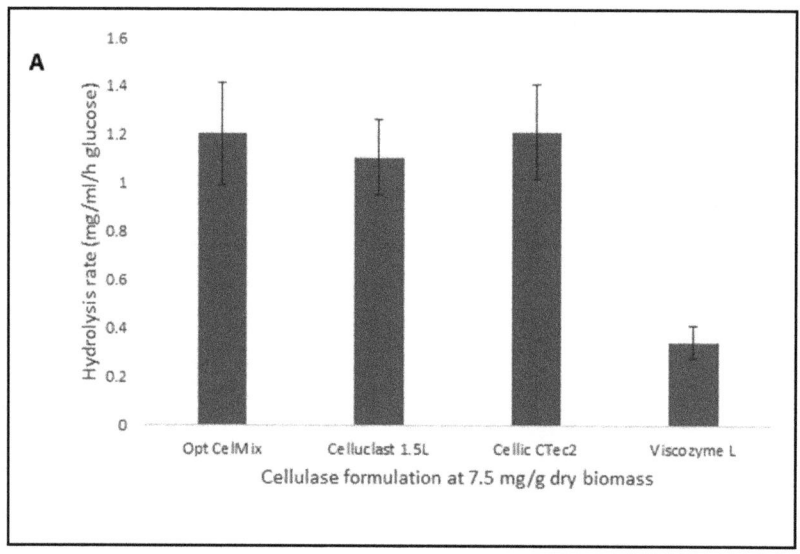

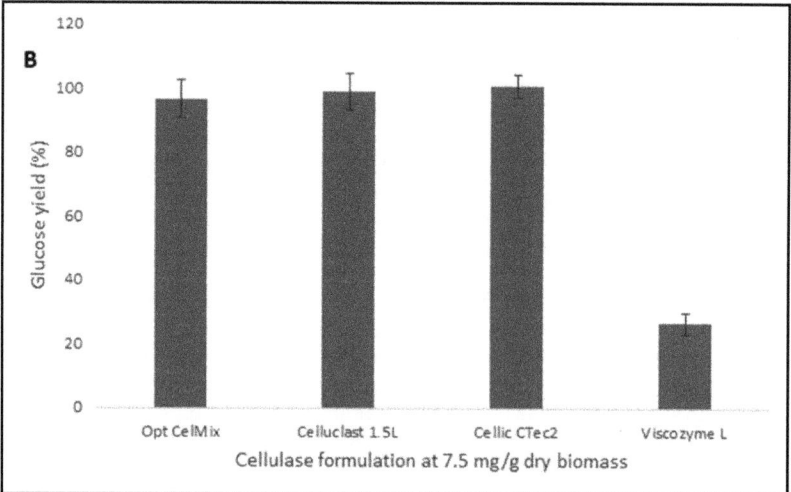

Figure 7. Comparison of (**A**) hydrolytic rates after 6 h and (**B**) glucose yields obtained after 48 h from the hydrolysis of 2% (*w/v*, dry mass basis) paper sludge for 48 h at 50 °C using the Opt CelMix, Celluclast® 1.5 L, Cellic® CTec2 and Viscozyme® L at a protein loading of 7.5 mg/g biomass. All experiments were performed in quadruplicate, and the mean values and error bars (±SD) were calculated. The statistical analysis showed that the differences were statistically significant (one-way ANOVA, $p < 0.05$).

3. Materials and Methods

3.1. Materials

Paper sludge was derived from Norway spruce (*Picea abies*) as residual fibre from a sulphite process (Domsjö Fabriker AB refinery, Örnsköldsvik, Sweden). All the chemicals and substrates (Carboxymethyl cellulose (CMC-Na), Avicel, *p*-nitrophenyl-glucopyranoside (*p*NPG) and *p*-nitrophenyl-cellobioside (*p*NPC)) were purchased from Sigma Aldrich (St. Louis, MI, USA) unless stated otherwise. The 1, 4-β-ᴅ-cellopentaitol was obtained from Megazyme™ (Ireland). The enzyme cocktails (Celluclast®

1.5 L, Cellic CTec2® and Viscozyme® L) were supplied by Novozymes A/S (Denmark). Amicon Pro Centrifugal filters (10 K) were purchased from MERCK (South Africa).

3.2. Paper Sludge Moisture Content Determination

Exactly 10 g of the paper sludge was added to a dry pre-weighed aluminium weighing boat and placed in an oven at 50 °C until a constant weight was achieved. The *% moisture content* of the paper sludge was calculated as follows:

$$\% \text{ moisture content} = \frac{\text{biomass wet weight} - \text{biomass dry weight}}{\text{biomass dry weight}} \times 100\% \qquad (1)$$

3.3. Paper Sludge Chemical Composition

The paper sludge was characterised using a modified sulphuric acid method by the National Renewable Energy Laboratory (NREL) (Golden, CO, USA) [48]. After the two-step acid hydrolysis, the biomass was filtered with the filtrate analysed with regards to acid-soluble lignin (ASL) and monosaccharide content using UV–V is spectroscopy and HPAEC, respectively, while the solid residue was set aside for Klason lignin and ash content estimation. The mass of the solid residue was measured after drying overnight at 105 °C to estimate the Klason lignin content. The mass of the solid residue used for Klason lignin was ignited in a muffle furnace at 525 °C for 6 h, cooled in a desiccator, and weighed to determine the ash content of the paper sludge.

3.4. Paper Sludge Crystallinity Index (CrI)

The crystallinity of paper sludge was determined by X-ray diffraction using Cu K radiation (1.5405 Å, nickel filter) on a Bruker D8® Discover (Bruker, United Kingdom) equipped with a proportional counter. The sample was scanned from 2θ of 10 to 40° with a step size of 0.02°. The determination time was 0.02° per second. The crystallinity index (*CrI*) was then defined as follows:

$$CrI = \frac{I_{002} - I_{am}}{I_{002}} \times 100 \qquad (2)$$

where I_{002} and I_{am} are the intensities of diffraction at 2θ 22.6° (crystalline portion) and 18.9° (amorphous portion), respectively.

3.5. Simon's Staining (SS) of Paper Sludge

The accessibility of cellulose contained in paper sludge to cellulases was assessed using a modified version of Simon's Stain technique with Direct Orange 15 (DO) dye as an adsorbent as described previously [49]. An increase in the DO dye absorption was proportional to increased cellulose accessibility [50]. SS (mg/g pulp) represents the amount of DO dye absorbed by paper sludge.

3.6. Media, Yeast Strains and Culture Conditions

The laboratory *S. cerevisiae* Y294 strain was used as a parental strain for the constitutive expression of the different cellulase genes. The *S. cerevisiae* Y294 [*Tr*EGII], Y294 [*Te*CBHI], Y294 [*Cl*CBHIIb] and Y294 [*Pc*BGLIIB] strains express the recombinant mono-component cellulase genes which produced the *Trichoderma reesei* endoglucanase II (*Tr*Cel5A) [12], the *Talaromyces emersonii* cellobiohydrolase I (*Te*Cel7A) with a carbohydrate binding module (CBM) attached to its carboxyl terminal (Tecbh1-TrCBM-C [11,12], the *Chrysosporium lucknowensis* cellobiohydrolase II (*Cl*Cel6A) [11] and *Phanerochaeta chrysospoprium* β-glucosidase (*Pc*BGL1B) [25], hereafter referred to as EGII, CBHI, CBHII and BGL, respectively.

SC^{-URA} growth medium was prepared in double strength (13.4 g/L yeast nitrogen base without amino acids (BD-Diagnostic Systems, Sparks, MD, USA), 20 g/L glucose, and 3 g/L yeast synthetic dropout medium supplements). The aerobic cultivation of *S. cerevisiae* Y294 strains was performed on a rotary shaker (200 rpm) at 30 °C in 500 mL Erlenmeyer flasks containing 100 mL medium. The supernatant of the individual strains was harvested after 72 h of cultivation through centrifugation for 10 min at 4000× g.

3.7. Preparation of Partially Purified Enzymes

Approximately 15 mL of crude cellulase supernatants were ultra-filtrated using an Amicon Ultra-15 Centrifugal device with a 10 kDa cut off membrane. The samples were centrifuged at 4000× g for 20 min at 4 °C using a swinging-bucket rotor in the Heraeus Megafuge 1.0R (ThermoFischer Scientific, Waltham, MA, USA). The 10 kDa retentate (approximately 0.5 mL) were diluted to 1.5 mL using 50 mM sodium citrate buffer (pH 5) and stored at 4 °C.

3.8. Protein Content and Purity Determination

The protein content of the yeast-derived cellulases was determined by the Bradford method with bovine serum albumin (BSA) used as a suitable protein standard [51]. To determine purity, protein samples were subsequently separated by electrophoresis on denaturing 12% SDS-PAGE gels according to a protocol described previously [52]. Coomassie Brilliant Blue R-250 staining was used to visualise the gels.

3.9. Enzyme Assays

3.9.1. Substrate Specificity Determination

The endo-glucanase, cellobiohydrolase II, beta-glucosidase and cellobiohydrolase I activities were determined using the CMC-Na, 1, 4-β-D-cellopentaitol, pNPG and pNPC substrates, respectively. Standard assay conditions (50 mM sodium citrate buffer (pH 5.0) at 50 °C with continuous agitation at 45 rpm) were followed as previously described [26]. One unit (U) of enzyme activity is defined as the amount of enzyme releasing 1 µmol of glucose per minute under the specified assay conditions [6,7].

3.9.2. Enzyme Cocktail Formulation

The experiments were carried out in quadruplicate at a paper sludge loading of 2% (w/v DW) in 50 mM sodium citrate buffer (pH 5.0) in a 400 µL total volume (with 0.1 mg/mL BSA as a stabiliser) using 1.5 mL safe-lock Eppendorf tubes. Hydrolysis took place at 50 °C with mixing at 45 rpm for up to 48 h. Unless otherwise specified, enzyme loadings were maintained at 1.875 mg/g paper sludge with BGL at 10% of the total protein loading (0.1875 mg/g paper sludge). The aforementioned enzyme loading was used for the synergy studies as it assured greater than 5% glucan conversion yields of paper sludge. Binary and ternary combinations between the cellobiohydrolases; CBHI and CBHII, and the endo-glucanase, EGII, were formulated (Table 3).

The optimal combination for glucose release was selected as the "core cellulase cocktail" (Opt CelMix). The hydrolysis was terminated by boiling for 5 min at 100 °C to inactivate the enzymes. Hydrolysis controls included substrate (without the enzyme) and enzyme controls (without substrate). The samples were stored at 4 °C until analysed.

3.9.3. Effect of Cellulase Cocktail Loading on the Hydrolysis Yields of Paper Sludge

The hydrolysis of 2% (w/v DM) paper sludge by the formulated cellulase enzyme cocktail was evaluated at a protein dosage range of 1.875–15 mg protein/g of paper sludge, following the procedure outlined in Section 3.9.2.

3.9.4. Effect of Paper Sludge Loading on Cellulase Cocktail Hydrolytic Efficiency

Hydrolysis experiments were carried out at 1%, 2%, 4%, 6%, 10%, 15% and 20% (w/v) DM consistency using paper sludge at a cellulase cocktail loading of 15 mg/g of paper sludge to evaluate the effect of paper sludge loading on cellulase enzyme cocktail hydrolytic efficiency, following the procedure outlined in Section 3.9.2.

Table 3. Enzyme combinations (%) at a total protein dosage of 1.875 mg/g biomass for paper sludge hydrolysis synergy studies with BGL added subsequently at 10% of the total protein dosage.

No	CBHI%	CBHII%	EGII%
1.	100	0	0
2.	75	25	0
3.	50	50	0
4.	25	75	0
5.	0	100	0
6.	0	75	25
7.	0	50	50
8.	0	25	75
9.	0	0	100
10.	25	0	75
11.	50	0	50
12.	75	0	25
13.	33.3	33.3	33.3
14.	33.5	33.5	25
15.	12.5	12.5	75
16.	18.75	56.25	25
17.	56.25	18.75	25

3.9.5. Comparison of the Formulated Cellulase Cocktail to Commercial Enzyme Preparations

The hydrolysis of 2% (w/v, DM) paper sludge was conducted using the Opt CelMix, Celluclast® 1.5 L, Cellic CTec2® and Viscozyme® L at a protein loading of 7.5 mg protein/g of paper sludge, following the procedure outlined in Section 3.9.2. Due to the low level of β-glucosidase activity detected in Celluclast® 1.5 L and Viscozyme® L (data not shown), these enzyme cocktails were supplemented with the β-glucosidase preparation, Novozyme® 188 (Novozymes A/S) at 10% of the total protein dosage.

3.10. Analytical Methods

Quantification of glucose in the hydrolysates was performed upon enzymatic hydrolysis using the glucose oxidase/peroxidase (GOPOD) method (K-GLUC, Megazyme, Ireland), while that of reducing sugars was performed using the DNS method with glucose as a suitable standard as described previously [26,53].

Paper sludge *saccharification yield* from all the enzymatic hydrolysis reactions was calculated using the following equation:

$$Saccharification\ yield\ (\%) = \frac{Glucose\ released \times 0.9}{Glucan\ content} \times 100 \qquad (3)$$

3.11. Scanning Electron Microscopy (SEM)

Prior to scanning electron microscopy (SEM), paper sludge samples before and after enzymatichydrolysis were mounted on a metal stub, dried using critical point-drying process and coated with a thin layer of gold prior to SEM analysis. The diameters and lengths of the paper sludge microfibrils from SEM images were quantified using the imaging software Image J. Approximately 30 microfibrils representing each sample were randomly and manually picked for diameter and length analysis.

3.12. Statistical Analysis

All statistical analyses were performed on Microsoft Excel 2013 software using one-way ANOVA. A *p*-value of less than 0.05 was considered to indicate statistically significant differences between compared data sets.

4. Conclusions

The study evaluated an optimised enzymatic cocktail that had been formulated with recombinant cellulases produced by *S. cerevisiae* Y294. The performance of the Opt CelMix was comparable to commercial preparations for paper sludge saccharification. At a protein loading of 7.5 mg/g biomass, over 80% glucose recovery was obtained using the formulated cellulase enzyme cocktail. When the enzyme cocktail was dosed at 15 mg/g biomass, it also showed consistent yields with a glucose recovery of 80% over a substrate loading range of 1 to 10% (*w/v*).

Author Contributions: Conceptualization, W.H.v.Z. and B.I.P.; methodology, S.M. and S.H.R.; software, S.M.; validation, S.M., S.H.R., W.H.v.Z. and B.I.P.; formal analysis, S.M., S.H.R., W.H.v.Z. and B.I.P.; investigation, S.M. and S.H.R.; resources, W.H.v.Z. and B.I.P.; data curation, S.M.; writing—original draft preparation, S.M.; writing—review and editing, S.M., S.H.R., W.H.v.Z. and B.I.P.; visualization, S.M.; supervision, W.H.v.Z. and B.I.P.; project administration, W.H.v.Z. and B.I.P.; funding acquisition, W.H.v.Z. and B.I.P. All authors have read and agreed to the published version of the manuscript.

Funding: This research was funded by the Department of Science and Technology (DST), the National Research Foundation of South Africa (Grant Number 86423-Senior SARChI Chair: Biofuels and other Clean Alternative Fuels), Rhodes University and Stellenbosch University for financial support. Any opinion, findings and conclusions or recommendations expressed in this material are those of the author(s) and therefore the NRF does not accept any liability in regard thereto.

Acknowledgments: The authors would like to thank Adnan Cavka (Sekab E-Technology AB, Örnsköldsvik, Sweden) for providing the paper sludge sample from Domsjö Fabriker AB refinery, Örnsköldsvik, Sweden.

Conflicts of Interest: The authors declare no conflict of interest.

References

1. Prasetyo, J.; Naruse, K.; Kato, T.; Boonchird, C.; Harashima, S.; Park, E.Y. Bioconversion of paper sludge to biofuel by simultaneous saccharification and fermentation using a cellulase of paper sludge origin and thermotolerant *Saccharomyces cerevisiae* TJ14. *Biotechnol. Biofuels* **2011**, *4*, 35. [CrossRef]
2. Gottumukkala, L.D.; Haigh, K.; Collard, F.X.; van Rensburg, E.; Görgens, J. Opportunities and prospects of biorefinery-based valorisation of pulp and paper sludge. *Bioresour. Technol.* **2016**, *215*, 37–49. [CrossRef]
3. Mendes, C.V.T.; Cruz, C.H.G.; Reis, D.F.N.; Carvalho, M.G.V.S.; Rocha, J.M.S. Integrated bioconversion of pulp and paper primary sludge to second generation bioethanol using *Saccharomyces cerevisiae* ATCC 26602. *Bioresour. Technol.* **2016**, *220*, 161–167. [CrossRef] [PubMed]
4. Kang, L.; Wang, W.; Lee, Y.Y. Bioconversion of kraft paper mill sludges to ethanol by SSF and SSCF. *Appl. Biochem. Biotechnol.* **2010**, *161*, 53–66. [CrossRef] [PubMed]
5. Gomes, D.G.; Serna-Loaiza, S.; Cardona, C.A.; Gama, M.; Domingues, L. Insights into the economic viability of cellulases recycling on bioethanol production from recycled paper sludge. *Bioresour. Technol.* **2018**, *267*, 347–355. [CrossRef]
6. Ganner, T.; Bubner, P.; Eibinger, M.; Mayrhofers, C.; Planks, H.; Nidetzky, B. Dissecting and reconstructing synergism: In situ visualization of cooperativity among cellulases. *J. Biol. Chem.* **2012**, *287*, 43215–43222. [CrossRef] [PubMed]
7. Kostylev, M.; Wilson, D. A distinct model of synergism between a processive endocellulase (TfCel9A) and an exocellulase (TfCel48A) from *Thermobifida fusca*. *Appl. Environ. Microbiol.* **2014**, *80*, 339–344. [CrossRef] [PubMed]
8. Malgas, S.; Thoresen, M.; van Dyk, S.J.; Pletschke, B.I. Time dependence of enzyme synergism during the degradation of model and natural lignocellulosic substrates. *Enzyme Microb. Technol.* **2017**, *103*, 1–11. [CrossRef]

9. Thoresen, M.; Malgas, S.; Pletschke, B.I. Enzyme adsorption-desorption and evaluation of various cellulase recycling strategies for steam pre-treated Eucalyptus enzymatic degradation. *Biomass Convers. Biorefinery* **2020**. [CrossRef]
10. Rizk, M.; Antranikian, G.; Elleuche, S. End-to-end gene fusions and their impact on the production of multifunctional biomass degrading enzymes. *Biochem. Biophys. Res. Commun.* **2012**, *428*, 1–5. [CrossRef]
11. Ilmén, M.; Den Haan, R.; Brevnova, E.; McBride, J.; Wiswall, E.; Froehlich, A.; Koivula, A.; Voutilainen, S.P.; Siika-Aho, M.; La Grange, D.C.; et al. High level secretion of cellobiohydrolases by *Saccharomyces cerevisiae*. *Biotechnol. Biofuels* **2011**, *4*, 30. [CrossRef] [PubMed]
12. Mhlongo, S.I.; den Haan, R.; Viljoen-Bloom, M.; van Zyl, W.H. Lignocellulosic hydrolysate inhibitors selectively inhibit/deactivate cellulase performance. *Enzyme Microb. Technol.* **2015**, *81*, 16–22. [CrossRef] [PubMed]
13. den Haan, R.; van Rensburg, E.; Rose, S.H.; Görgens, J.F.; van Zyl, W.H. Progress and challenges in the engineering of non-cellulolytic microorganisms for consolidated bioprocessing. *Curr. Opin. Biotechnol.* **2015**, *33*, 32–38. [CrossRef] [PubMed]
14. Davison, S.A.; den Haan, R.; van Zyl, W.H. Heterologous expression of cellulase genes in natural *Saccharomyces cerevisiae* strains. *Appl. Microbiol. Biotechnol.* **2016**, *100*, 8241–8254. [CrossRef]
15. Voutilainen, S.P.; Murray, P.G.; Tuohy, M.G.; Koivula, A. Expression of Talaromyces emersonii cellobiohydrolase Cel7A in *Saccharomyces cerevisiae* and rational mutagenesis to improve its thermostability and activity. *Protein Eng. Des. Sel.* **2010**, *23*, 69–79. [CrossRef]
16. Kroukamp, H.; den Haan, R.; van Zyl, J.-H.; van Zyl, W.H. Rational strain engineering interventions to enhance cellulase secretion by *Saccharomyces cerevisiae*. *Biofuels Bioprod. Biorefining* **2018**, *12*, 108–124. [CrossRef]
17. Davison, S.A.; den Haan, R.; van Zyl, W.H. Exploiting strain diversity and rational engineering strategies to enhance recombinant cellulase secretion by Saccharomyces cerevisiae. *Appl. Microbiol. Biotechnol.* **2020**. [CrossRef]
18. Park, S.; Baker, J.O.; Himmel, M.E.; Parilla, P.A.; Johnson, D.K. Cellulose crystallinity index: Measurement techniques and their impact on interpreting cellulase performance. *Biotechnol. Biofuels* **2010**, *3*, 10. [CrossRef]
19. Segal, L.; Creely, J.; Martin, A.; Conrad, C. An empirical method for estimating the degree of crystallinity of native cellulose using the x-ray diffractometer. *Text. Res. J.* **1959**, *29*. [CrossRef]
20. Malgas, S.; Kwanya Minghe, V.M.; Pletschke, B.I. The effect of hemicellulose on the binding and activity of cellobiohydrolase I, Cel7A, from *Trichoderma reesei* to cellulose. *Cellulose* **2020**, *27*, 781–797. [CrossRef]
21. Thygesen, A.; Oddershede, J.; Lilholt, H.; Thomsen, A.B.; Ståhl, K. On the determination of crystallinity and cellulose content in plant fibres. *Cellulose* **2005**, *12*, 563–576. [CrossRef]
22. den Haan, R.; van Zyl, J.M.; Harms, T.M.; van Zyl, W.H. Modeling the minimum enzymatic requirements for optimal cellulose conversion. *Environ. Res. Lett.* **2013**, *8*, 025013. [CrossRef]
23. Tsukada, T.; Igarashi, K.; Yoshida, M.; Samejima, M. Molecular cloning and characterization of two intracellular β-glucosidases belonging to glycoside hydrolase family 1 from the basidiomycete *Phanerochaete chrysosporium*. *Appl. Microbiol. Biotechnol.* **2006**, *73*, 807–814. [CrossRef]
24. Rath, A.; Glibowicka, M.; Nadeau, V.G.; Chen, G.; Deber, C.M. Detergent binding explains anomalous SDS-PAGE migration of membrane proteins. *Proc. Natl. Acad. Sci. USA* **2009**, *106*, 1760–1765. [CrossRef] [PubMed]
25. Njokweni, A.P.; Rose, S.H.; Van Zyl, W.H. Fungal β-glucosidase expression in *Saccharomyces cerevisiae*. *J. Ind. Microbiol. Biotechnol.* **2012**, *39*, 1445–1452. [CrossRef] [PubMed]
26. Malgas, S.; Chandra, R.; Van Dyk, J.S.; Saddler, J.N.; Pletschke, B.I. Formulation of an optimized synergistic enzyme cocktail, HoloMix, for effective degradation of various pre-treated hardwoods. *Bioresour. Technol.* **2017**, *245*, 52–65. [CrossRef]
27. Nakamura, A.; Tsukada, T.; Auer, S.; Furuta, T.; Wada, M.; Koivula, A.; Igarashi, K.; Samejima, M. The tryptophan residue at the active site tunnel entrance of *Trichoderma reesei* cellobiohydrolase Cel7A is Important for initiation of degradation of crystalline cellulose. *J. Biol. Chem.* **2013**, *288*, 13503–13510. [CrossRef]

28. Zhong, L.; Matthews, J.F.; Hansen, P.I.; Crowley, M.F.; Cleary, J.M.; Walker, R.C.; Nimlos, M.R.; Brooks, C.L.; Adney, W.S.; Himmel, M.E.; et al. Computational simulations of the *Trichoderma reesei* cellobiohydrolase I acting on microcrystalline cellulose Iβ: The enzyme-substrate complex. *Carbohydr. Res.* **2009**, *344*, 1984–1992. [CrossRef]
29. Igarashi, K.; Koivula, A.; Wada, M.; Kimura, S.; Penttila, M.; Samejima, M. High speed atomic force microscopy visualizes processive movement of *Trichoderma reesei* cellobiohydrolase I on crystalline cellulose. *J. Biol. Chem.* **2009**, *284*, 36186–36190. [CrossRef]
30. Hoshino, E.; Shiroishi, M.; Amano, Y.; Nomura, M.; Kanda, T. Synergistic actions of exo-type cellulases in the hydrolysis of cellulose with different crystallinities. *J. Ferment. Bioeng.* **1997**, *84*, 300–306. [CrossRef]
31. Väljamäe, P.; Sild, V.; Pettersson, G.; Johansson, G. The initial kinetics of hydrolysis by cellobiohydrolases I and II is consistent with a cellulose surface-erosion model. *Eur. J. Biochem.* **1998**, *253*, 469–475. [CrossRef] [PubMed]
32. Wood, T.M.; McCrae, S.I. Synergism Between Enzymes Involved in the Solubilization of Native Cellulose. In *Hydrolysis of Cellulose: Mechanisms of Enzymatic and Acid Catalysis*; Advances in Chemistry; American Chemical Society: Washington, DC, USA, 1979; Volume 181, pp. 181–209. ISBN 0-8412-0460-8.
33. Jalak, J.; Kurašin, M.; Teugjas, H.; Väljamä, P. Endo-exo synergism in cellulose hydrolysis revisited. *J. Biol. Chem.* **2012**, *287*, 28802–28815. [CrossRef] [PubMed]
34. Boisset, C.; Fraschini, C.; Schülein, M.; Henrissat, B.; Chanzy, H. Imaging the enzymatic digestion of bacterial cellulose ribbons reveals the endo character of the cellobiohydrolase Cel6A from *Humicola insolens* and its mode of synergy with cellobiohydrolase Cel7A. *Appl. Environ. Microbiol.* **2000**, *66*, 1444–1452. [CrossRef] [PubMed]
35. Boisset, C.; Pétrequin, C.; Chanzy, H.; Henrissat, B.; Schlein, M. Optimized mixtures of recombinant *Humicola insolens* cellulases for the biodegradation of crystalline cellulose. *Biotechnol. Bioeng.* **2001**, *72*, 339–345. [CrossRef]
36. Woodward, J.; Lima, M.; Lee, N.E. The role of cellulase concentration in determining the degree of synergism in the hydrolysis of microcrystalline cellulose. *Biochem. J.* **1988**, *255*, 895–899. [CrossRef] [PubMed]
37. Kallioinen, A.; Puranen, T.; Siika-Aho, M. Mixtures of thermostable enzymes show high performance in biomass saccharification. *Appl. Biochem. Biotechnol.* **2014**, *173*, 1038–1056. [CrossRef]
38. Väljamäe, P.; Sild, V.; Nutt, A.; Pettersson, G.; Johansson, G. Acid hydrolysis of bacterial cellulose reveals different modes of synergistic action between cellobiohydrolase I and endoglucanase I. *Eur. J. Biochem.* **1999**, *266*, 327–334. [CrossRef]
39. Zhang, Y.H.P.; Lynd, L.R. A functionally based model for hydrolysis of cellulose by fungal cellulase. *Biotechnol. Bioeng.* **2006**, *94*, 888–898. [CrossRef]
40. Zhang, Y.H.P.; Lynd, L.R. Determination of the number-average degree of polymerization of cellodextrins and cellulose with application to enzymatic hydrolysis. *Biomacromolecules* **2005**, *6*, 1510–1515. [CrossRef]
41. Zhou, J.; Wang, Y.H.; Chu, J.; Luo, L.Z.; Zhuang, Y.P.; Zhang, S.L. Optimization of cellulase mixture for efficient hydrolysis of steam-exploded corn stover by statistically designed experiments. *Bioresour. Technol.* **2009**, *100*, 819–825. [CrossRef]
42. Przybysz Buzała, K.; Kalinowska, H.; Przybysz, P.; Małachowska, E. Conversion of various types of lignocellulosic biomass to fermentable sugars using kraft pulping and enzymatic hydrolysis. *Wood Sci. Technol.* **2017**, *51*, 873–885. [CrossRef]
43. Hu, J.; Arantes, V.; Saddler, J.N. The enhancement of enzymatic hydrolysis of lignocellulosic substrates by the addition of accessory enzymes such as xylanase: Is it an additive or synergistic effect? *Biotechnol. Biofuels* **2011**, *4*, 36. [CrossRef] [PubMed]
44. Huang, W.D.; Percival Zhang, Y.H. Analysis of biofuels production from sugar based on three criteria: Thermodynamics, bioenergetics, and product separation. *Energy Environ. Sci.* **2011**, *4*, 784–792. [CrossRef]
45. Mukasekuru, M.R.; Hu, J.; Zhao, X.; Sun, F.F.; Pascal, K.; Ren, H.; Zhang, J. Enhanced high-solids fed-batch enzymatic hydrolysis of sugar cane bagasse with accessory enzymes and additives at low cellulase loading. *ACS Sustain. Chem. Eng.* **2018**, *6*, 12787–12796. [CrossRef]
46. Gomes, D.; Gama, M.; Domingues, L. Determinants on an efficient cellulase recycling process for the production of bioethanol from recycled paper sludge under high solid loadings. *Biotechnol. Biofuels* **2018**, *11*, 111. [CrossRef]

47. Steffen, F.; Janzon, R.; Saake, B. Enzymatic treatment of deinking sludge–effect on fibre and drainage properties. *Environ. Technol.* **2018**, *39*, 2810–2821. [CrossRef]
48. Sluiter, J.B.; Ruiz, R.O.; Scarlata, C.J.; Sluiter, A.D.; Templeton, D.W. Compositional analysis of lignocellulosic feedstocks. 1. Review and description of methods. *J. Agric. Food Chem.* **2010**, *58*, 9043–9053. [CrossRef]
49. Chandra, R.; Ewanick, S.; Hsieh, C.; Saddler, J.N. The characterization of pretreated lignocellulosic substrates prior to enzymatic hydrolysis. *Biotechnol. Prog.* **2008**, *24*, 1178–1185. [CrossRef]
50. Chandra, R.P.; Saddler, J.N. Use of the Simons' staining technique to assess cellulose accessibility in pretreated substrates. *Ind. Biotechnol.* **2012**, *8*, 230–237. [CrossRef]
51. Bradford, M.M. A rapid and sensitive method for the quantitation of microgram quantities of protein utilizing the principle of protein-dye binding. *Anal. Biochem.* **1976**, *72*, 248–254. [CrossRef]
52. Laemmli, U.K. Cleavage of Structura l Proteins during the Assembly of the Head of Bacteriop hage T4. *Nature* **1970**, *227*, 680–685. [CrossRef]
53. Miller, G.L. Use of Dinitrosalicylic Acid Reagent for Determination of Reducing Sugar. *Anal. Chem.* **1959**, *31*, 426–428. [CrossRef]

© 2020 by the authors. Licensee MDPI, Basel, Switzerland. This article is an open access article distributed under the terms and conditions of the Creative Commons Attribution (CC BY) license (http://creativecommons.org/licenses/by/4.0/).

Article

Study of Curing Characteristics of Cellulose Nanofiber-Filled Epoxy Nanocomposites

Mohan Turup Pandurangan * and Krishnan Kanny

Department of Mechanical Engineering, Durban University of Technology, Durban 4001, South Africa; kannyk@dut.ac.za
* Correspondence: mohanp@dut.ac.za; Tel.: +2-73-1373-2831

Received: 24 June 2020; Accepted: 13 July 2020; Published: 24 July 2020

Abstract: In recent years, much attention was focused on developing green materials and fillers for polymer composites. This work is about the development of such green nanofiller for reinforcement in epoxy polymer matrix. A cellulose nanofiber (CNF)-filled epoxy polymer nanocomposites was prepared in this work. The effect of CNF on curing, thermal, mechanical, and barrier properties of epoxy polymer is evaluated in this study. CNF were extracted from banana fiber using acid hydrolysis method and then filled in epoxy polymer at various concentration (0–5 wt.%) to form CNF-filled epoxy nanocomposites. The structure and morphology of the CNF-filled epoxy nanocomposites were examined by scanning electron microscopy (SEM) and transmission electron microscopy (TEM) analysis. Curing studies shows CNF particles acts as a catalytic curing agent with increased cross-link density. This catalytic effect of CNF particles has positively affected tensile, thermal (thermogravimetry analysis and dynamic mechanical analysis) and water barrier properties. Water uptake test of nanocomposites was studied to understand the barrier properties. Overall result also shows that the CNF can be a potential green nanofiller for thermoset epoxy polymer with promising applications ahead.

Keywords: thermoset polymer; epoxy; cellulose nanofiber; curing characteristics; thermal properties; mechanical properties

1. Introduction

Over the past two decades, much research attention focused on developing synthetic nanofillers such as carbon nanotubes (CNT), nanoclays, TiO_2, SiO_2, $CaCO_3$, and Al_2O_3, etc., for polymeric matrices. These nanofillers were reinforced in a range of polymeric matrices such as thermosets, thermoplastics, elastomers and their blends to develop polymer nanocomposites with nanofillers as reinforcement material. Dramatic improvements in thermal, mechanical, chemical, and physical properties were observed in nanoparticle filled in polymeric matrix composites [1–5]. However, in recent years because of the environmental pollution, rapid depletion in natural resources, recycling, and greenhouse carbon gas emission effects, researchers started looking for alternative green nanofillers for polymeric materials [6–8].

Some of the green nanofillers such as eggshell-based nanoparticles [9,10], plant-based materials [11–14] such as nanocellulose, lignin and hemicellulose, polylactic acid (PLA) [15,16], chitin/chitosan [17–19], etc., were the focus in recent past. Considerable breakthrough had been obtained with comparable thermo-mechanical properties using these green nanofillers to that of synthetic nanofillers.

In particular, plant-based cellulosic particles have gained a significant interest in the research communities because of their abundant availability and cost effective processing methods. Cellulose is abundantly available worldwide from plant source. Their availability is higher than that of commonly produced synthetic materials such as polyethylene, polypropylene, PET, glass fibers, and carbon fibers. One observation suggests that cellulose fibril has a comparable tensile property to that of plain carbon steel material. Cellulose is an important component in the plant source which induces structural properties such as strength, stiffness, thermal, and physical stability. In general, cellulose content varies from 20 wt.% to 90 wt.% in plant source depending on the type of the plant [20–25].

Plants have three major chemical constituents, namely, cellulose, hemicellulose, lignin and other chemicals at minor level. Successful attempts have been made in the recent past regarding extraction of cellulosic material from plant source by chemical and physical methods [26–28]. Cellulose is available in the form of microfibril, nanocrystal, or nanofibers depending upon the obtainable size and processing methods [29,30].

These cellulosic particles were filled in polymer matrix to develop micro or nanocomposites, depending upon the size of the cellulosic particle fillers. In general, most of the literature focused on the addition of cellulosic particles in thermoplastic polymeric matrices, such as PP (polypropylene), LDPE (low density polyethylene), PLA (Polylactic acid), etc., [31–34]. The literature on cellulosic particles filled in thermoset polymers are relatively scarce or negligibly available [35]. It is important to understand the effect of these green nanoparticles on thermoset polymers such as epoxy, polyester, and urethanes. Thermoset epoxy and epoxy composites are widely used from bicycle industry to aerospace industry, because of their high thermal, mechanical, and barrier properties than that of polymeric materials [36,37]. A possible replacement of synthetic fillers with cellulosic nanofiber (CNF) fillers in thermosets will have huge impact on economic and social benefits.

Therefore, understanding the effect of CNF on thermal, mechanical, and barrier properties of the epoxy polymer matrix is vital as the study has significant industrial application. This paper focuses on the effect of CNF in epoxy polymer matrix on curing characteristics, structure, thermal, and mechanical properties. The sub-micron sized CNF were extracted from the stem section of banana fibers using acid hydrolysis method and the fibers were reinforced in the epoxy polymer to form epoxy-CNF nanocomposites.

2. Results and Discussion

Figure 1a shows the photograph image of raw banana fiber and CNF extracted from banana fiber. The raw banana fibers appear as a brown colored long fiber extracted from stem section of the banana plant. The CNF appears as a bright white colored particle, with above 95% cellulose content as obtained from our earlier study, with the CNF yield of ~28% from initial banana fiber mass content [38]. Figure 1b shows SEM image of cellulose nanofibers (CNF). The diameter of CNF varies from 50 nm to 100 nm and length 100 nm to 1000 nm. Figure 2 shows the TEM image of 2 wt.% CNF-filled epoxy nanocomposite (Figure 2a) and 5 wt.% CNF-filled epoxy nanocomposite (Figure 2b) observed under bright field mode. In this mode the CNF phase is represented by dark discontinuous needle like phase and the matrix phase is represented by bright continuous phase. Dispersion of CNF particles on the matrix was uniform at lower (2 wt.%) and higher concentration (5 wt.%). However, CNF particles tend to form agglomeration at higher concentration (5 wt.%).

Figure 1. Fiber and cellulose nanofiber (CNF) images shows (**a**) photography image of raw banana fiber and extracted CNF; (**b**) SEM image of CNF fibrils.

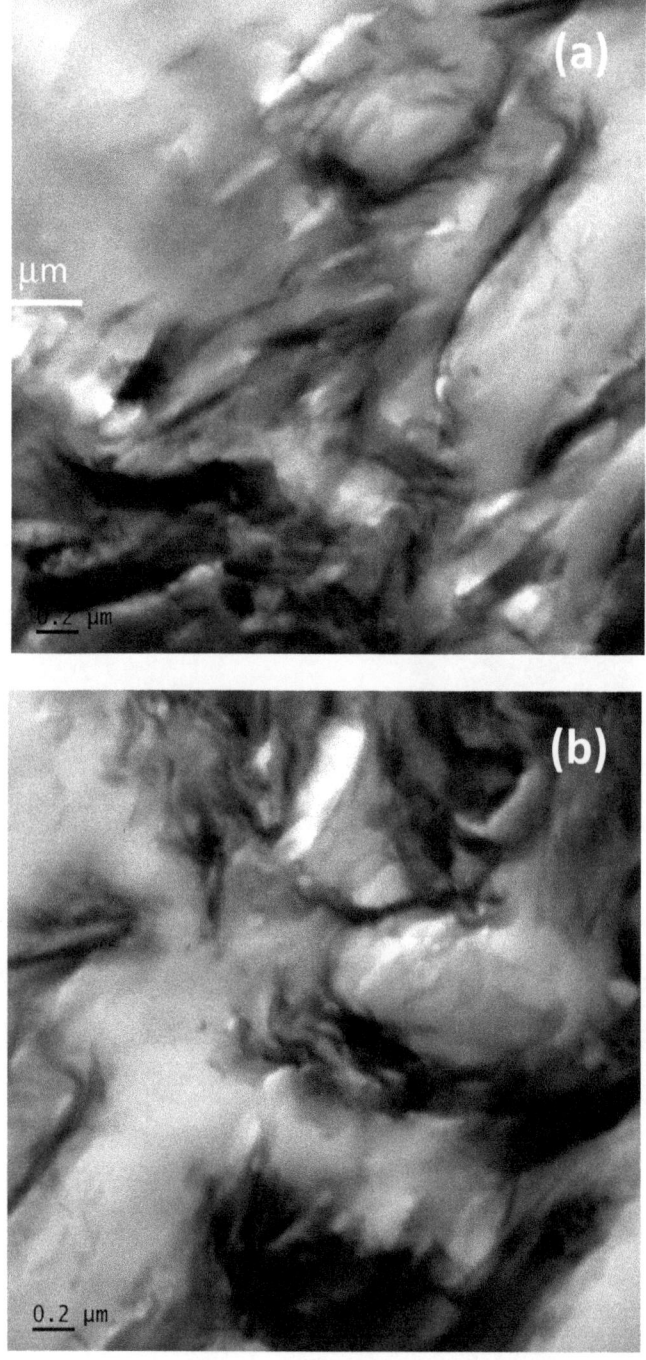

Figure 2. TEM image of epoxy with (**a**) 2 wt.% CNF; (**b**) 5 wt.% CNF-filled nanocomposites.

Figure 3 shows the measurement of heat release during gelation of epoxy and epoxy-CNF series during gelation. Unfilled neat epoxy resin shows the maximum heat release of 76 °C at around 96 min of gelation, whereas, epoxy with CNF up to 1 wt.% shows maximum heat release at relatively lower time around 90 min. The reduction in gelation time suggest the catalytic effect of CNF particles during epoxy curing. CNF with above 1 wt.% shows the gelation peak similar to that of neat epoxy resin, however, with reduced peak width as function of CNF concentration. This suggests the higher concentration may not act as a catalytic agent, however, it may aid faster curing. This curing characteristics of the of nanocomposite series was further examined using the FTIR study. Figure 4 shows the FTIR spectrum of CNF, unfilled neat cured epoxy polymer, 0.5 wt.% and 3 wt.% CNF-filled epoxy nanocomposites. CNF shows a broad band around 3400 cm^{-1} suggesting presence of –OH group. Presence of a shoulder at around 1660 cm^{-1} suggests oxidation of carbohydrate during CNF extraction. A peak around 1102 cm^{-1} is due to C–H stretching of the cellulosic group. A peak at 1102 cm^{-1} suggest changes induced in hydrogen bonds indicating transition from cellulose I to cellulose II structure during chemical treatment. A band at around 1034 cm^{-1} suggests a fraction of xyloglucans associated with non-hydrolyzed hemicellulose which is strongly bound within cellulosic fibrils. A sharp peak at around 888 cm^{-1} suggests a typical cellulosic structure [39,40]. Unfilled neat epoxy polymer shows a characteristics peak of a typical epoxy group (3500 cm^{-1} OH stretching vibration; shoulder at 2950 cm^{-1}–2775 cm^{-1} of CH stretching of CH_2 and CH aromatic or aliphatic vibration; 1608 cm^{-1} of aromatic C=C stretching; 1504 cm^{-1} of C-C stretching; 1031 cm^{-1} C–O–C of ether stretching; 830 cm^{-1} of oxirane group) [41,42]. The nanocomposites show almost all characteristics peaks of epoxy and CNF and suggest the presence of both these phases. However, in nanocomposites the intensity of oxirane peak (830 cm^{-1}) was reduced with the disappearance of carbohydrate oxidation peak of CNF (1660 cm^{-1}) in 0.5 wt.% CNF and 3 wt.% CNF-filled epoxy nanocomposites. This suggests that CNF might induce a catalytic curing of oxirane rings of epoxy polymer. The reduced oxirane peak shows the higher level of cross-link density of the epoxy polymer in nanocomposites due to CNF particles, possibly the oxidized cellulose in presence of amine curing agent induced amide bonding. This "amidated" cellulose molecules might have incorporated covalent bonding with epoxide structure, showing the presence of amide group's band at around 1660 cm^{-1} in nanocomposites. This effect suggests catalytic effect of CNF in epoxy polymer and may also affect the properties of composite.

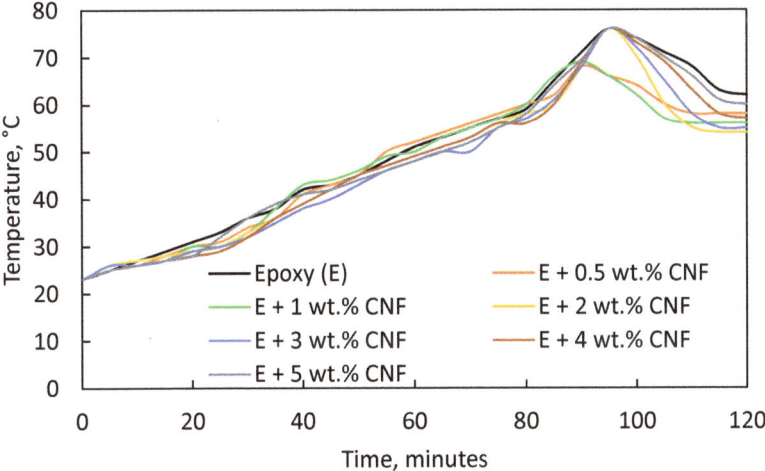

Figure 3. Time–temperature gelation of unfilled and CNF-filled epoxy nanocomposite series.

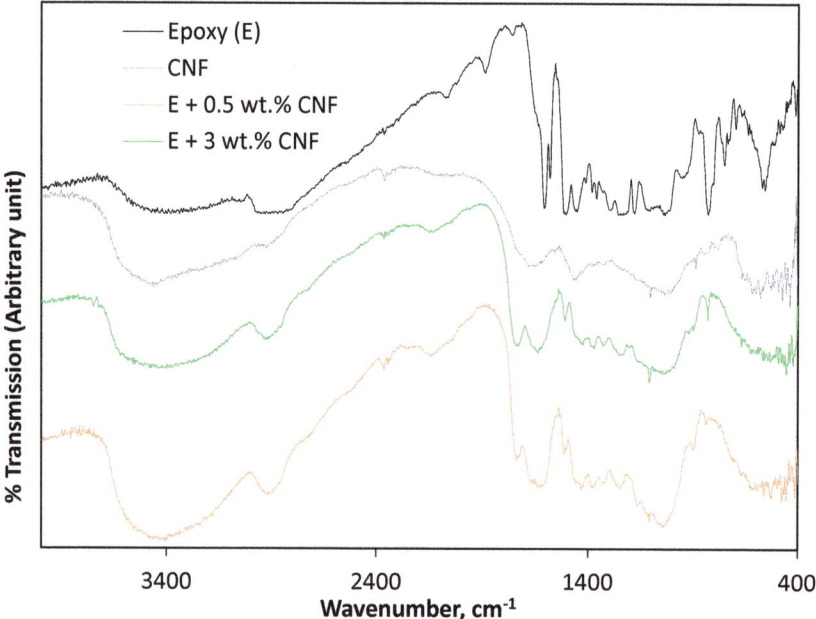

Figure 4. FTIR spectrum of CNF, epoxy and epoxy filled with 0.5 wt.% CNF and 3 wt.% CNF-filled nanocomposites.

Tensile properties of nanocomposites were examined and the tensile stress–strain curves are shown in Figure 5 with their values in Table 1. In general, tensile properties are positively affected because of the addition of CNF particles. Maximum improvement of 86% increased modulus at 5 wt.% CNF, 11% increased strength at 5 wt.% CNF, and an 26% increased elongation at 3 wt.% CNF was obtained than that of unfilled neat epoxy polymer. Although the general trend shows increased tensile properties of nanocomposites, Table 1 shows the tensile values are fluctuating as the function of CNF content. This could be due to the few factors, namely, inconsistent particle size, cellulose concentration within and among the particles, and other non-cellulosic phases presenting in the particles. Since the material was extracted from natural source such level inconsistence properties are expected. Similar results were also obtained elsewhere [8,31,33].

TGA of nanocomposites were examined and their plot is shown in Figure 6 with their values in Table 2. The thermal decomposition of CNF was lower than that of neat banana fiber. This could be due to the larger heat energy exposed by CNF due to its high surface area when compared with banana fiber [43]. Whereas, the thermal stability of the nanocomposite is higher than that of neat epoxy polymer, possible that the nanolevel dispersion of CNF particles acted as a thermal barrier to the epoxy polymer matrix and resulted in improved thermal stability. Similar improved thermal properties were also obtained in CNF-filled thermoplastic polymer nanocomposites [32,44]. Relatively lesser improvement at thermal property at higher content CNF (>3 wt.%) could be due to the higher amount of surface area exposure of particles. Much of the improvement in thermal stability is obtained at 2 to 3 wt.% CNF-filled epoxy nanocomposites.

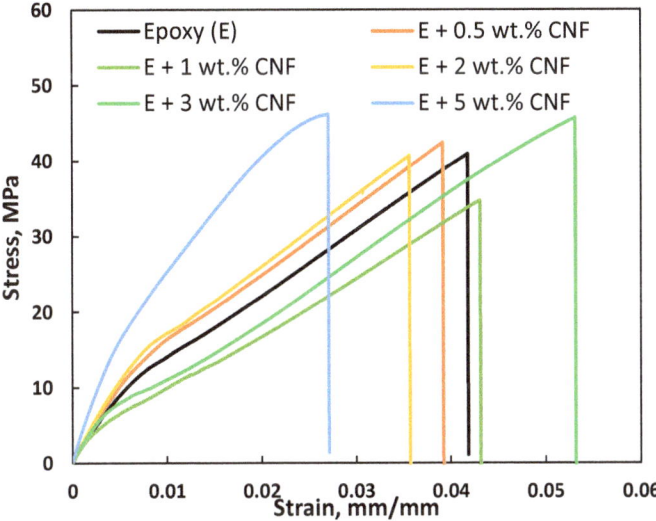

Figure 5. Tensile stress–strain curves of unfilled and CNF-filled epoxy nanocomposite series.

Table 1. Tensile properties of unfilled epoxy and CNF-filled epoxy nanocomposites.

Material	Modulus, GPa	Ultimate Tensile Strength, MPa	Elongation at Break, %
Epoxy (E)	2.2	41.6	4.2
E + 0.5 wt.% CNF	2.2	42.3	3.9
E + 1 wt.% CNF	2.0	34.4	4.3
E + 2 wt.% CNF	2.4	40.6	3.5
E + 3 wt.% CNF	2.1	45.6	5.3
E + 5 wt.% CNF	4.1	46.2	2.7

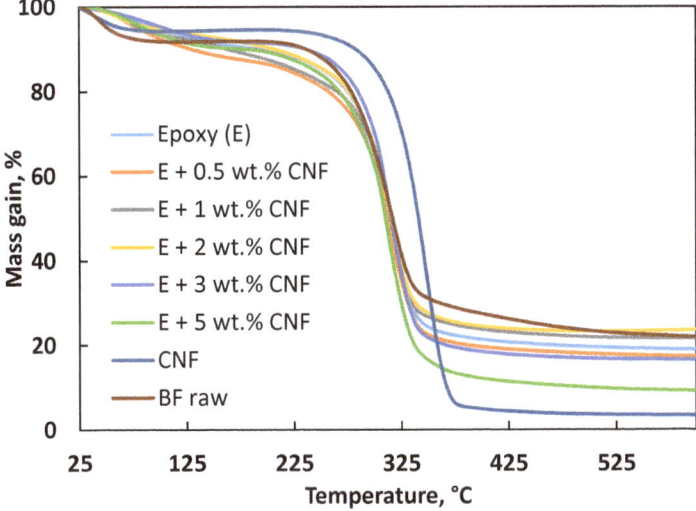

Figure 6. TGA curves of unfilled and CNF-filled epoxy nanocomposite series.

Table 2. TGA properties of unfilled epoxy and CNF-filled epoxy nanocomposites.

Material	Mass Loss at 125 °C, %	Mass Loss at 200 °C, %	Mass Gain at 425 °C, %	Mass Gain at 525 °C, %	Onset Decomposition Temperature, °C	Mass Loss at Onset Decomposition, %	Endset Decomposition Temperature, °C	Mass Gain at Endset Decomposition, %
CNF	5.7	5.4	4.4	3.6	295	12.7	380	5.8
Banana fiber	8.2	8.2	26.2	22.9	255	12.8	350	31.7
Epoxy (E)	7.1	10.1	20.8	19.5	260	16.1	345	25.2
E + 0.5 wt.% CNF	9.5	13.5	19.1	17.8	270	22.8	345	23.3
E + 1 wt.% CNF	7.7	12.4	23.0	21.8	275	22.2	340	28.1
E + 2 wt.% CNF	6.2	9.6	24.0	23.3	265	16.7	340	29.5
E + 3 wt.% CNF	6.4	8.5	17.9	16.8	270	14.8	345	22.7
E + 5 wt.% CNF	8.4	10.6	11.5	9.8	265	19.1	350	16.6

Figure 7 shows the DMA properties of CNF-filled epoxy nanocomposite series with their properties in Table 3. The CNF addition increases the storage modulus of epoxy at room and elevated temperatures (Figure 7a). This improvement could be due to the hard and stiff properties of the CNF particles when compared with epoxy polymer matrix. Figure 7b shows the Tanδ versus temperature values of nanocomposite series. The temperature at which maximum Tanδ value obtained is called glass transition temperature (T_g). T_g is a temperature at which molecular relaxation occurs in the amorphous phase of the polymer and changes toward rubbery phase. Marginal increase in T_g value is obtained in nanocomposites due to the addition of CNF particles. The CNF particles possibly resisted the molecular relaxation movement during heating and increased the T_g temperature. The Tan δ peak value was reduced in composites (>1 wt.% CNF) than that of neat epoxy polymer. Similar results on the reduced Tan δ peak values were observed elsewhere, suggesting effective reinforcement of particles in the polymer matrix [35,45,46]. Tan δ peak value relates to damping and impact characteristics of the material. The reduction in Tan δ peak value in composites (>1 wt.% CNF) shows improved damping/impact characteristics due to increased packing efficiency of the CNF particles in the matrix [45,46]. DMA results show optimized improvement was observed at 2–3 wt.% nanocomposite.

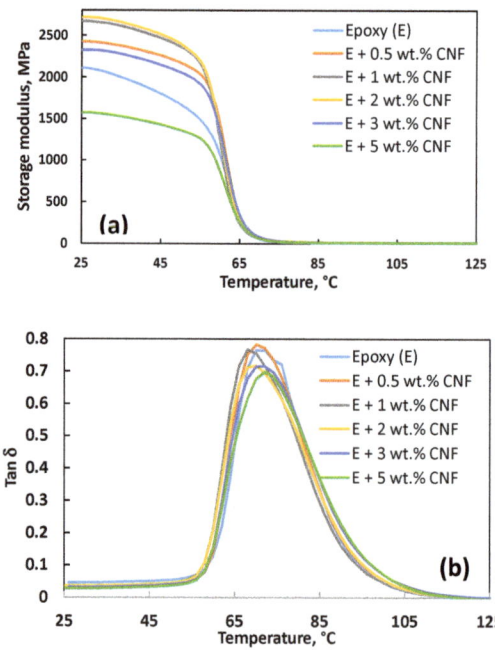

Figure 7. Dynamic mechanical analysis (DMA) of (a) storage modulus; (b) Tan δ of unfilled and CNF-filled nanocomposite series.

Table 3. DMA properties of unfilled epoxy and CNF-filled epoxy nanocomposites.

Material	Storage Modulus at 25 °C, MPa	Storage Modulus at 80 °C, MPa	Storage Modulus at 100 °C, MPa	T_g, °C	Tan δ Peak
Epoxy	2107	11.84	7.31	70	0.765
E + 0.5 wt.% CNF	2424	14.51	8.91	70	0.782
E + 1 wt.% CNF	2656	15.27	9.97	68	0.766
E + 2 wt.% CNF	2708	18.51	10.85	70	0.717
E + 3 wt.% CNF	2323	16.56	9.47	72	0.716
E + 5 wt.% CNF	1570	12.24	6.36	72	0.698

Water uptake properties of nanocomposites is shown in Figure 8 with their values in Table 4. When compared with neat epoxy polymer, water uptake is reduced in CNF-filled epoxy nanocomposites. The reduction in water uptake is proportional to CNF content in epoxy polymer matrix. Maximum reduction of ~47% decreased water uptake was observed in 5 wt.% CNF-filled epoxy nanocomposite than that of neat epoxy polymer. This reduction may be due to the hydrophobic characteristics of CNF particles than that of banana fiber [26–30] and possibly this effect would have resulted in improved water barrier properties of nanocomposites. Moreover, large surface of CNF particles might have exposed a larger area of water medium and served as a barrier medium and protected the matrix polymer against water uptake.

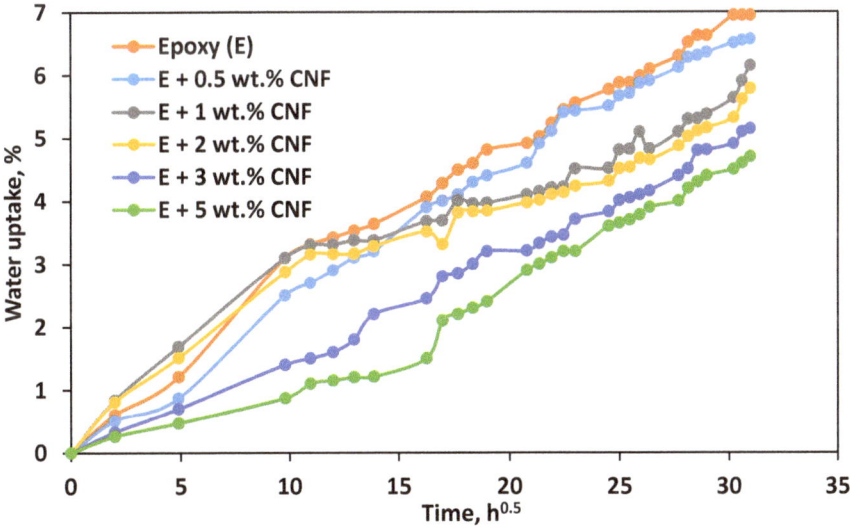

Figure 8. Water uptake of unfilled and CNF-filled epoxy nanocomposite series.

Table 4. Water uptake properties of unfilled epoxy and CNF-filled epoxy nanocomposites.

Material	Equilibrium Water Uptake, %
Epoxy (E)	6.94
E + 0.5 wt.% CNF	6.57
E + 1 wt.% CNF	6.15
E + 2 wt.% CNF	5.78
E + 3 wt.% CNF	5.15
E + 5 wt.% CNF	4.10

3. Materials and Methods

3.1. Raw Materials

The matrix material is a cured thermoset epoxy polymer made up of resin and hardener.

Epoxy resin (Diglycidyl ether of bisphenol-A) and hardener (unmodified cyclic aliphatic amine) supplied under trade name LR-20 and LH-281 respectively were obtained from AMT Composites (Mumbai, South Africa). Banana fibers were obtained from Richmond traders, Mumbai, India. The fibers were mechanically extracted from the stem part of the banana plant. All other chemicals used for the extraction of cellulose nanofibers were obtained from Merck Chemicals, Durban, South Africa.

3.2. Extraction of Cellulose Nanofiber (CNF)

CNF were extracted from banana fibers using acid hydrolysis process. Initially the banana fibers were cut into lengths of 3 cm. CNF extraction process involved three steps, namely, alkaline treatment, bleaching, and acid hydrolysis.

In alkaline treatment, the chopped banana fibers were soaked in 5 wt.% NaOH solution at the weight ratio of 1:10 of fiber to solution content. The fibers were soaked for 24 h after which the fibers were extracted and dried in an oven at 60 °C for 4 h.

In the bleaching process, alkaline-treated fibers were soaked and rinsed in concentrated sodium hypochlorite solution for 1 h. 3 wt.% fraction of fibers were soaked in a solution and bleached in running tap water. The fibers were removed and then placed in an oven at 60 °C for 4 h.

In the acid hydrolysis process, 10% of diluted sulfuric acid was prepared using distilled water in a beaker. Thereafter, bleached fibers were then soaked into the acid solution at 7 wt.% fiber weight fraction. The fiber-solution was stirred using a mechanical stirrer for 30 min. Thereafter, the excess liquid solution was extracted by a vacuum process. The acid-hydrolyzed fibers were then washed with distilled water and dried in an oven at 60 °C for 4 h.

In the above processes, cellulosic phase in the banana fiber was extracted and all other non-cellulosic phases such as lignin, hemicellulose, and other non-cellulosic phases were eliminated.

3.3. Processing of CNF-Filled Epoxy Nanocomposites

Synthesis of nanocomposites were carried out in two steps. First step involved mixing of resin with desired concentration of CNF and the second step involved casting of resin-hardener-CNF mixture in a glass mold. An electric shear mixer (Heidolph MR Hei: Standard, Labotec, South Africa) was used to mix resin and CNF particles. Initially, 100 g epoxy resin was heated in a glass beaker at 80 °C and CNF were added followed by shear mixing at 500 rpm for 0.5 h at 80 °C. The resin and CNF particle mixture were then cooled to room temperature (RT) and 30% weight fraction of hardener to that of epoxy resin (as per supplier's manual) was added into the mixture for curing purpose. The resin, hardener, and CNF together were gently stirred using a glass rod for ~3 min and then cast into glass molds. The casting process involved pouring of resin-hardener-CNF mixture in between two glass mold plates (30 cm × 30 cm × 3 cm) separated by a rubber gasket running along the three sides of the plates. The pouring of the resin mixture was assisted by a runner attached to the top side of the glass plate molds. To facilitate easier removal of the cast product, wax was used as a mold release agent and applied at the inner face of the glass plates and rubber gasket before the pouring process. The nanocomposite cast sample (with dimension of ~27 cm × 27 cm × 3 mm) was obtained 24 h after it was poured inside the mold cavity. The testing and characterization was carried out after seven days of curing (i.e., fully cured state of the epoxy polymer under experimental condition, though 100% curing cannot be achieved). The cured sheets were further cut and sized as per the standard dimension required for conducting testing.

3.4. Characterisation

A scanning electron microscope (SEM) was used to analyze the surface morphology of CNF particles. CNF surfaces were examined by Zeiss Environmental SEM (ESEM: model EVO HD 15 operating at controlled pressure conditions at 20 kV). Before conducting the actual SEM surface analysis, CNF specimens were gold surface-coated using Quorum-150R ES model thin film coating equipment.

A high-resolution transmission electron microscope (TEM) was used to study the dispersion of CNF particles in the epoxy polymer matrix. TEM was carried out on an ultrathin microtomed nanocomposite specimen using JEOL HR-TEM (JEM-2100 series), operating at 120 kV.

Thermogravimetric analysis (TGA) and dynamic mechanical analysis (DMA) were used to examine the thermal properties of nanocomposites using TA instruments SDT Q600 model and Q800 model respectively. In TGA, thermal properties of nanocomposites such as weight loss, decomposition temperature, and degradation were evaluated. In TGA analysis, ~5 mg of the sample was placed in an alumina crucible of TA apparatus and the sample was scanned from RT to 600 °C at a scanning rate of 10 °C/min under atmospheric condition.

DMA testing on nanocomposites was carried out at a frequency of 10 Hz using a 3-point bending mode of TA instrument from 25 °C to 125 °C under atmospheric conditions. In DMA testing, specimen dimension of 5.5 cm × 1 cm × 0.3 cm was used. DMA parameters such as storage modulus, Tanδ (damping factor) and T_g (glass transition temperature) were measured using DMA.

FTIR (Nicolet) analysis was carried out for the cured epoxy and epoxy-CNF nanocomposite series using an attenuated total reflectance (ATR) mode to study the functional group of epoxy and curing characteristics of nanocomposites. Moreover, curing characteristics of epoxy and epoxy-CNF nanocomposite were studied by directly monitoring the exothermic cure temperature at regular time intervals. The temperature of the curing reaction was recorded at regular intervals as soon as the hardener was mixed into the epoxy resin. The time–temperature graph was plotted for all the curing sample series.

3.5. Testing

The tensile test of nanocomposite series was conducted on the cured samples to study the tensile modulus, strength, and elongation properties of CNF-filled epoxy composites. The tensile test was conducted as per ASTM D3039 standard test, with specimen dimension of 5 cm gauge length × 1 cm width × 0.3 cm thick. The tensile test was conducted using an MTS UTM Tensile Tester (Model LPS 304—424708 series) with a crosshead speed of 1 mm/min and 1kN load cell. A mean value of tensile property of three specimens was selected and considered for analysis.

Barrier properties of nanocomposites was studied by a water immersion test method, as per ASTM D570-98 (2005) test procedure at 25 °C. Three test specimens, each of dimension 3 cm × 3 cm × 0.3 cm, were chosen for this study. In order to eliminate surface or subsurface entrapped moisture and retain actual solid mass in the specimen, test samples were dried at 60 °C for 4 h using an oven. During the 4 h drying, samples were taken out of the oven at an interval of 1 h and immediately transferred into the airtight desiccator. The RT-cooled sample was weighed and ensured that the mass loss remained constant until the 4 h heating cycle.

The actual solid mass sample was then immediately taken out of the desiccator and fully immersed in a distilled water medium which was placed in a temperature-controlled water bath set-up. The bath temperature was constantly maintained at 25 °C for entire duration of the water immersion test. The water soaked specimen was taken from the water bath at different time intervals and wiped using a paper towel to eliminate surface water. The sample was then weighed in an electronic balance and then immediately transferred back into the water bath set-up. This weighing procedure was repeated until the water soaked sample showed no or negligible increase in the water mass uptake (i.e., equilibrium water uptake content, W_e). The W_e was measured as per Equation (1)

$$W_e = \frac{(W_t - W_i)}{(W_i)} \times 100 \tag{1}$$

where W_i and W_t are initial dry solid mass of the test specimen and the water mass uptake of soaked sample at time t of testing was done respectively. The barrier property was examined by selecting the mean test specimen water uptake result.

4. Conclusions

In this work, polymer nanocomposites consisting of CNF particles as reinforcement filler and thermoset epoxy polymer were produced by shear mixing process. The prime objective of this study is to prepare a CNF-filled thermoset polymer-based composites. CNF particles were filled up to 5 wt.% in epoxy polymer matrix. The effect of CNF concentration on curing tensile, DMA, TGA, and water uptake properties was evaluated. CNF addition shows positive effect in these properties. FTIR and curing studies shows that CNF may act as a curing catalyst during epoxy gelation and increases the cross-link density of the epoxy polymer and reduces with curing time. Well-dispersed CNF particles were obtained in this processing method. Almost comparable modulus, 10% increased tensile strength and 26% increased elongation were observed in 3 wt.% CNF-filled epoxy nanocomposite. An optimized improved onset and endset decomposition temperature was observed at 2–3 wt.% CNF-filled epoxy nanocomposite. In the DMA study, storage modulus of 2 wt.% CNF-filled epoxy nanocomposite was increased by 28%, 56%, and 48% respectively at 25 °C, 80 °C, and 100 °C respectively when compared with unfilled epoxy polymer. Water uptake results suggest that the water uptake proportionally reduces in nanocomposites as concentration of CNF particles increases in matrix polymer. Maximum 47% reduction of water mass uptake was seen in 5 wt.% CNF-filled epoxy nanocomposite. Overall results suggest that an optimum level of improvement is obtained at 2–3 wt.% CNF-filled epoxy nanocomposite. The result suggests that the CNF can be successfully incorporated in thermoset epoxy polymer matrix with improved properties and serve as a promising green nanofiller for the epoxy matrix.

Author Contributions: M.T.P. conducted the study, analyzed the result and wrote the paper. K.K. was project supervision and fact checking of scientific claims in the manuscript. Both the authors together formulated the objective of the manuscript. All authors have read and agreed to the published version of the manuscript.

Funding: "This research was funded by South African National Research Foundation (NRF), grant number 119779" and "The Research & Postgraduate Directorate of Durban University of Technology, South Africa."

Conflicts of Interest: The authors declare no conflict of interest.

References

1. Valentino, O.; Melanie, M.; George, P.S.; Landi, G.; Neitzert, H.-C. The effect of the nanotube oxidation on the rheological and electrical properties of CNT/HDPE nanocomposites. *Polym. Eng. Sci.* **2017**, *57*, 665–673.
2. Fei, H.; Wenjian, Z.; Armin, T.R.; Nieh, M.P.; Cornelius, C.J. SiO_2-TiO_2-PBC nanocomposite film morphology, solvent swelling, estimated parameter, and liquid transport. *Polymer* **2017**, *123*, 247–257.
3. He, H.; Li, K.; Wang, J.; Sun, G.; Li, Y.; Wang, J. Study on thermal and mechanical properties of nano-calcium carbonate/epoxy composites. *Mater. Des.* **2011**, *32*, 4521–4527. [CrossRef]
4. Yang, Q.; Lin, Y.; Li, M.; Shen, Y.; Nan, C.-W. Characterization of mesoporous silica nanoparticle composites at low filler content. *J. Compos. Mater.* **2016**, *50*, 715–722. [CrossRef]
5. Velmurugan, R.; Mohan, T.P. Epoxy–Clay Nanocomposites and Hybrids: Synthesis and Characterization. *J. Reinf. Plast. Compos.* **2009**, *28*, 17–37. [CrossRef]
6. Xu, Q.; Wang, Y.; Chi, M.; Hu, W.; Zhang, N.; He, W. Porous polymer-titanium dioxide/copper composite with improved photocatalytic activity toward degradation of organic pollutants in wastewater: Fabrication and characterization as well as photocatalytic activity evaluation. *Catalysts* **2020**, *10*, 310. [CrossRef]

7. Giuseppina, L.; Claudio, I.; Giuseppe, V. Photosensitive hybrid nanostructured materials: The big challenges for sunlight capture. *Catalysts* **2020**, *10*, 103.
8. Kargarzadeh, H.; Huang, N.; Lin, I.; Ahmad, M.; Mariano, A.; Dufresne, A.; Thomas, S.; Galeski, A. Recent developments in nanocellulose-based biodgradable polymers, thermoplastic polymers, and porous nanocomposites. *Prog. Polym. Sci.* **2018**, *87*, 197–227. [CrossRef]
9. Punyanich, I.; Aroon, K.; Kavichat, K. The potential of chicken eggshell waste as a bio-filler filled epoxidized natural rubber (ENR) composite and its properties. *J. Polym. Environ.* **2013**, *21*, 245–258.
10. Munlika, B.; Kaewta, K. Biodegradation of thermoplastic starch/eggshell powder composites. *Carbohydr. Polym.* **2013**, *97*, 315–320.
11. Gallo, E.; Schartel, B.; Acierno, D.; Cimino, F.; Russo, P. Tailoring the flame retardant and mechanical performances of natural fibre-reinforced biopolymer by multi-component laminate. *Compos. B Eng.* **2013**, *44*, 112–119. [CrossRef]
12. Zuhri, M.Y.M.; Guan, Z.W.; Cantwell, W.J. The mechanical properties of natural fibre based honeycomb core materials. *Compos. B Eng.* **2014**, *58*, 1–9. [CrossRef]
13. Sanjay, M.R.; Suchart, S.; Jyotishkumar, P.; Jawaid, M.; Pruncu, C.I.; Khan, A. A comprehensive review of techniques for natural fibers as reinforcement in composites: Preparation, processing and characterization. *Carbohydr. Polym.* **2019**, *207*, 108–121.
14. Engin, S.; Hasan, C.; Hakan, D. Production of epoxy composites reinforced by different natural fibers and their mechanical properties. *Compos. B Eng.* **2019**, *167*, 461–466.
15. Mohamadreza, N.; Dilara, S.; Pierre, J.C.; Kamal, M.R.; Heuzey, M.-C. Poly (lactic acid) blends: Processing, properties and applications. *Int. J. Biol. Macromol.* **2019**, *125*, 307–360.
16. Jin, F.-L.; Hu, R.-R.; Park, S.J. Improvement of thermal behaviors of biodegradable poly (lactic acid) polymer: A review. *Compos. B Eng.* **2019**, *164*, 287–295. [CrossRef]
17. Nguyen, M.A.; Wyatt, H.; Susser, L.; Geoffrion, M.; Rasheed, A.; Duchez, A.-C.; Cottee, M.L.; Afolayan, E.; Farah, E.; Kahiel, Z.; et al. Delievery of microRNAs by chitosan nanoparticles to functionally alter macrophage cholesterol efflux in vitro and in vivo. *ACS Nano* **2019**, *13*, 6491–6505. [CrossRef]
18. Nooshin, V.; Farhad, G.; Behjat, T.; Cacciotti, I.; Jafari, S.M.; Omidi, T.; Zahedi, Z. Biodegradable zein film composites reinforced with chitosan nanoparticles and cinnamon essential oil: Physical, mechanical, structural and antimicrobial attributes. *Colloids. Surf. B* **2019**, *177*, 25–32.
19. Monika, Y.; Priynshi, G.; Kunwar, P.; Kumar, M.; Pareek, N.; Vivekanand, V. Seafood waste: A source for preparation of commercially employable chitin/chitosan materials. *Bioresour. Bioprocess.* **2019**, *6*, 8.
20. Mondher, H.; Abderrahim, E.M.; Zouhaier, J.; Akrout, A.; Haddar, M. Static and fatigue characterization of flax fiber reinforced thermoplastic composites by acoustic emission. *Appl. Acoust.* **2019**, *147*, 100–110.
21. Amy, L.; William, P.; Shannon, B.; Frantz, D.; Burholder, J.; Kiziltas, A.; Mielewski, D. Heat-treated blue agave fiber composites. *Compos. B Eng.* **2019**, *165*, 712–724.
22. Dang, C.-Y.; Shen, X.-J.; Nie, H.-J.; Yang, S.; Shen, J.-X.; Yang, X.-H.; Fu, S.-Y. Enchanced interlaminar shear strength of ramie fiber/polypropylene composites by optimal combination of graphene oxide size and content. *Compos. B Eng.* **2019**, *168*, 448–495. [CrossRef]
23. Abraham, E.; Thomas, M.S.; John, C.; Pothan, L.A.; Shoseyov, O.; Thomas, S.; Smith, R.H. Green nanocomposites of natural rubber/nanocellulose: Membrane transport, rheological and thermal degradation characterisations. *Ind. Crop. Prod.* **2013**, *51*, 415–424. [CrossRef]
24. Codispoti, R.; Oliveira, D.V.; Olivito, R.S.; Paulo, B.L.; Fangueiro, R. Mechanical performance of natural fiber-reinforced composites for the strengthening of masonry. *Compos. B Eng.* **2015**, *77*, 74–83. [CrossRef]
25. Mohan, T.P.; Kanny, K. Nanoclay infused banana fiber and its effects on mechanical and thermal properties of composites. *J. Compos. Mater.* **2016**, *50*, 261–276. [CrossRef]
26. Roni, M.d.S.; Wilson, P.F.N.; Hudson, A.S.; Dantas, N.O.; Neto, W.P.F. Cellulose nanocrystals from pineapple leaf, a new approach for thereuse of this agro-waste. *Ind. Crop. Prod.* **2013**, *50*, 707–714.
27. Ligia, M.M.C.; Gabriel, M.d.O.; Bibin, M.C.; Leao, A.L.; de Souza, S.F.; Ferreira, M. Bionanocomposites from electrospun PVA/pineapple nanofibers/Stryphnodendron adstringens bark extract for medical applications. *Ind. Crop. Prod.* **2013**, *41*, 198–202.
28. Li, J.; Wei, X.; Wang, Q.; Chen, J.; Chang, G.; Kong, L.; Su, L.; Liu, Y. Homogeneous isolation of nanocellulose from sugarcane bagasse by high pressure homogenization. *Carbohydr. Polym.* **2012**, *90*, 1609–1613. [CrossRef]

29. Erika, M.; Riccardo, R.; Marco, A.O.; Simone, G.B.; Luciano, P. Comparison of cellulose nanocrystals obtained by sulphuric acid hydrolysis and ammonium persulfate, to be used as coating on flexible food-packaging materials. *Cellulose* **2016**, *23*, 779–793.
30. Fleur, R.; Mohamed, N.B.; Alessandro, G.; Bras, J. Recent advances in surface-modified cellulose nanofibrils. *Prog. Polym. Sci.* **2019**, *88*, 241–264.
31. Maiju, H.; Kristiina, O. Pelletized cellulose fibres used in twin-screw extrusion for biocomposite manufacturing: Fibre breakage and dispersion. *Compos. Part A Appl. Sci. Manuf.* **2018**, *109*, 538–545.
32. Xu, K.; Liu, C.; Kang, K.; Zheng, Z.; Wang, S.; Tang, Z.; Yang, W. Isolation of nanocrystalline cellulose from rice straw and preparation of its biocomposites with chitosan: Physicochemical characterization and evaluation of interfacial compatibility. *Compos. Sci. Technol.* **2018**, *154*, 8–17. [CrossRef]
33. Goeun, S.; Yanqing, L.; van de Ven, T. Transparent composite films prepared from chemically modified cellulose fibers. *Cellulose* **2016**, *23*, 2011–2024.
34. Sinke, H.O.; Christina, D.; Sven, F.; Andres, B.; Engstrand, P.; Norgren, S.; Engström, A.-C. Nanofibrillated cellulose/nanographite composite films. *Cellulose* **2016**, *23*, 2487–2500.
35. Saba, N.; Ahmad, S.; Sanyang, M.L.; Andres, B. Thermal and dynamic mechanical properties of cellulose nanofibers reinforced epoxy composites. *Int. J. Biol. Macromol.* **2017**, *102*, 822–828. [CrossRef] [PubMed]
36. Ziaullah, K.; Yousif, B.F.; Mainul, I. Fracture behaviour of bamboo fiber reinforced epoxy composites. *Compos. B Eng.* **2017**, *116*, 186–199.
37. Zheng, N.; Huang, Y.; Liu, H.-Y.; Gao, J.; Mai, Y.W. Improvement of interlaminar fracture toughness in carbon fiber/epoxy composites with carbon nanotubes/polysulfone interleaves. *Compos. Sci. Technol.* **2017**, *140*, 8–15. [CrossRef]
38. Sifiso, P.N.; Krishnan, K. Extraction of hemicellulose and lignin from sugarcane bagasse for biopolymer films: Green process. *J. Adv. Mater. Process.* **2018**, *6*, 57–65.
39. Heloisa, T.; Franciele, M.P.; Florencia, C.M. Cellulose nanofibers produced from banana peel by chemical and enzymatic treatment. *LWT Food Sci. Technol.* **2014**, *59*, 1311–1318.
40. Alemdar, A.; Sain, M. Biocomposites from wheat straw nanofibers: Morphology, thermal and mechanical properties. *Compos. Sci. Technol.* **2008**, *68*, 557–565. [CrossRef]
41. Patterson, W.A. Infrared absorption bands characteristic of oxirane rings. *Anal. Chem.* **1954**, *26*, 823–835. [CrossRef]
42. Chike, K.E.; Myrick, M.L.; Lyon, R.E.; Angel, S.M. Raman and near infrared studies of an epoxy resin. *Appl. Spectrosc.* **1993**, *47*, 1631–1635. [CrossRef]
43. Ticiane, T.; Luana, A.S.; Simone, M.L.R.; Bica, C.I.D.; Nachtigall, S.M.B. Cellulose nanocrystals from acacia bark–Influence of solvent extraction. *Int. J. Biol. Macromol.* **2017**, *101*, 553–561.
44. Abeer, A.; Amira, E.; Atef, I.; El-Shafei, A.M.; Kamar, M. Extraction of oxidized nanocellulose from date palm (Phoenix Dactylifera, L.) sheath fibers: Influence of CI and CII polymorphs on the properties of chitosan/bionanocomposite films. *Ind. Crop. Prod.* **2018**, *124*, 155–165.
45. George, T.M.; Abraham, E.; Jyotishkumar, P.; Maria, H.J.; Pothen, L.A.; Thomas, S. Nanocelluloses from jute fibers and their nanocomposites with natural rubber: Preparation and characterization. *Int. J. Biol. Macromol.* **2015**, *81*, 768–777.
46. Joseph, S.; Sreekala, M.S.; Oommen, Z.; Koshy, P.; Thomas, S. A comparison of the mechanical properties of phenol formaldehyde composites reinforced with banana fibres and glass fibres. *Compos. Sci. Technol.* **2002**, *62*, 1857–1868. [CrossRef]

© 2020 by the authors. Licensee MDPI, Basel, Switzerland. This article is an open access article distributed under the terms and conditions of the Creative Commons Attribution (CC BY) license (http://creativecommons.org/licenses/by/4.0/).

Article

Conversion of Xylose to Furfural over Lignin-Based Activated Carbon-Supported Iron Catalysts

Annu Rusanen [1,*], Riikka Kupila [2], Katja Lappalainen [1,2], Johanna Kärkkäinen [1], Tao Hu [1] and Ulla Lassi [1,2]

[1] Research Unit of Sustainable Chemistry, University of Oulu, P.O. Box 4300, FIN-90014 Oulu, Finland; katja.lappalainen@oulu.fi (K.L.); johanna.karkkainen@oulu.fi (J.K.); tao.hu@oulu.fi (T.H.); ulla.lassi@oulu.fi (U.L.)
[2] Unit of Applied Chemistry, Kokkola University Consortium Chydenius, University of Jyväskylä, FI-67100 Kokkola, Finland; riikka.kupila@chydenius.fi
* Correspondence: annu.rusanen@oulu.fi; Tel.: +358-294-48-1636

Received: 30 June 2020; Accepted: 20 July 2020; Published: 22 July 2020

Abstract: In this study, conversion of xylose to furfural was studied using lignin-based activated carbon-supported iron catalysts. First, three activated carbon supports were prepared from hydrolysis lignin with different activation methods. The supports were modified with different metal precursors and metal concentrations into five iron catalysts. The prepared catalysts were studied in furfural production from xylose using different reaction temperatures and times. The best results were achieved with a 4 wt% iron-containing catalyst, 5Fe-ACs, which produced a 57% furfural yield, 92% xylose conversion and 65% reaction selectivity at 170 °C in 3 h. The amount of Fe in 5Fe-ACs was only 3.6 µmol and using this amount of homogeneous $FeCl_3$ as a catalyst, reduced the furfural yield, xylose conversion and selectivity. Good catalytic activity of 5Fe-ACs could be associated with iron oxide and hydroxyl groups on the catalyst surface. Based on the recycling experiments, the prepared catalyst needs some improvements to increase its stability but it is a feasible alternative to homogeneous $FeCl_3$.

Keywords: furfural; carbon-supported catalyst; xylose conversion; iron; heterogeneous catalysts

1. Introduction

Furfural is an important biomass-based high-value chemical with numerous applications [1]. While it is widely used directly as a solvent and a fungicide, it is most commonly converted into pharmaceuticals, chemicals and biopolymers, many of which are used as substitutes for petrochemical-derived analogs [2,3].

Approximately 300,000 tons of furfural are produced annually and the biggest manufacturer is China [4]. It is formed by the dehydration of pentoses, mainly xylose, which can be obtained through the hydrolysis of different agricultural residues (corn stover, wheat straw, sugarcane bagasse, rice husk, oat hull) or forest industry wastes (birch or poplar sawdust). Current industrial processes use mineral acids, such as sulfuric, phosphoric or hydrochloric acid, as catalysts, with an approximate furfural yield of 50% [5,6]. The relatively low yield, high energy consumption (caused by the high, 150–240 °C, reaction temperature) and environmental concerns related to acidic process wastes are driving scientists to develop catalysts with high selectivity for furfural formation. A number of studies have used inexpensive water-soluble inorganic salts (mainly chlorides) as catalysts instead of mineral acids for xylose conversion to furfural [7–10]. Many of those studies show that $FeCl_3$ results in the highest furfural yields compared to other metal chlorides [7,10].

Compared to water-soluble salts, solid catalytic systems have fewer environmental impacts than liquid ones, reduce operational costs and are technically feasible alternatives for industry [11]. A broad range of different heterogeneous catalysts have been studied for xylose conversion to furfural, the most common of which are acidic zeolites and mesoporous silicas, such as SBA-15 and MCM-41 [11,12]. Additionally, sulfonated metal oxides (especially TiO_2, ZrO_2) have been used, either as such or combined with mesoporous supports [13–16]. Many studies have shown that some Brønsted acidity is needed, either from metal oxide or support, in addition to metal oxides with a Lewis character, to achieve good furfural yields [17,18]. The need for Brønsted acidity is associated with the furfural production mechanism; while Lewis acid can isomerize xylose to xylulose, Brønsted acid is needed for the dehydration step (Scheme 1) [19].

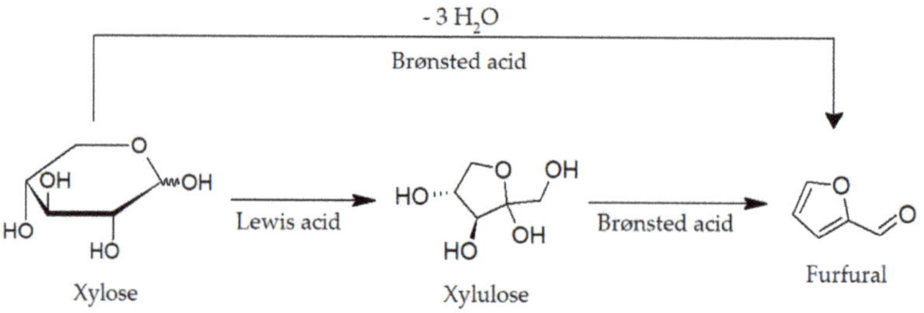

Scheme 1. Dehydration of xylose to furfural by Brønsted acid and Lewis acid catalysts.

Carbon-based catalysts are attractive since the carbon supports are low-cost materials with high surface area and good thermal stability and they are easily modified with functional groups [20]. Carbon surface groups containing heteroatoms, such as oxygen, can act as anchoring sites for metal particles and generate high metal dispersion [21]. The diverse surface also enables Brønsted acidity in addition to metal's Lewis acidity, which can further promote xylose conversion to furfural. To the best of our knowledge, few papers have been published on carbon-supported metal oxide catalysts. Mazzotta et al. used a sulfonated carbonaceous material with TiO_2 sites, while Russo et al. used TiO_2/carbon black in the conversion of xylose to furfural [22,23]. Barroso-Bogeat et al. prepared a class of different activated carbon-metal oxide catalysts (Fe, Al, Zn, Sn, Ti and W) but did not use them in any reaction [24–26]. Therefore, cheap metal oxides, such as iron oxide, have not been utilized in xylose conversion to furfural, even though iron is common in homogeneous catalysts. The aim of this study was to create iron oxide sites on a carbon-supported catalyst and apply the catalyst in the conversion of xylose to furfural. Activated carbon support was derived from hydrolysis lignin, which is a side stream of cellulosic ethanol production. Reactions were performed in biphasic media to increase reaction selectivity.

2. Results and Discussion

2.1. Preliminary Studies

The purpose of the preliminary studies was to optimize the reaction media and select the best catalytic metal for heterogeneous catalysts.

2.1.1. Furfural Partitioning in Biphasic Reaction System

The reaction system was optimized in terms of the appropriate organic solvent and solvent:water ratio. Experiments were carried out using a 4.7 wt% furfural solution in water as feed and toluene or methyl isobutyl ketone (MIBK) as an organic solvent. The purpose was to compare furfural partitioning from water into these two solvents. Solvent ratios of 1:1, 1:2 and 1:3 were used based on the literature [17,27,28]. The results of partitioning experiments are shown in Figure 1, and, as expected, the more organic solvent was added, the better furfural was extracted to the organic phase (Figure 1a). MIBK showed a better ability to extract furfural than toluene, as the furfural content in water was relatively higher in the toluene experiments than in the MIBK experiments. In the best case, when 1:3 water:MIBK was used as a solvent, 96% of furfural was extracted to the MIBK phase. Similarly, MIBK was reported to extract furfural better than toluene in the literature [27,29]. MIBK has a polar carbonyl group structure that may interact better with furfural, which is an aldehyde, compared to nonpolar toluene. The partitioning coefficient of MIBK has been reported to be 7 [29], while in our study it was 8 using a similar solvent ratio (1:1) and 8.5 when the solvent ratio was 1:3 (Figure 1b). For toluene, the partitioning coefficient has been reported to be 3 [29] but in our study it was significantly higher (6) with a 1:1 solvent ratio. Differences in reported and measured partitioning coefficients are likely due to different experimental setups. In the cited study, the partitioning coefficients were determined without any heating, using only shaking as an extraction technique. In our study, 5 min microwave heating (at 160 °C) was used, as the higher temperature corresponds to the reaction conditions in conversion. Based on the partitioning results, MIBK was considered most suitable for the furfural removal from water and it was used as an organic solvent in further experiments.

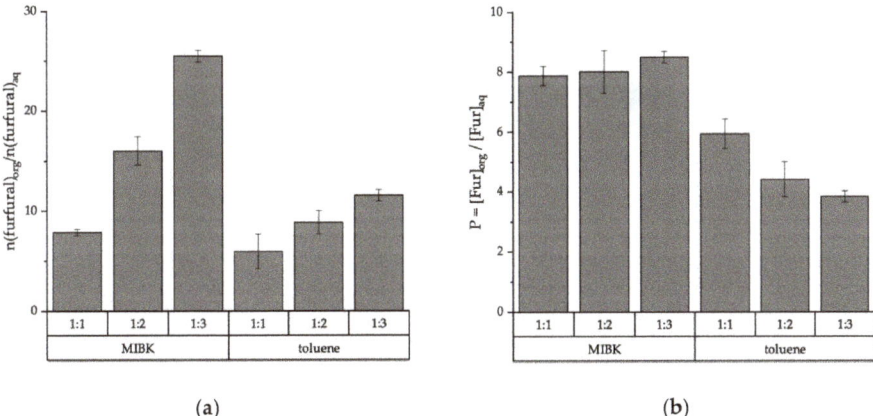

Figure 1. Furfural partitioning between water and methyl isobutyl ketone (MIBK) and water and toluene using different solvent ratios. (**a**) Partitioning calculated as the ratio of moles of furfural in the organic phase and in the water phase; (**b**) Partitioning calculated as the partitioning coefficient (based on concentrations).

2.1.2. Furfural Production Using Homogeneous Catalysts

After choosing the reaction media, homogeneous metal salts ($ZnCl_2$, $AlCl_3·6H_2O$, $CrCl_3·6H_2O$, $SnCl_2·2H_2O$ and $FeCl_3·6H_2O$) were used as catalysts to determine the best metal for furfural production. Different metal salts have been studied in the literature but it was important to perform tests in our reaction system, using microwave heating and water/MIBK media. Reactions were performed at a temperature of 160 °C, a reaction time of 1.5 h and a catalyst amount of 0.05 mmol, based on the literature [10]. Almost full xylose conversion (98%–99%) was achieved using all catalysts other than zinc chloride, which yielded only 91% conversion (Figure 2, asterisks). In the furfural yields, there was variation between catalysts and the yields varied between 33% and 68% (Figure 2, bars). Chromium chloride and aluminum chloride produced the lowest furfural yields, although the xylose conversion was almost complete. Therefore, their selectivities (34% and 38%, respectively) were poorest among all catalysts (Figure 2, squares). Tin chloride and zinc chloride produced slightly higher furfural yields compared to chromium or aluminum and their selectivities were also higher (46% and 48%, respectively). However, the significantly highest furfural yield (68%) and selectivity (70%) were achieved using iron chloride as a catalyst. Our result agrees with the literature, as many studies have shown that $FeCl_3$ produces the highest furfural yield in water media compared to other metal halides [7,10]. The 68% furfural yield achieved in our study with FeCl$_3$ catalyst is also comparable to that found in other studies. For example, Ershova et al. achieved a 60.3% furfural yield in water at 180 °C, while Zhang et al. reported a 77% furfural yield in gamma-valerolactone at 160 °C [10,30]. However, in most studies NaCl has been used as a phase modifier, and, because Cl$^-$ has been shown to catalyze furfural formation, it is difficult to compare those results to our study without NaCl [31–34].

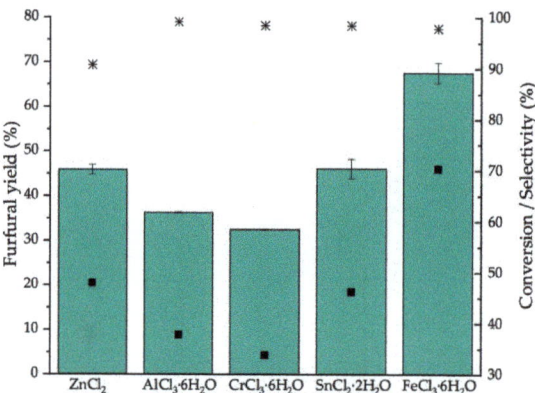

Figure 2. Furfural yield (bar), respective conversion (asterisk) and reaction selectivity (square) during xylose dehydration at 160 °C for 1.5 h using different metal chlorides (0.05 mmol) as catalysts.

2.2. Preparation and Characterization of Heterogeneous Catalysts

After the appropriate reaction media and catalytic metal were chosen, three different catalyst supports (ACz, ACz$_N$, ACs) and five different catalysts (5Fe$_{NO3}$-ACz, 5Fe-ACz, 5Fe-ACz$_N$, 5Fe-ACs and 10Fe-Acs) were prepared (Table 1). In the preparations, different activation methods, metal precursors, metal concentrations and additional treatments were used. ACz support was prepared applying common ZnCl$_2$ chemical activation [35]. ACz-based catalysts 5Fe$_{NO3}$-ACz and 5Fe-ACz were prepared using 5 wt% of either FeNO$_3$ or FeCl$_3$ as an iron precursor, respectively. ACz$_N$ support was prepared using HNO$_3$ treatment after ZnCl$_2$ chemical activation and ACz$_N$-based catalyst 5Fe-ACz$_N$ by further impregnation with FeCl$_3$. The third support type, ACs, was prepared using the physical activation method. Compared to chemical activation, physical activation is performed at a higher temperature but only using water steam as an activation agent [21,36]. ACs-based catalysts 5Fe-ACs and 10Fe-ACs were prepared using FeCl$_3$ as a metal precursor and 5 or 10 wt% as the metal concentration.

Table 1. Key factors in the preparation of various supports and catalysts and the metal contents (wt%) of the prepared catalysts measured by inductively coupled plasma optical emission spectrometry (ICP-OES).

Sample	Activation Method	Other Treatment	Metal Precursor	Initial Fe (wt%)	Measured Metal (wt%)	
					Zn	Fe
ACs	Steam (H$_2$O)	-	-	-	0.01	0.06
5Fe-ACs	Steam (H$_2$O)	-	FeCl$_3$	5	-	4.0
10Fe-ACs	Steam (H$_2$O)	-	FeCl$_3$	10	-	9.2
ACz	ZnCl$_2$	-	-	-	8.2	0.08
5Fe$_{NO3}$-ACz	ZnCl$_2$	-	FeNO$_3$	5	4.6	5.0
5Fe-ACz	ZnCl$_2$	-	FeCl$_3$	5	3.8	4.5
ACz$_N$	ZnCl$_2$	HNO$_3$	-	-	0.07	0.06
5Fe-ACz$_N$	ZnCl$_2$	HNO$_3$	FeCl$_3$	5	0.08	5.5

The metal contents (according to inductively coupled plasma optical emission spectrometry (ICP-OES)) of all supports and catalysts are listed in Table 1. ACs support was analyzed most comprehensively since it was the least treated and therefore provided some indication about the metal content of the biomass-based lignin. According to the results, ACs contained minor amounts of metals, such as Ca (0.47 wt%) and Na (0.57 wt%) but its Fe content was very low (0.06 wt%) (Table S1). ACs-based iron-impregnated catalysts 5Fe-ACs and 10Fe-ACs contained 4.0 and 9.2 wt% iron, respectively. Based on these values, iron impregnation was considered successful, as the target amounts were 5 and 10 wt%. The other support, ACz, contained a significant amount of zinc (8.2 wt%), which originated from the chemical activation step in the preparation process. Zinc was also naturally present in ACz-supported 5Fe-ACz and $5Fe_{NO3}$-ACz (3.8 and 4.6 wt%, respectively) but the amount decreased during the impregnation step. The iron contents for 5Fe-ACz and $5Fe_{NO3}$-ACz were 4.5 and 5.0 wt%, respectively. Because of the remaining zinc, a third catalyst support ACz_N was prepared similarly to ACz but afterwards it was treated with HNO_3 in order to remove remaining zinc. According to the ICP-OES results, this treatment did remove zinc because the remaining amount was only 0.07 wt%. Meanwhile, the iron content of 5Fe-ACz_N was 5.5 wt%.

Additionally, the surface area (SA) according to Brunauer–Emmett–Teller (BET) theory, average pore diameter and pore volume distributions according to density functional theory (DFT) were determined for all supports and catalysts using N_2-physisorption analysis (Table 2). As can be seen from the results, the surface area and pore volume of the chemically activated ACz and ACz_N supports were higher (1470/1091 m^2g^{-1} and 0.72/0.49 cm^3g^{-1}) than for the ACs support prepared with steam activation (760 m^2g^{-1} and 0.47 cm^3g^{-1}) (entries 1, 4 and 7). Further, the total pore volume and the relative amount of micropores were higher in ACz and ACz_N than in ACs. Treatment with HNO_3 after chemical activation decreased the support surface area, average pore diameter and total pore volume but the quantities were still greater than with steam activated ACs.

Table 2. N_2-physisorption analysis from different activated carbon (AC) supports and catalysts.

Entry	Sample	BET			DFT	
		BET SA (m^2/g)	Avg. Pore Diam. (nm)	Total Pore Volume (cm^3/g)	Mesopores (cm^3/g)	Micropores (cm^3/g)
1	ACs	760	2.90	0.47	0.26	0.21
2	5Fe-ACs	455	3.44	0.34	0.23	0.11
3	10Fe-ACs	380	3.15	0.26	0.16	0.10
4	ACz	1470	2.29	0.72	0.31	0.41
5	5Fe-ACz	1000	2.28	0.48	0.19	0.29
6	$5Fe_{NO3}$-ACz	948	2.16	0.45 *	0.15	0.29
7	ACz_N	1091	2.15	0.49	0.16	0.33
8	5Fe-ACz_N	790	2.07	0.35	0.10	0.25

* also contained macro-pores 0.01 cm^3/g (3%).

The surface area and total pore volume of all catalysts decreased while the iron was impregnated on the carbon surface. This is reasonable, as metal is deposited on the surface and goes into the pores (Table 2). For ACs-based catalysts, the surface area decreased from 860 to 455 or 380 m^2/g and total pore volume from 0.47 to 0.34 or 0.26 cm^3/g, depending on how much metal was impregnated on the catalyst surface (entries 1–3). For ACz-based catalysts, the surface area decreased from 1470 to 1000 or 948 m^2/g and the total pore volume from 0.72 to 0.48 or 0.45 cm^3/g, depending on which iron precursor was used (entries 4–6). For ACz_N-based catalysts, the surface area decreased from 1091 to 790 m^2/g and the total pore volume decreased from 0.49 to 0.35 (entries 7–8).

Absolute volumes of the meso- and micropores decreased with all catalysts when iron was impregnated but there were differences in the final pore volume distributions between meso- and micropores, partly because of different initial pore volume distributions of the different supports.

For steam-activated ACs, the micropore volume decreased in 5Fe-ACs and 10Fe-ACs catalysts by 48%–52% compared to ACs. With mesopores, the decrease was smaller, only 12%–38%, meaning that the micropores were primarily filled in ACs when iron was impregnated. Because micropores were more filled with iron than mesopores, the average pore diameter also increased with ACs-based catalysts when compared to plain ACs. Comparing 5Fe-ACs and 10Fe-ACs, the difference in mesopore volume was notable, which meant that 5% iron addition did not affect the mesopores significantly while 10% iron addition filled the mesopores considerably. For chemically activated ACz and ACz$_N$, the micropore volume decreased only 29% or 24%, respectively, while the volume of the mesopores was decreased 39%–52% for ACz and 38% for ACz$_N$. Different precursors did not have a significant effect on surface area or pore volumes. Because the volume of the micropores decreased less than the volume of the mesopores, the relative amount of micropores was increased in chemically activated iron catalysts. In addition, the average pore diameter decreased compared to the supports.

Other surface properties, such as functional groups, acidity and detailed metal composition, were studied using X-ray photoelectron spectroscopy (XPS), the Boehm titration method and X-ray diffraction (XRD). These analyses were only conducted for ACs, 5Fe-ACs and 10Fe-ACs, which were determined to be the most promising ones in the conversion of xylose to furfural (see Section 2.3). XPS results show that plain ACs support already contains some oxygen functionalities (hydroxyl and carboxyl groups) on the surface but iron addition increases the oxygen–carbon ratio of the catalysts compared to ACs support (Table 3, Table S2). The increase is also dependent on how much iron is added, as the percentage of carbon atoms decreases from 96.8 to 93.6 or 86.3% and that of oxygen atoms increases from 2.9 to 4.1 or 9.5% when the amount of iron increased from 0.06 to 4.0 or to 9.2 wt%, respectively. In addition to C, O and Fe, Cl was detected from iron-impregnated catalysts. An increase of oxygen functionalities was detected from an O1s scan at 531.0eV and at 532.5eV (Table S2). The former can originate from carbonyl groups or metal oxides and the latter, for example, from hydroxyl groups or O-Fe bonds [37,38]. From the XPS Fe2p spectra, a peak at 711.3 eV was detected from both 5Fe-ACs and 10Fe-ACs catalysts, indicating the presence of oxidized iron, Fe_2O_3 or FeOOH [37,39]. According to XRD, the iron was mostly present as oxides (Fe_3O_4 and Fe_2O_3, Figure S1), so it is proposed that oxygen content was increased together with iron content as iron oxide. According to XRD, 10Fe-ACs contained mostly Fe_2O_3 (hematite, 01-080-5405) and only small amounts of Fe_3O_4 (magnetite, 04-015-9120). Conversely, 5Fe-ACs contained clearly more Fe_3O_4 than Fe_2O_3. However, no iron chlorides were detected with XRD measurements.

Table 3. Surface analysis of prepared carbon support and catalysts according to X-ray photoelectron spectroscopy (XPS) and Boehm titration.

Sample	XPS [a]				
	Total C-% from C1s	Total O-% from O1s	Total Fe-% from Fe2p	Total Cl-% from Cl2p	Total Acidic Groups (mmol/g) [b]
ACs	96.8	3.0	nd	nd	0.07
5Fe-ACs	93.7	4.1	0.6	1.5	1.77
10Fe-ACs	86.2	9.6	2.4	1.6	1.95

[a] as atom-%, [b] by Boehm titration, nd = not detected.

According to Boehm titration, plain ACs contained a small amount of acidic oxygen functionalities (0.07 mmol/g, Table 3, Figure S2.). The amount of acidic oxygen groups increased when the metal was added to 1.77 or 1.95, depending on the metal amount. Acidic oxygen functionalities are probably formed during iron impregnation as a consequence of HCl formation from $FeCl_3$ hydration in water solution. Barroso-Bogeat et al. showed that metal ions markedly influence the pH of the impregnation solution and thereby the oxidizing power of this solution toward the activated carbon support [40]. Since XPS revealed the potential for the presence of iron hydroxides, it is possible that Brønsted acidity of 5Fe-ACs and 10Fe-ACs is induced by iron hydroxides (e.g., FeOOH) [25,41]. Regardless of the specific nature of the acidic oxygen groups, they favor metal adsorption [42].

The morphology of the physically activated support (ACs) and catalysts (5Fe-ACs and 10Fe-ACs) was observed using a scanning electron microscope (SEM) and a scanning transmission electron microscope (STEM). SEM images clearly revealed particles on the carbon surface for 5Fe-ACs and 10Fe-ACs, while plain ACs did not contain any visible particles (Figure 3). The support and both catalysts showed a very porous structure in SEM as well as in STEM. Figure 4 (and Figure S3) shows chemical mapping of the elements C, O, Fe and Cl using energy-dispersive x-ray spectroscopy in scanning transmission electron microscopy (STEM-EDS). Comparing the distribution of Fe and O, it is clear that both elements appear at the same location, which indicates the presence of iron oxide. Therefore, the results obtained from XPS and XRD showing that the iron particles were oxides were confirmed. Mapping also showed that significant amount of residual chlorine was evenly distributed on the surface. Notable chlorine remains have been also detected in the literature when $FeCl_3$ has been used as a metal precursor [43].

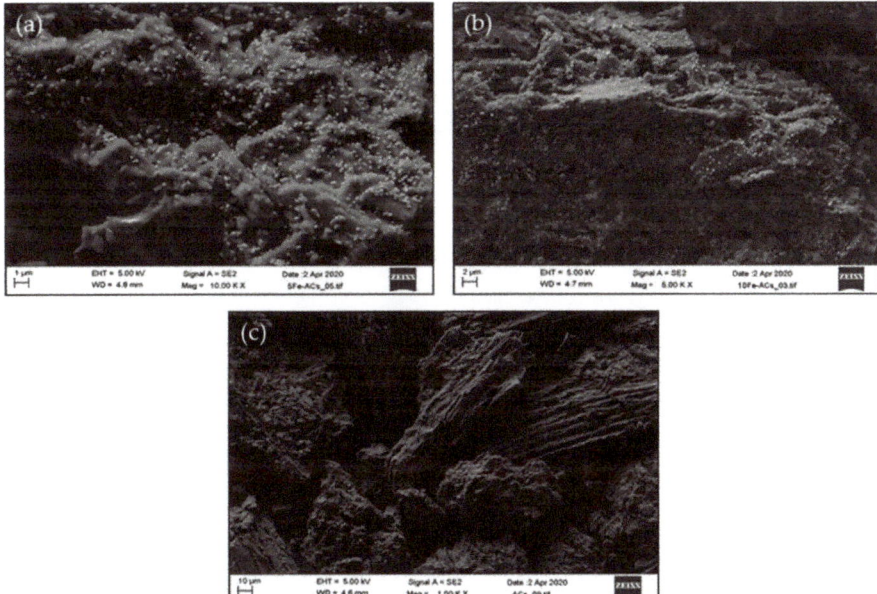

Figure 3. Scanning electron microscopy (SEM) images of 5Fe-ACs (**a**), 10Fe-ACs (**b**) and ACs (**c**). Iron particles are clearly visible in 5Fe-ACs and 10Fe-ACs while ACs shows a porous structure.

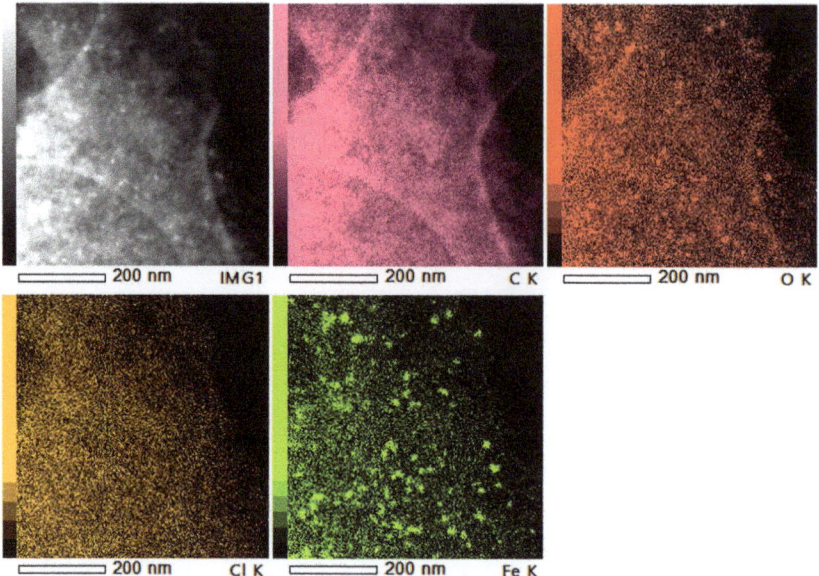

Figure 4. Chemical mapping of 5Fe-ACs using energy-dispersive x-ray spectroscopy in scanning transmission electron microscopy (STEM-EDS). The figure reveals that iron is most likely present as iron oxide on the carbon surface, as iron and oxygen appear in the same locations.

2.3. Furfural Production Using Heterogeneous Catalysts

Conversion studies were started with the control experiment without any support or catalyst (Table 4). This so-called autocatalysis was able to produce a 12% furfural yield and 18% xylose conversion at 160 °C in 1.5 h. Autocatalysis is based on high temperature and pressure, where the dissociation constant of water is increased [44]. In addition, formation of organic acids (e.g., formic and lactic acid) during the reaction might occur, which then further catalyzes the hydrolysis reaction [45]. However, a 160 °C reaction temperature and 1.5-h reaction time represent rather mild conditions and therefore only low conversion and yield were achieved. All supports and catalysts were able to produce higher furfural yields and conversions than the control experiment. First, chemically activated ($ZnCl_2$) carbon support (ACz) was tested and it resulted in good conversion (82%, Table 4, entry 2). The furfural yield was also considerably high (28%). The high conversion is most likely connected to the high zinc content of ACz (Table 1), which originated from chemical activation. Iron was impregnated to the support using an $FeNO_3$ precursor and $FeCl_3$ precursor (Table 4, entries 3¨C4, respectively). NO_3-based salts are commonly favored as precursors since they leave no residue on the catalyst [46,47]. With a nitrate precursor, conversion increased further (from 82 to 91%) compared to ACz support but the yield decreased from 28 to 23% (Table 4, entries 2–3). Therefore, the reaction selectivity toward furfural also decreased (from 36 to 27%). With a chloride precursor, the yield increased (from 28 to 32%) compared to the ACz support and the reaction selectivity also increased (from 36 to 51%, Table 4, entries 2 and 4). Based on the higher furfural yield and reaction selectivity, the $FeCl_3$ precursor was determined to be more suitable than $FeNO_3$ and was used in further catalysts. Similar results were obtained by Chareonlimkun et al., who discovered that chloride-based precursors resulted in higher reactivity compared to nitrate-based precursors in ZrO and TiO catalysts [14]. As mentioned in Section 2.1.2, chloride ions have been shown to enhance furfural formation from xylose by favoring the 1,2-enediol formation before dehydration [34]. This is most probably reason why chlorine-based precursors increase furfural yield and reaction selectivity compared to nitrate-based precursors.

Table 4. Furfural yield, xylose conversion and reaction selectivity using various chemically activated catalysts at 160 °C with a 1.5-h reaction time.

Entry	Catalyst	Yield (%)	Conversion (%)	Selectivity (%)
1	-	12	18	12
2	ACz	28	82	36
3	5Fe$_{NO3}$-ACz	23	91	27
4	5Fe-ACz	32	66	51

Based on the results so far (Table 4), the 5Fe-ACz catalyst appeared to be the most promising catalyst. Therefore, it was used in conversion at different temperatures using various reaction times (Table 5, graphical presentation Figure S4). Temperature and reaction time are highly dependent on each other, so the time was increased at each temperature until the conversion reached 98%. At 160 °C, a 5 h reaction time was needed to achieve full xylose conversion (98%). In these conditions, the furfural yield was 48%. However, with a shorter 3-h reaction time, a similar furfural yield of 47% was achieved with lower conversion (89%). At a higher reaction temperature of 170 °C, the same 98% conversion was achieved in 2.5 h. However, the furfural yield was lower at this temperature with a 2.5 h reaction time (44%) compared to the yield at 160 °C but the highest 50% furfural yield was achieved with a shorter 2 h reaction time at 170 °C. At 180 °C, 99% conversion was obtained in 1.5 h but the furfural yield was low (39%). The highest reaction selectivities (55–56%) were achieved at 160 °C with 4 and 3 h reaction times and at 170 °C with a 2 h reaction time. At 180 °C, reaction selectivity was lowest because of the increasing amount of side reactions. Side reaction products were visible in chromatograms produced by high performance liquid chromatography (HPLC) (Figure S5).

Table 5. Effect of reaction temperature and time on furfural yield, xylose conversion and reaction selectivity using 5Fe-ACz as the catalyst. Y = furfural yield, C = xylose conversion, S = reaction selectivity.

Time	160 °C			170 °C			180 °C		
	Y (%)	C (%)	S (%)	Y (%)	C (%)	S (%)	Y (%)	C (%)	S (%)
1	25	50	54	36	82	46	38	97	41
1.5	32	66	51	44	93	49	39	99	41
2	38	79	51	50	96	55	-	-	-
2.5	38	85	48	44	98	48	-	-	-
3	47	89	55	-	-	-	-	-	-
4	47	95	56	-	-	-	-	-	-
5	48	98	54	-	-	-	-	-	-

Even though good furfural yields and reaction selectivities were achieved with 5Fe-ACz catalyst, there was some problems related to its stability. That is, all zinc already leached out of the catalyst at 160 °C in 3 h. In addition, a significant amount of iron (68–78% from initial, depending on reaction conditions) leached out of the catalyst. Therefore, more catalyst supports and catalysts were prepared and tested (Table 6). ACz$_N$ was chemically activated similarly to ACz but it was treated with HNO$_3$ after activation in order to remove remaining zinc. It produced a 14% furfural yield and 21% xylose conversion in 1.5 h at 160 °C (Table 6, entry 1), which is significantly less compared to ACz. However, the support was now zinc free (Table 1). When iron was impregnated on ACz$_N$, the furfural yield increased from 14 to 22% and the conversion increased from 21 to 47% (Table 6, entries 1–2). The selectivity of 5Fe-ACz$_N$ was similar to that of 5Fe-ACz but the furfural yield and conversion were lower (Table 6, entry 2 and Table 4, entry 4). The third AC support, ACs, was physically steam activated with H$_2$O instead of chemical activation. Its surface area was lower than that of ACz or ACz$_N$ because of the activation method but the conversion and furfural yields were comparable to ACz$_N$ (Table 6, entries 1 and 3). After iron impregnation to ACs, the furfural yield increased from 14 to 25% and the conversion increased from 19 to 36%. The reaction selectivity with 5Fe-ACs was significantly higher than with any other iron catalyst: 72% compared to 27/50/51% (Tables 4 and 6).

Table 6. Furfural yield, xylose conversion and reaction selectivity using various catalysts at 160 °C with a 1.5-h reaction time.

Entry	Catalyst	Yield (%)	Conversion (%)	Selectivity (%)
1	ACz_N	14	21	71
2	5Fe-ACz_N	22	47	50
3	ACs	14	19	81
4	5Fe-ACs	25	36	72

To summarize the experiments so far, the chemical activation method with the AC support produced the best furfural yields before and after iron impregnation (Tables 4 and 6). However, catalyst characterization revealed that zinc chloride activation left some zinc remains in the catalyst support, which further affected its catalytic activity. Furthermore, all the zinc leached out of the catalyst during the first use and therefore the catalyst was suitable for a single use only. HNO_3-treated chemically activated catalyst did not contain zinc remains but its catalytic activity in furfural production was comparable to that of physically activated ACs and its selectivity was lower than that of ACs. In addition, the chemical activation of ACz_N demands significant amounts of $ZnCl_2$, which is toxic to the environment. Therefore, the most reasonable carbon support for iron impregnation would be ACs, which does not require any chemicals other than water for preparation. As a precursor, $FeCl_3$ was more selective than $FeNO_3$. The reaction temperature of 170 °C led to the highest furfural yield and thus it was chosen for further experiments.

Conversion studies were continued with ACs-based catalysts—5Fe-ACs and 10Fe-ACs. In addition, control experiments with plain ACs, without any catalyst and with a similar amount of homogeneous iron (as in 5Fe-ACs) were carried out. Reactions without any catalyst were able to produce at most a 37% furfural yield at 170 °C in 3.5 h (Figure 5, black squares). Conversion increased with time and reached 88% at highest. Conversion was notably higher at 170 °C than at the lower 160 °C reaction temperature and with a shorter 1.5-h reaction time (Table 4). The pH of the water phase also clearly changed to acidic during the reaction. In autocatalyzed reactions, furfural yields are strongly dependent on temperature and time but the highest furfural yields have been found around 50% [6]. Our 37% yield at 170 °C in 3.5 h is in good agreement with the study of Ershova et al., who achieved a 42% furfural yield at 180 °C in 3.75 h [10].

A plain activated carbon support (ACs) produced a similar furfural yield as water with all reaction times (Figure 5, red circles). Conversion was slightly higher, probably due to oxygen functionalities on the carbon surface (Table 3), until after 3 h reaction time it decreased slightly. Additionally, the selectivity was higher with ACs than with plain water. A 40% furfural yield and 77% conversion were the highest achieved with ACs in 3.5 h. When the activated carbon support was impregnated with iron, the furfural yield clearly increased. As expected, iron promoted furfural formation. Yields with both iron catalysts (5Fe-ACs and 10Fe-ACs) were similar with time (Figure 5, blue triangles up and green triangles down, respectively). The yields increased until a 3 h reaction time, after which they leveled off to 55–57% depending on the catalyst. The only exception was the 2.5 h reaction with 10Fe-ACs, in which the furfural yield did not increase when compared to the 2 h reaction. Conversion was similar with both the 5Fe-ACs and 10Fe-ACs catalysts and it increased with time from 59 to 96%. Conversion was clearly higher than with ACs or without a catalyst, which indicated that iron impregnation increased the catalysts' activities. Because the furfural yields and xylose conversions with 10Fe-ACs and 5Fe-ACs were so similar, it was concluded that 4 wt% iron was already enough to increase activated carbon catalyst activity and no benefit was obtained with a higher metal content. In fact, the reaction selectivity was lower with 10Fe-ACs than with 5Fe-ACs. The best reaction selectivity

(67%) was achieved with the 5Fe-ACs catalyst in 2 h but it did not decrease notably when the reaction time was increased to 3 h, which resulted in the highest furfural yield.

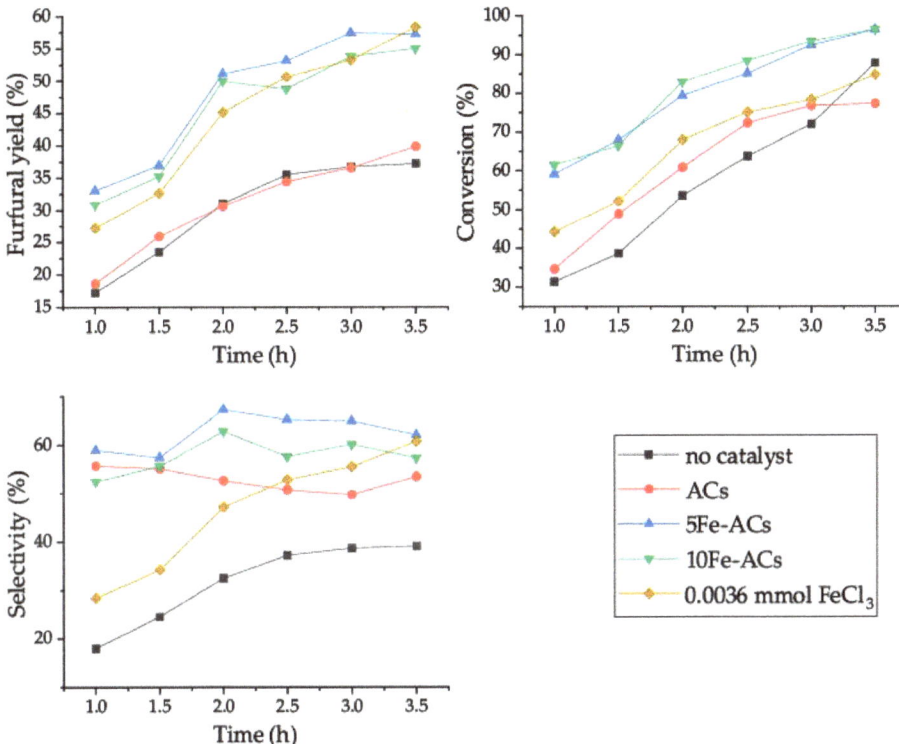

Figure 5. Furfural yield, xylose conversion and reaction selectivity using various catalysts at 170 °C.

Based on the previous results, 5Fe-ACs was the most promising catalyst studied. Therefore, it was compared to homogeneous $FeCl_3 \cdot 6H_2O$ in similar conditions and using similar amounts of catalytic iron (Figure 5, brown rhombuses). A total of 0.0036 mmol of homogeneous $FeCl_3 \cdot 6H_2O$ was able to produce a 27–58% furfural yield in 1–3.5 h. The yield was lower than with 5Fe-ACs until the reaction time reached 3.5 h. Even then, the increase in furfural yield was marginal compared to that with 5Fe-ACs, only one percentage unit. Conversion increased with time from 44 to 85% but was always clearly lower than with 5Fe-ACs. Reaction selectivity also increased with time from 28 to 61% but again it was always lower than with 5Fe-ACs. Better results with carbon-supported 5Fe-ACs than with homogeneous $FeCl_3 \cdot 6H_2O$ demonstrated that activated carbon support is a promising option for furfural production. Based on catalyst characterization, iron was oxidized in a heterogeneous catalyst, which may have affected its catalytic activity positively compared to $FeCl_3$. Moreover, hydroxyl groups were detected on the surface of 5Fe-ACs, which increases the catalyst's Brønsted acid sites and therefore can increase furfural production.

The results with 5Fe-ACs (57% yield at 170 °C in 3 h) are comparable with studies using carbon-supported titanium catalysts—Mazzotta et al. reported a 51% furfural yield (180 °C, 30 min) in a biphasic MeTHF/water system, while Russo et al. achieved a 69% furfural yield (170 °C, 3.5 h) in toluene/water [22,23]. The results were higher than with sulfonated SBA- and MCM-supported metal oxide catalysts (SBA-15/ZrO_2-Al_2O_3/SO_4^{2-} and MCM-41/ZrO_2/SO_4^{2-}), which resulted in 53 and 50% furfural yields, respectively, at 160 °C with a 4-h reaction time [15,16].

2.4. Recycling Experiments

Catalyst recycling experiments were performed with 5Fe-ACs catalyst at 170 °C with a 3 h reaction time. After the reaction, the catalyst was filtered, washed with methanol and water, dried and then used again in the same conditions for three cycles. As a result, the furfural yield decreased from 48% to 30% after the first run and then remained constant (Figure 6). Conversion and reaction selectivity behaved similarly; conversion decreased from 76% to 53% and reaction selectivity from 66% to 59% after the first run and then they remained constant (Figure 6). The decrease in furfural yield, conversion and selectivity may be explained by iron leaching. Leaching was observed by measuring the iron content of the water phase after every reaction. After the first run, 66% of the initial iron amount was leached out but after second and third runs leaching was only 3% and 1%, respectively. It has been reported that hot acidic water promotes the solubility of some metal oxides and in particular the solubility of iron oxides is strongly influenced by the solution pH [48,49]. In our experiments, the pH changed from 6 to 2 during the reaction with 5Fe-ACs, so the reaction media was clearly acidic. Chang et al. were able to decrease iron leaching by post-treating the catalyst after impregnation, first with NaOH and then with HCl [43]. Post treatment was designated to convert iron to ferric hydroxide, which is poorly soluble in water. This treatment also decreased iron leaching at a low pH (pH = 2) to 21%, while the loss of iron during the post-treatment process was very minor (from 5.2 to 4.5 wt%). In future work, this must be taken into consideration.

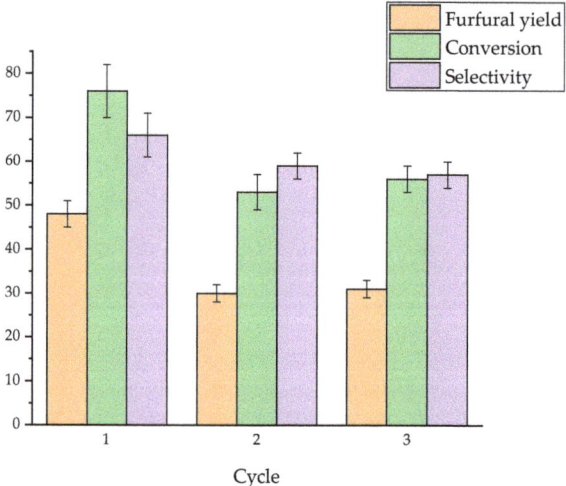

Figure 6. Furfural yield, xylose conversion and reaction selectivity in catalyst recycling experiments after 1–3 cycles.

The stability of the catalyst was also observed by monitoring changes at the catalyst surface with SEM and STEM. The images revealed that the iron content was decreased and leaching was more connected to larger iron particles (approx. 15–40 nm) than to small ones (approx. 5 nm) (see Figure S6). Small iron particles were present throughout the recycled catalysts, while large ones occurred only randomly. However, some large particles were also left on the catalyst surface after the third cycle, indicating that they were not all leached out. Leaching of large particles (or agglomerates of small particles) is reasonable since they are more vulnerable to leaching because they have poorer interaction with the support surface than smaller particles. In future work, it would be important to optimize the catalyst preparation to obtain less-agglomerated metal particles.

Based on catalyst weighing, the mass of the catalyst was increased from 5 to 6.7 mg in each cycle. This indicates that the catalyst was adsorbing some reaction products, such as humins. Humins are black carbonaceous side-products generated either by the cross-polymerization of furfural by itself, between the just-formed furfural and free xylose present in solution or between furfural and intermediate products [50–52]. Solid humins can plug the pores of the catalyst surface, block the access of xylose to metal surface sites or totally encapsulate a metal particle [53]. It is also possible that this blocking of the catalytic sites on the porous structure of the catalyst caused the decrease in furfural yields. Indeed, pore blocking caused by carbon deposition is a common problem in heterogeneous porous catalysts; for example, with MCM-41-SO$_3$H, the surface area and pore volume decreased by 50%–60% after the first run and the furfural yield decreased from 54% to 37% [54]. BET analysis of the used catalysts could not be performed because a very small amount of the catalysts was used in the reactions. In the SEM and STEM images, there was not any clear coating visible on the surface of the carbon, as in Reference [55] but some slight changes were observed in the SEM images (Figure S7). However, based on the figures, no differences were found between catalysts that were used one, two or three times.

3. Materials and Methods

3.1. Materials

For catalyst preparation, hydrolysis lignin was obtained through a biomass hydrolysis process from Sekab Oy, Sweden. Other chemicals used in catalyst preparation, conversion reactions, partitioning experiments and analyses were used as received, without any purification. For high pressure liquid chromatography (HPLC) sample preparation, regenerated cellulose (RC) syringe filters (0.45 µm, 25 mm, Phenomenex) were used for the organic samples and polytetrafluoroethylene (PTFE) syringe filters (0.45 µm, 4 mm, Phenomenex) were used for the water samples.

3.2. Furfural Partitioning in Biphasic Reactor System

Furfural partitioning experiments were carried out using a 4.7 wt% furfural solution in water as feed and toluene or MIBK as an organic solvent. The furfural solution and organic solvent were measured with alternative ratios (1:1, 1:2 and 1:3) into a microwave reactor tube with a magnetic stirring bar. The tube was closed, heated five minutes at 160 °C and cooled to room temperature. Samples from both layers were analyzed by HPLC to calculate the partitioning for furfural in different solvent systems.

3.3. Catalyst Preparation and Characterization

Activated carbon (AC) supports were prepared from hydrolysis lignin, dried in oven at 105 °C and crushed to a particle size of <425 µm. Activation was performed using either a chemical or physical activation method. Chemical activation was done by impregnation of zinc chloride into the dried lignin using a 2:1 mass ratio of $ZnCl_2$:biomass. $ZnCl_2$ dissolved into H_2O was mixed with the biomass for 3 h at 85 °C and then dried in the oven at 105 °C until achieving a constant weight. The carbonization and activation of the dried $ZnCl_2$-impregnated lignin was done in a stainless-steel tube in a tube furnace (Nabertherm RT200/13) (Nabertherm GmbH, Lilienthal, Germany) at 600 °C for 2 h using a heating ramp of 10 °C/min. During the thermal heating process, the reactor was flushed continuously with N_2 (flow 10 mL/min). Alternatively, carbonization followed by physical activation was performed in in one-step process in a stainless-steel tube in a tube furnace using a heating ramp of 10 °C/min to 800 °C. At the target temperature, steam was added by feeding water at 0.5 mL/min into the reactor for 2 h. During the thermal heating process, the reactor was flushed continuously with N_2 (flow 10 mL/min). Both resulting activated carbons were washed with hot water, dried overnight at 105 °C, crushed and sieved to a fraction size of <100 µm. The supports were named ACz (AC zinc chloride-activated and water washed) and AC$_S$ (AC steam-activated and water washed). In addition, a support with chemical activation and HNO$_3$ treatment was prepared (ACz$_N$). This was performed in a round bottom flask with a 10:1 mass ratio of 3 M HNO$_3$ per support and heated for 4 h at 85 °C. After the acid treatment, the support was filtrated and washed with hot distilled water until neutral pH was obtained and finally it was dried in the oven at 105 °C.

In order to modify the carbon supports with iron, metal salts (FeCl$_3$·6H$_2$O or Fe(NO$_3$)$_3$·9H$_2$O) were added by incipient wetness impregnation on the support, aiming that the targeted concentration of iron in the catalyst was 5 or 10 wt% of the total catalyst mass. The metal salts were dissolved in distilled water equal to the pore volume of the support and mixed with the support, matured for 5 h at room temperature and finally dried in an oven at 105 °C for 16 h. Finally, the catalysts were calcined at 400 °C for 2 h with a continuous flush of N$_2$ (flow 10 mL/min). The iron-impregnated catalysts were named 5Fe-ACs, 10Fe-ACs, 5Fe$_{NO3}$-ACz, 5Fe-ACz and 5Fe-ACz$_N$ according to the targeted iron concentration, type of support and type of iron precursor (FeNO$_3$ if mentioned, otherwise FeCl$_3$).

Specific SAs and pore size distributions were determined from the physisorption adsorption isotherms using nitrogen as the adsorbate. Determinations were performed with a Micromeritics ASAP 2020 instrument (Micromeritics Instrument, Norcross, GA, USA). Portions of each sample (100–200 mg) were degassed at low pressure (0.27 kPa) at a temperature of 140 °C for 3 h in order to remove adsorbed gas. Adsorption isotherms were obtained by immersing sample tubes in liquid nitrogen (-196 °C) to achieve constant temperature conditions. Gaseous nitrogen was added to the samples in small doses and the resulting isotherms were obtained. SAs were calculated from adsorption isotherms according to the BET (Brunauer–Emmett–Teller) method [56]. The percentual distribution of pore volumes (vol%) was calculated from the individual volumes of the micropores (pore diameter <2 nm), mesopores (pore diameter 2–50 nm) and macro-pores (diameter >50 nm) using the DFT (Density Functional Theory) model [57]. The instrumental setup enabled the measurement of micropores down to 1.5 nm in diameter, even if there might have been some contribution from smaller pores. The SAs were measured with a precision of ~5%.

The metal contents of the catalysts and supports were measured by ICP-OES using the 5110 VDV instrument (Agilent Technologies, Santa Clara, CA, USA). Zn, Fe, Ca, K, Mn, S, Na and Mg were measured from the ACs support; Zn and Fe contents were measured from ACz, 5Fe$_{NO3}$-ACz, 5Fe-ACz, ACz$_N$ and 5Fe-ACz$_N$; while only the Fe content was determined from 5Fe-ACs and 10Fe-ACs. For determination, samples of 0.1–0.2 g were first digested in a microwave oven (MARS, CEM Corporation) using the EPA 3051A method with 9 mL of HNO$_3$ and 3 mL of HCl [58]. Subsequently, the solution was diluted to 50 mL with water and the former elements were analyzed with the ICP-OES.

XPS analyses were performed using the ESCALAB 250Xi XPS System (Thermo Fisher Scientific, Waltham, MA, USA). With a pass energy of 20 eV and a spot size of 900 μm, the accuracy of the reported binding energies (BEs) was ±0.3 eV. Fe, C, O and Cl were measured for all samples. The measurement data were analyzed using Avantage software. The monochromatic AlKα radiation (1486.6 eV) was operated at 20 mA and 15 kV. Charge compensation of the BEs was performed by applying the C1s line at 284.8 eV as a reference.

XRD was used to study the phases of 5Fe-ACs and 10Fe-ACs utilizing PANalytical X'Pert Pro X-ray diffraction equipment (Malvern Panalytical, Almelo, Netherlands). The diffractograms were collected in the 2θ range of 5–90°, with a step size of 0.017° and a scan speed of 1.06°/min using monochromatic CuKα1 radiation (λ = 1.5406 Å) at 45 kV and 40 mA. The crystalline phases and structures were analyzed with HighScore Plus software and the peaks were identified using International Centre for Diffraction Data ICDD (PDF-4 + 2020).

The morphology of the catalyst particles was studied using SEM and STEM. A JEOL JEM-2200FS energy-filtered transmission electron microscope equipped with a scan generator (EFTEM/STEM) (JEOL Ltd., Tokyo, Japan) was used for STEM analysis. The catalyst samples were dispersed in pure ethanol and pretreated in an ultrasonic bath for several minutes to create a microemulsion. A small drop of the microemulsion was deposited on a copper grid pre-coated with carbon (Lacey/Carbon 200 Mesh Copper) and evaporated in air at room temperature. The accelerating voltage in the measurements was 200 kV, while the resolution of the STEM image was 0.2 nm. The metal particle sizes were estimated visually from the STEM high-angle annular dark field (HAADF) images. The SEM was performed with a Zeiss Sigma Field emission scanning electron microscope (FESEM). In the sample preparation, a powder sample was placed on a conductive glue tape. The SEM images were taken at a voltage of 5 kV and a working distance around 5 mm.

Catalyst surface acidity was characterized by applying the Boehm titration method [59–63]. A total of 100 mg of catalyst was weighed and mixed with 50 mL 0.01 M NaOH. Samples were shaken (300 rpm) in sealed tubes for 72 h at room temperature and then filtered using a syringe and syringe filter (0.45 µm, regenerated cellulose). Titration was carried out using a back-titration method by taking 10 mL of filtrate, mixing it with 20 mL of 0.01 M HCl and finally back-titrating with 0.01 M NaOH. Acidic groups were calculated using Equation (2), based on the theory that NaOH neutralizes all acidic oxygen groups (including phenols, lactonic groups and carboxylic acids) present on carbon. Nonconsumed base content was neutralized with acid and then nonconsumed acid was quantified through simple acid-base titration.

3.4. Furfural Production from Xylose

In a conversion reaction, 0.25 mmol (37.6 mg) of xylose and 0.0036/0.050 mmol of homogeneous metal salt ($AlCl_3·6H_2O$, $ZnCl_2$, $CrCl_3·6H_2O$, $SnCl_2·2H_2O$ or $FeCl_3·6H_2O$) or 5 mg heterogeneous carbon-based catalyst were placed into a 5 mL reaction tube. A magnetic stirring bar, water (1 mL) and MIBK (3 mL) were added and the tube was sealed. The reaction was carried out in a Biotage Initiator microwave reactor (Biotage, Uppsala, Sweden) at 160/170/180 °C for 30 min to 3.5 h. After the reaction, approximate 1 mL samples from both layers were filtered with a syringe filter (an RC filter for the organic layer and a PTFE filter for the water layer) and then analyzed with HPLC.

3.5. Catalyst Recycling

Catalyst recycling experiments were carried out with 5Fe-ACs catalyst with a 3 h reaction time and a reaction temperature of 170 °C. After the reaction, liquid samples were taken normally for furfural and xylose analyses. In addition, metal leaching (Fe and Zn) was monitored by measuring the metal content of the water phase by AAS (see Section 3.6) and the catalyst was collected using a PALL Easy Pressure Syringe Filter Holder and hydrophilic polypropylene membrane (GHP). The catalyst was first washed with methanol (4 + 10 mL) and water (3*10 mL) and then dried and weighed. Used catalysts were analyzed with SEM and TEM (see Section 3.3).

3.6. Analytical Methods for Conversion Studies

Two different HPLC analyses were used to detect furfural and xylose in the samples. In the analyses, calibrations were performed with commercial furfural or xylose. HPLC analysis for furfural was carried out using a Waters 2695 separation module fitted with an Atlantis T3 (3 µm, 4.6 × 150 mm) column and a Waters 996 photodiode array (PDA) detector (Waters Corp., Milford, MA, USA). A mixture of water (0.1% TFA) and methanol (0.1% TFA) (90:10) was used as the mobile phase, with a flow rate of 1 mL/min. The column temperature was kept constant at 30 °C and the UV detection for furfural was performed at 277 nm. HPLC analysis for xylose was carried out using a Shimadzu LC-20AT liquid chromatograph instrument fitted with an SIL-20A TH autosampler, RID-20A refractive index detector, SUGAR SH-G pre-column and Shodex SUGAR SH1821 column (8.0 × 300 mm). Sulfuric acid (5 mM) was used as a mobile phase with a flow rate of 0.8 mL/min and the column temperature was kept constant at 60 °C.

Atomic absorption spectroscopy (AAS) was used to determine iron and zinc leaching from 5Fe-ACs and 5Fe-ACz catalysts. First, water phase samples of the reactions were diluted with water to a minimum 10 mL. Then, determinations were made using Varian AA240FS equipment (Varian Inc., Palo Alto, CA, USA), air-acetylene fuel, a Varian SpectrAA lamp (Cu/Mn/Zn/Fe) and flame emission wavelengths of 372.0 nm for Fe and 213.9 nm for Zn.

3.7. Equations

The partitioning coefficient (P) was calculated using the following formula:

$$P = [furfural]_{org}/[furfural]_{aq}, \qquad (1)$$

where $[furfural]_{org}$ is the concentration (g/l) of furfural in the organic layer and $[furfural]_{aq}$ is the concentration of furfural in the water layer.

The total amount of acid sites according to Boehm titration was calculated as follows:

$$n_{total\ acids} = [(C_{NaOH} * V_{NaOH\ added} - (C_{HCl} * V_{HCl\ added} - C_{NaOH} * V_{NaOH\ titration})/\tfrac{1}{5}) - n_{total\ acids\ in\ reference}]/m, \qquad (2)$$

where c (NaOH and HCl) are concentrations in mol/L, V (NaOH and HCl) are added volumes in mL, V (NaOH titration) is the volume of NaOH in ml needed to achieve equilibrium in titration, m is the mass of carbon weighed and n (total acids in reference) represents a blank solution without carbon. The factor $\tfrac{1}{5}$ is due to the measurement of the 10 mL aliquots representing $\tfrac{1}{5}$ of the reaction base.

The yield of furfural was calculated as follows:

$$Y_{furfural}\ (\%) = [c_{furf\ meas\ org}/c_{furf\ max}] \times 100\%, \qquad (3)$$

where $c_{furf.\ meas.\ org}$ is the measured furfural concentration in the organic phase of the sample and $c_{furf\ max}$ is the theoretical maximum concentration of furfural in the sample.

The conversion of xylose was calculated as follows:

$$C_{xylose}\ (\%) = [n_{xyl\ initial}/n_{xyl\ final}] \times 100\%, \qquad (4)$$

where $n_{xyl\ initial}$ is the initial amount of xylose (in moles) fed to the reaction and $n_{xyl\ final}$ is the amount of xylose left in the reaction mixture after the reaction.

The selectivity of the xylose to furfural conversion was calculated as follows:

$$S\ (\%) = [(c_{furf\ meas\ total}/c_{furf\ max})/conversion] \times 100\%, \qquad (5)$$

where $c_{furf.\ meas\ total.}$ is the measured total furfural concentration in the organic and aqueous phases of the sample and $c_{furf\ max}$ is the theoretical maximum concentration of furfural in the sample.

4. Conclusions

In this study, the conversion of xylose to furfural was studied using lignin-based activated carbon-supported iron oxide catalysts. Three different activated carbon supports and five different catalysts were prepared and studied in furfural production. Different activation methods, metal precursors and metal concentrations were used for the catalysts and different temperatures and reaction times were studied in the conversion reactions. Chemical activation resulted in a higher surface area and pore volume than physical activation but in conversion reactions, physically activated catalysts produced better reaction selectivity. $FeNO_3$ precursor yielded higher xylose conversion than $FeCl_3$ precursor but the furfural yield and selectivity were higher with $FeCl_3$ precursor. The best results for xylose conversion to furfural were achieved with a 4 wt% iron-containing catalyst (5Fe-ACs), which produced a 57% yield, 92% conversion and 65% selectivity at 170 °C in 3 h. The results with a catalyst containing more iron (9.2 wt%) were lower (54% yield, 93% conversion and 60% selectivity) in similar conditions. The catalytic amount of Fe in 5Fe-ACs was only 3.6 µmol and using this amount of homogeneous $FeCl_3$ as a catalyst, reduced the furfural yield, xylose conversion and selectivity. Based on catalyst characterization, iron was in the form of iron oxide on the surface of the heterogeneous catalyst, which may have affected to its catalytic activity positively compared to $FeCl_3$. Moreover, hydroxyl groups were detected on the surface of 5Fe-ACs, which increases catalyst Brønsted acid sites and therefore can increase furfural production. The recycling experiments revealed that part of the iron is easily leached out of the catalyst at a high temperature and in acidic conditions and the catalyst adsorbed some reactions products. These factors decreased the furfural yield and xylose conversion after the first round of recycling but then they remained constant. Although the activated carbon-supported iron oxide catalyst needs some improvements for better stability, it is a feasible alternative to homogeneous $FeCl_3$.

Supplementary Materials: The following are available online at http://www.mdpi.com/2073-4344/10/8/821/s1, Table S1: Metal analysis from ACs by ICP-OES, Table S2: XPS results of ACs, 5-Fe-ACs and 10Fe-ACs, Figure S1: XRD results of 5Fe-ACs and 10Fe-ACs, Figure S2: Boehm titration curves, Figure S3: EDS spectra of area shown in Figure 4, Figure S4: Graphical presentation of the results presented in Table 5, Figure S5: HPLC chromatogram of water (a) and organic (b) phase of reaction solution using 5Fe-ACz catalyst. Grams show increasing side product peak at 3.1 min and furfural shoulder at 8.5 min, when 180 °C was used as reaction temperature, Figure S6: STEM HAADF image of three times used 5Fe-ACs, which shows large agglomerated iron particles (diameter approx. 15˜C40 nm) as well as small single particles (diameter approx. 5 nm), Figure S7: SEM images of unused 5Fe-ACs (a,c) and used 5Fe-ACs (b,d) catalysts.

Author Contributions: Conceptualization, A.R.; methodology, A.R. and R.K.; formal analysis, A.R., R.K. and T.H.; investigation, A.R. and R.K.; data curation, A.R.; writing—original draft preparation, A.R.; writing—review and editing, K.L., R.K., J.K., T.H. and U.L.; visualization, A.R.; supervision, K.L., J.K. and U.L.; funding acquisition, A.R., U.L. and K.L. All authors have read and agreed to the published version of the manuscript.

Funding: This research was funded by Fortum Foundation, grant numbers 201800022 and 20190005, the EU/Interreg Botnia-Atlantica, grant number 20201508, the Foundation of Tauno Tönning, grant number 20190154 and Nessling Foundation, grant number 201800070.

Acknowledgments: Sari Tuikkanen is acknowledged for completing part of HPLC measurements and Riina Hemmilä for AAS-measurements.

Conflicts of Interest: The authors declare no conflict of interest.

References

1. Werpy, T.; Petersen, G. *Top Value Added Chemicals from Biomass: Volume I–Results of Screening for Potential Candidates from Sugars and Synthesis Gas*; U.S. Department of Energy: Washington, DC, USA, 2004. [CrossRef]
2. Kottke, R.H. Furan derivatives. In *Encyclopedia of Chemical Technology*; Othmer, K., Ed.; John Wiley & Sons, Inc.: New York, NY, USA, 2000; Volume 12, pp. 259–286. [CrossRef]
3. Eseyin, A.E.; Steele, P.H. An overview of the applications of furfural and its derivatives. *Int. J. Adv. Chem.* **2015**, *3*, 42–47. [CrossRef]
4. Win, D. Furfural–Gold from Garbage. *Au J. Technol.* **2005**, *8*, 185–190.
5. Zeitsch, K. *The Chemistry and Technology of Furfural and its Many By–Products*; Elsevier: Amsterdam, The Netherlands, 2000; pp. 36–74.
6. Dashtban, M.; Gilbert, A.; Fatehi, P. Production of furfural: Overview and challenges. *J. Sci. Technol. For. Prod. Process* **2012**, *2*, 44–53.
7. Yemis, O.; Mazza, G. Catalytic Performances of Various Solid Catalysts and Metal Halides for Microwave–Assisted Hydrothermal Conversion of Xylose, Xylan, and Straw to Furfural. *Waste Biomass Valori.* **2019**, *10*, 1343–1353. [CrossRef]
8. Le Guenic, S.; Delbecq, F.; Ceballos, C.; Len, C. Microwave–assisted dehydration of D–xylose into furfural by diluted inexpensive inorganic salts solution in a biphasic system. *J. Mol. Catal. A Chem.* **2015**, *410*, 1–7. [CrossRef]
9. Liu, C.; Wyman, C.E. The enhancement of xylose monomer and xylotriose degradation by inorganic salts in aqueous solutions at 180 °C. *Carbohydr. Res.* **2006**, *341*, 2550–2556. [CrossRef]
10. Ershova, O.; Nieminen, K.; Sixta, H. The Role of Various Chlorides on Xylose Conversion to Furfural: Experiments and Kinetic Modeling. *Chem. Cat. Chem.* **2017**, *9*, 3031–3040. [CrossRef]
11. Romo, J.E.; Bollar, N.V.; Zimmermann, C.J.; Wettstein, S.G. Conversion of Sugars and Biomass to Furans Using Heterogeneous Catalysts in Biphasic Solvent Systems. *Chem. Cat. Chem.* **2018**, *10*, 4819–4830. [CrossRef]

12. Da Costa Lopez, A.M.; Morais, A.R.C.; Łukasik, R.M.; Fang, Z.; Smith, R.L., Jr.; Qi, X. Sustainable Catalytic Strategies for C5-Sugars and Biomass Hemicellulose Conversion Towards Furfural Production. In *Production of Platform Chemicals from Sustainable Resources*; Springer Nature: Singapore, 2017; Volume 7, pp. 45–80. [CrossRef]
13. Zhang, J.; Lin, L.; Liu, S. Efficient Production of Furan Derivatives from a Sugar Mixture by Catalytic. *Process. Energ. Fuel* **2012**, *26*, 4560–4567. [CrossRef]
14. Chareonlimkun, A.; Champreda, V.; Shotipruk, A.; Laosiripojana, N. Reactions of C5 and C6-sugars, cellulose, and lignocellulose under hot compressed water (HCW) in the presence of heterogeneous acid catalysts. *Fuel* **2010**, *89*, 2873–2880. [CrossRef]
15. Shi, X.; Wu, Y.; Li, P.; Yi, H.; Yang, M.; Wang, G. Catalytic conversion of xylose to furfural over the solid acid SO42-/ZrO2–Al2O3/SBA-15 catalysts. *Carbohydr. Res.* **2011**, *346*, 480–487. [CrossRef]
16. Dias, A.S.; Pillinger, M.; Valente, A.A. Mesoporous silica–supported 12–tungstophosphoric acid catalysts for the liquid phase dehydration of d-xylose. *Micropor. Mesopor. Mater.* **2006**, *94*, 214–225. [CrossRef]
17. Weingarten, R.; Cho, J.; Conner, W.C., Jr.; Huber, G.W. Kinetics of furfural production by dehydration of xylose in a biphasic reactor with microwave heating. *Green Chem.* **2010**, *12*, 1423–1429. [CrossRef]
18. Choudhary, V.; Pinar, A.B.; Sandler, S.I.; Vlachos, D.G.; Lobo, R.F. Xylose Isomerization to Xylulose and its Dehydration to Furfural in Aqueous Media. *ACS Catal.* **2011**, *1*, 1724–1728. [CrossRef]
19. Choudhary, V.; Sandler, S.I.; Vlachos, D.G. Conversion of Xylose to Furfural Using Lewis and Brønsted Acid Catalysts in Aqueous Media. *ACS Catal.* **2012**, *2*, 2022–2028. [CrossRef]
20. Rodríguez-Reinoso, F. The role of carbon materials in heterogeneous catalysis. *Carbon* **1998**, *36*, 159–175. [CrossRef]
21. Rodríguez-Reinoso, F.; Molina-Sabio, M. Activated carbons from lignocellulosic materials by chemical and/or physical activation: An overview. *Carbon* **1992**, *30*, 1111–1118. [CrossRef]
22. Mazzotta, M.G.; Gupta, D.; Saha, B.; Patra, A.K.; Bhaumik, A.; Abu-Omar, M.M. Efficient Solid Acid Catalyst Containing Lewis and Brønsted Acid Sites for the Production of Furfurals. *ChemSusChem* **2014**, *7*, 2342–2350. [CrossRef]
23. Russo, P.A.; Lima, S.; Rebuttini, V.; Pillinger, M.; Willinger, M.; Pinna, N.; Valente, A.A. Microwave-assisted coating of carbon nanostructures with titanium dioxide for the catalytic dehydration of d-xylose into furfural. *RSC Adv.* **2013**, *3*, 2595–2603. [CrossRef]
24. Barroso-Bogeat, A.; Alexandre-Franco, M.; Fernández-González, C.; Gómez–Serrano, V. Preparation of activated carbon-metal oxide hybrid catalysts: Textural characterization. *Fuel Process. Technol.* **2014**, *126*, 95–103. [CrossRef]
25. Barroso-Bogeat, A.; Alexandre-Franco, M.; Fernández-González, C.; Gómez–Serrano, V. Preparation and Microstructural Characterization of Activated Carbon–Metal Oxide Hybrid Catalysts: New Insights into Reaction Paths. *J. Mater. Sci. Technol.* **2015**, *31*, 806–814. [CrossRef]
26. Barroso-Bogeat, A.; Alexandre-Franco, M.; Fernández-González, C.; Macías-García, A.; Gómez-Serrano, V. Preparation of Activated Carbon-SnO$_2$, TiO$_2$, and WO$_3$ Catalysts. Study by FT-IR Spectroscopy. *Ind. Eng. Chem. Res.* **2016**, *55*, 5200–5206. [CrossRef]
27. Mittal, A.; Black, S.K.; Vinzant, T.B.; O'Brien, M.; Tucker, M.P.; Johnson, D.K. Production of Furfural from Process-Relevant Biomass-Derived Pentoses in a Biphasic Reaction System. *ACS Sustain. Chem. Eng.* **2017**, *5*, 5694–5701. [CrossRef]
28. Moreau, C.; Durand, R.; Peyron, D.; Duhamet, J.; Rivalier, P. Selective preparation of furfural from xylose over microporous solid acid catalysts. *Ind. Crops Prod.* **1998**, *7*, 95–99. [CrossRef]
29. Brouwer, T.; Blahusiak, M.; Babic, K.; Schuur, B. Reactive extraction and recovery of levulinic acid, formic acid and furfural from aqueous solutions containing sulphuric acid. *Sep. Purif. Technol.* **2017**, *185*, 186–195. [CrossRef]
30. Zhang, L.; Yu, H.; Wang, P.; Li, Y. Production of furfural from xylose, xylan and corncob in gamma-valerolactone using FeCl$_3$·6H$_2$O as catalyst. *Bioresour. Technol.* **2014**, *151*, 355–360. [CrossRef]
31. Vom Stein, T.; Grande, P.M.; Leitner, W.; Dominguez de Maria, P. Iron-Catalyzed Furfural Production in Biobased Biphasic Systems: From Pure Sugars to Direct Use of Crude Xylose Effluents as Feedstock. *ChemSusChem* **2011**, *4*, 1592–1594. [CrossRef]
32. Zhang, Y.; Chen, M.; Wang, J.; Hu, Q. Furfural production from dehydration of xylose catalyzed by chromium chloride in biphasic system. *Chem. Eng.* **2014**, *3*, 54–58. [CrossRef]

33. Yang, Y.; Hu, C.; Abu–Omar, M.M. Conversion of carbohydrates and lignocellulosic biomass into 5–hydroxymethylfurfural using AlCl$_3$·6H$_2$O catalyst in a biphasic solvent system. *Green Chem.* **2012**, *14*, 509–513. [CrossRef]
34. Marcotullio, G.; Jong, W.D. Chloride ions enhance furfural formation from D–xylose in dilute aqueous acidic solutions. *Green Chem.* **2010**, *12*, 1739–1746. [CrossRef]
35. Varila, T.; Bergna, D.; Lahti, R.; Romar, H.; Hu, T.; Lassi, U. Activated carbon production from peat using ZnCl2: Characterization and applications. *Bioresources* **2017**, *12*, 8078–8092. [CrossRef]
36. Lahti, R.; Bergna, D.; Romar, H.; Tuuttila, T.; Hu, T.; Lassi, U. Physico–chemical properties and use of waste biomass-derived activated carbons. *Chem. Eng. Trans.* **2017**, *57*. [CrossRef]
37. Brion, D. Etude par spectroscopie de photoelectrons de la degradation superficielle de FeS2, CuFeS2, ZnS et PbS a l'air et dans l'eau. *Appl. Surf. Sci.* **1980**, *5*, 133–152. [CrossRef]
38. Moulder, J.F.; Stickle, W.F.; Sobol, P.E.; Bomben, K.D. *Handbook of X-ray Photoelectron Spectroscopy*; Perkin-Elmer Corporation: Eden Prairie, MN, USA, 1992; p. 45.
39. Li, J.; Liu, J.; Zhou, H.; Fu, Y. Catalytic Transfer Hydrogenation of Furfural to Furfuryl Alcohol over Nitrogen-Doped Carbon-Supported Iron Catalysts. *ChemSusChem* **2016**, *9*, 1339–1347. [CrossRef]
40. Barroso-Bogeat, A.; Alexandre-Franco, M.; Fernández-González, C.; Gómez-Serrano, V. Activated carbon surface chemistry: Changes upon impregnation with Al(III), Fe(III) and Zn(II)-metal oxide catalyst precursors from NO3− aqueous solutions. *Arab. J. Chem.* **2019**, *12*, 3963–3976. [CrossRef]
41. Xiong, Y.; Tong, Q.; Shan, W.; Xing, Z.; Wang, Y.; Wen, S.; Lou, Z. Arsenic transformation and adsorption by iron hydroxide/manganese dioxide doped straw activated carbon. *Appl. Surf. Sci.* **2017**, *416*, 618–627. [CrossRef]
42. Pakuła, M.; Biniak, S.; Świątkowski, A. Chemical and Electrochemical Studies of Interactions between Iron(III) Ions and an Activated Carbon Surface. *Langmuir* **1998**, *14*, 3082–3089. [CrossRef]
43. Chang, Q.; Lin, W.; Ying, W. Preparation of iron-impregnated granular activated carbon for arsenic removal from drinking water. *J. Hazard. Mater.* **2010**, *184*, 515–522. [CrossRef]
44. Dallinger, D.; Kappe, C.O. Microwave–Assisted Synthesis in Water as Solvent. *Chem. Rev.* **2007**, *107*, 2563–2591. [CrossRef]
45. Antal, M.J.; Leesomboon, T.; Mok, W.S.; Richards, G.N. Mechanism of formation of 2-furaldehyde from d-xylose. *Carbohydr. Res.* **1991**, *217*, 71–85. [CrossRef]
46. Hutchings, G.J.; Védrine, J.C. Heterogeneous Catalyst Preparation. In *Basic Principles in Applied Catalysis*; Baerns, M., Ed.; Springer: Berlin, Germany, 2004; Volume 75, pp. 215–257. [CrossRef]
47. Jüntgen, H. Activated carbon as catalyst support: A review of new research results. *Fuel* **1986**, *65*, 1436–1446. [CrossRef]
48. Matatov-Meytal, Y.I.; Sheintuch, M. Catalytic Abatement of Water Pollutants. *Ind. Eng. Chem. Res.* **1998**, *37*, 309–326. [CrossRef]
49. Kraemer, S.M. Iron oxide dissolution and solubility in the presence of siderophores. *Aquat. Sci.* **2004**, *66*, 3–18. [CrossRef]
50. Cheng, B.; Wang, X.; Lin, Q.; Zhang, X.; Meng, L.; Sun, R.; Xin, F.; Ren, J. New Understandings of the Relationship and Initial Formation Mechanism for Pseudo–lignin, Humins, and Acid–Induced Hydrothermal Carbon. *J. Agric. Food Chem.* **2018**, *66*, 11981–11989. [CrossRef] [PubMed]
51. Van Zandvoort, I.; Wang, Y.; Rasrendra, C.B.; van Eck, E.R.H.; Bruijnincx, P.C.A.; Heeres, H.J.; Weckhuysen, B.M. Formation, Molecular Structure, and Morphology of Humins in Biomass Conversion: Influence of Feedstock and Processing Conditions. *ChemSusChem* **2013**, *6*, 1745–1758. [CrossRef] [PubMed]
52. Sumerskii, I.V.; Krutov, S.M.; Zarubin, M.Y. Humin-like substances formed under the conditions of industrial hydrolysis of wood. *Russ. J. Appl. Chem.* **2010**, *83*, 320–327. [CrossRef]
53. Bartholomew, C.H. Mechanisms of catalyst deactivation. *Appl. Catal. A Gen.* **2001**, *212*, 17–60. [CrossRef]
54. Dias, A.S.; Pillinger, M.; Valente, A.A. Dehydration of xylose into furfural over micro–mesoporous sulfonic acid catalysts. *J. Catal.* **2005**, *229*, 414–423. [CrossRef]
55. Jeong, G.H.; Kim, E.G.; Kim, S.B.; Park, E.D.; Kim, S.W. Fabrication of sulfonic acid modified mesoporous silica shells and their catalytic performance with dehydration reaction of d–xylose into furfural. *Micropor. Mesopor. Mater.* **2011**, *144*, 134–139. [CrossRef]
56. Brunauer, S.; Emmett, P.H.; Teller, E. Adsorption of Gases in Multimolecular Layers. *J. Am. Chem. Soc.* **1938**, *60*, 309–319. [CrossRef]

57. Seaton, N.A.; Walton, J.P.R.B.; Quirke, N. A new analysis method for the determination of the pore size distribution of porous carbons from nitrogen adsorption measurements. *Carbon* **1989**, *27*, 853–861. [CrossRef]
58. U.S. EPA. *Method 3051A (SW–846): Microwave Assisted Acid Digestion of Sediments, Sludges, and Oils*; U.S. EPA: Washington, DC, USA, 2007.
59. Boehm, H.P. Chemical Identification of Surface Groups. In *Advances in Catalysis*; Eley, D.D., Pines, H., Weisz, P.B., Eds.; Elsevier Academic Press: San Diego, CA, USA, 1966; Volume 16, pp. 179–274. [CrossRef]
60. Boehm, H.P. Some aspects of the surface chemistry of carbon blacks and other carbons. *Carbon* **1994**, *32*, 759–769. [CrossRef]
61. Schönherr, J.; Buchheim, J.R.; Scholz, P.; Adelhelm, P. Boehm titration revisited (part i): Practical aspects for achieving a high precision in quantifying oxygen-containing surface groups on carbon materials. *Carbon* **2018**, *4*, 21. [CrossRef]
62. Goertzen, S.L.; Theriault, K.D.; Oickle, A.M.; Tarasuk, A.C.; Andreas, H.A. Standardization of the Boehm titration. Part, I. CO_2 expulsion and endpoint determination. *Carbon* **2010**, *48*, 1252. [CrossRef]
63. Oickle, A.M.; Goertzen, S.L.; Hopper, K.R.; Abdalla, Y.O.; Andreas, H.A. Standardization of the Boehm titration: Part II. Method of agitation, effect of filtering and dilute titrant. *Carbon* **2010**, *48*, 3313–3322. [CrossRef]

© 2020 by the authors. Licensee MDPI, Basel, Switzerland. This article is an open access article distributed under the terms and conditions of the Creative Commons Attribution (CC BY) license (http://creativecommons.org/licenses/by/4.0/).

MDPI
St. Alban-Anlage 66
4052 Basel
Switzerland
Tel. +41 61 683 77 34
Fax +41 61 302 89 18
www.mdpi.com

Catalysts Editorial Office
E-mail: catalysts@mdpi.com
www.mdpi.com/journal/catalysts

www.ingramcontent.com/pod-product-compliance
Lightning Source LLC
LaVergne TN
LVHW070155120526
838202LV00013BA/1146